MODELS FOR LIFE

MODELS FOR LIFE: AN INTRODUCTION TO DISCRETE MATHEMATICAL MODELING WITH MICROSOFT® OFFICE EXCEL®

JEFFREY T. BARTON
Birmingham-Southern College,
Birmingham,
Alabama, USA

WILEY

Published by John Wiley & Sons, Inc., Hoboken, New Jersey
Published simultaneously in Canada

For general information on our other products and services or for technical support, please contact our Customer Care
Department within the United States at (800) 762-2974, outside the United States at (317) 572-3993 or fax (317) 572-4002.

Wiley also publishes its books in a variety of electronic formats. Some content that appears in print may not be
available in electronic formats. For more information about Wiley products, visit our web site at www.wiley.com.

Library of Congress Cataloging-in-Publication data applied for

ISBN: 9781119039754

Set in 10/12pt Times by SPi Global, Pondicherry, India

Printed in the United States of America

10   9   8   7   6   5   4   3   2   1

1   2016

For Nora and Tess.

CONTENTS

EXCEL TABLE OF CONTENTS

PREFACE

Over the past 15 years I have had the pleasure of teaching mathematics at a small, student-centered liberal arts college. During that time I have taught students of all interests and abilities, and a considerable amount of my time and energy has been devoted to teaching mathematics to nonmathematics majors. These students are typically taking mathematics to satisfy a graduation requirement, and for most of them the class they take with me will be the last mathematics class they will ever take. For such students I feel a special responsibility to make their one mathematics class a satisfying, engaging experience. Over the years of trying many different texts and approaches, I came to realize that a text for the kind of course I really wanted to share with these students did not yet exist. Thus, I began writing my own in hopes of creating a text with the qualities I believe are necessary to provide the kind of experience I want for my students, one based on mathematics and applications that are relevant, authentic, and accessible.

The topics presented in the text have been carefully selected to be of relevance and interest to a general audience. Furthermore, problems throughout the text are personalized in that they invite readers to supply their own numbers when making a model projection. The questions below are examples of the kinds of questions readers will find:

- How much do you need to save each month to reach your retirement goal?
- Is a lease takeover for the car you want a good deal?
- What will the mortgage payment be for your dream house?
- What will your blood alcohol concentration be after two of your favorite drinks?
- How much will you weigh in 6 months if you follow a given diet?
- What will your body fat percentage be 1 month from today if you become more active?

- Who was the best team in your favorite sport last season?
- What would happen if there were an outbreak of Ebola in your hometown?
- How many cups of Starbucks coffee could you have and still be under the NCAA competition limit for caffeine?

Questions like these and others are found throughout the text. It is my hope and my belief that readers will find their solutions interesting and their answers truly useful.

As a mathematics book this text is somewhat nontraditional. In a traditional mathematics text, the mathematical content comes first, sometimes followed by a few cursory applications. As readers can tell by a quick glance at the table of contents, this text is very much applications driven. Here there are no chapters on "Solving Systems of Equations" or "The Quadratic Formula." Instead there are chapters on blood alcohol concentration, body weight, and infectious diseases. Within those chapters the necessary mathematics for creating and analyzing the models is presented as it is needed. In organizing the text this way, there is no opportunity to ask of the mathematics, "What is this good for?" because the mathematics only comes up if it is needed to answer a real question. We will, incidentally, solve a few systems of equations, and we will occasionally need the quadratic formula, but these are the "means," not the "ends."

The models presented in this text are authentic— that is, they are real mathematical models developed to answer real questions, not artificial or forced applications of the mathematical material. Most of the models can be found in the literature of the relevant discipline and, with few exceptions, parameter values, constants, data, etc., are all taken from research on the subject at hand. Sources for all models, parameters, and data are cited so that the reader may verify them or pursue further information as desired. The few numbers that are estimated by the author or that serve as stand-in values are so noted in the text.

To keep the text accessible to a general audience, the mathematical content and technology requirements have been selected to keep prerequisites to a minimum. Thus, the text focuses on only one type of model: discrete dynamical systems. Though the name may sound daunting to those unfamiliar with it, at its heart a discrete dynamical system is simply a way of describing how something changes from one time step to the next. Once that change is understood, its repeated application allows us to make long-term predictions. Thus, we do not need sophisticated equations to generate results, and only high school algebra is assumed in the text.

The idea of a discrete dynamical system has been around for a very long time. After all, observing how a quantity changes from one time to the next is a natural kind of thing to do. However, if one attempts to use a discrete dynamical system to project a quantity very far into the future, it quickly becomes impractical to do the calculations by hand. The structure of discrete dynamical systems and how they are calculated make the use of spreadsheet software like Microsoft® Office Excel® a perfect match. With such a tool we can focus on setting up and understanding a model, leaving the tedious calculations to Excel. Excel is also a widely used tool throughout the business world, and knowing how to use it is a valuable skill on its own. To ensure that the use of Excel is really a benefit to the reader and not an additional barrier to learning, no prior knowledge of Excel is assumed. The text employs step-by-step screenshots of Excel techniques as an integral part of the presentation with the implicit assumption that the reader has never opened the program before.

HOW TO USE THIS BOOK

Any engaging and satisfying mathematical experience must involve doing mathematics as opposed to just witnessing it. Thus, I encourage readers to follow along by creating their own spreadsheet models as they are presented in the text. It will also save much time later in the exercises if these spreadsheets are saved in an easily accessible way, with folders organized by chapter and section. The use of descriptive file names such as "Widmark Blood Alcohol Concentration Model" will also make models easier to locate when it is time to work the exercises.

As with any mathematics text, the exercises are an integral part of the learning experience. They serve to reinforce, deepen, and extend the reader's knowledge and skill. Exercises are suggested at the end of each section. These will go most smoothly if the reader has already created the spreadsheets for the models developed in the section as many exercises require the reader to apply those models. The exercises range in difficulty from routine calculations to more difficult conceptual questions. Some exercises are tagged as *Extension* problems. These problems require something not discussed explicitly in the text. Some are purely mathematical in nature, but most of them involve making modifications to the models in the text in order to improve the model or apply it in a slightly different context. Thus, these are exercises where the reader will be engaged in the *process* of modeling as opposed to the *use* of existing models. Both are useful skills. The *Extension* exercises are generally open ended with more than one reasonable answer, and they also range in difficulty. They are meant to be stimulating and fun, and it is my hope that the reader will attempt many of them.

The mathematical modeling material is organized by chapter, section, and subsection. The Excel material is marked by sections numbered E1, E2, etc., in order to make it easier for the reader to reference specific Excel topics.

Unlike many mathematics texts, the material does not "build on itself" as the text progresses. The book is semimodular in structure in that once Chapter 1 has been digested, it is possible to read the remaining chapters in any order according to the reader's interests. If chapters are skipped, it will only occasionally be necessary for the reader to refer to a previous section for a relevant definition or Excel technique. The sequence of topics as presented in the table of contents is only a suggestion.

JEFFREY T. BARTON

ACKNOWLEDGMENTS

I have been very fortunate to work at a place like Birmingham-Southern College where faculty have the freedom to develop and teach courses like the one that inspired this book and where faculty scholarship continues to be supported through funding for sabbaticals, travel, and professional development. This text has certainly benefited from this very tangible kind of support, and I have personally benefited even more from being part of a faculty that is so dedicated to learning and teaching.

I am particularly appreciative of my colleagues in the mathematics department—a more dedicated, professional, and supportive group of colleagues I could not hope to have. Their help, during the last few months especially, was freely offered, very much needed, and deeply appreciated. Thanks Anne, Bernie, Maria, and Doug!

Special thanks go to those who devoted their time and energy to reading drafts of the manuscript, including the anonymous reviewers who I hope will see that their suggestions are reflected in the final product. Thanks to Dr. Doug Riley, Dr. Bernadette Mullins, Dr. Anne Yust, and Dr. Barry Spieler for their willingness to read and suggest improvements for the book. It is a much better book than it would have been without their input. Thanks also to Sandeep Kumar at SPi Global and to everyone at John Wiley & Sons who helped make this book happen: Susanne Steitz-Filler, Senior Editor; Sari Friedman, Senior Editorial Assistant; Allison McGinniss, Project Editor; and Nomita Swaminathan, Production Editor. I am indebted to all for their professionalism and patience in working with me on this project. While a great many people worked very hard to make this book as good as it can be, any errors that remain in the text are solely the responsibility of the author.

This book would not have been possible without the understanding and support of my family. To my parents, A.K. and Shirley Barton, thank you for all of your encouragement and for the writer's retreat. To two wonderful daughters, Nora and Tess, thank you for your excitement and curiosity about the book and for your patience along the way. Finally, to my wife, Gabrielle, whose enthusiasm for the project and continual optimism were much appreciated, thank you for being so willing to discuss ideas for the book, for offering feedback throughout the process from the proposal to the finished product, and for making it easy for me to find the time necessary to write. I am very grateful to you all.

1

DENSITY-INDEPENDENT POPULATION MODELS

This chapter is our introduction to **discrete dynamical systems**, which are mathematical models that involve the repeated application of relatively simple equations. In this chapter we set the stage by developing the language, notation, and tools that will be fundamental to our model building and analysis. In particular, we show how to represent a model graphically using a flow diagram, and we show how to implement models using the spreadsheet software Microsoft Excel. We begin our discussion of modeling in the context of population growth, but we will soon see that the mathematics we develop is immediately applicable to other situations as well.

1.1 EXPONENTIAL GROWTH

When a biologist, ecologist, or wildlife manager studies a population, certain fundamental, quantitative questions immediately arise:

- How many are there in the population?
- How many will there be in the future?
- How fast is the population growing or declining?
- If a population is declining, is it due to a low birth rate or excess mortality?
- What will be the effect of human efforts to manage the population?
- What will be the effects of natural disasters on the population?

Models for Life: An Introduction to Discrete Mathematical Modeling with Microsoft® Office Excel®,
First Edition. Jeffrey T. Barton.
© 2016 John Wiley & Sons, Inc. Published 2016 by John Wiley & Sons, Inc.

The material in this chapter describes some of the attempts that mathematicians and scientists have made to answer these and similar questions through **mathematical modeling**. A mathematical model is a mathematical description of a situation whose purpose is to help us understand it or predict how it will change.

The mathematical models that we consider first are models of populations that are said to be **density independent**. A density-independent population is one whose rate of growth or decline does not depend on its size. For example, a population that always grows by 10% per year whether the population is 5 or 5,000,000 would be considered density independent because the growth rate does not change with the size. Similarly, a population that declines by 20 members per year regardless of its size would also be considered density independent. Many real populations exhibit this property, though usually over relatively short time intervals.

A population is said to exhibit exponential growth if it increases by the same percentage each year. In 1798 the influential English economist Thomas Malthus suggested that the world's human population was growing exponentially. He further argued that the growth of the human population would outstrip the growth of the world's food supply, a situation that would of course lead to a stark and difficult existence (Malthus, 1798). Malthus was in fact not the first to make this claim; he was preceded in this hypothesis by the Swiss mathematician Leonhard Euler (1707–1783) (Murray, 1993). In any case, due to Malthus's pioneering work, exponential growth is sometimes referred to as **Malthusian growth**.

As will be our habit throughout the book, we introduce our mathematical model by using real data from a real situation. In this first case we study the population of grizzly bears in Yellowstone National Park.

1.1.1   Modeling Yellowstone Grizzlies

The population of grizzly bears in Yellowstone National Park is an example of the successful management and subsequent recovery of an endangered species. Through the problems and discussion that follow, we will learn about the history of the bear population and make predictions about its future using an exponential population model.

A theme that we will emphasize as we go along is that in order to create a mathematical model, we must make simplifying assumptions about the situation we are modeling. We must also continually ask ourselves whether the assumptions we have made are reasonable or whether we have simplified the situation so much that our model is no longer useful. Here we assume that the grizzly population exhibits exponential growth during the time period of interest. This assumption is reasonable based on data presented in the sources cited below.

Example 1.1: In 1993, the National Biological Service estimated that the population of grizzly bears in Yellowstone National Park was 197 and that it was growing at a rate of 1% per year (Mattson, Wright, Kendall, & Martinka, 1993). Based on these estimates, what would you predict the population to be in 2002?

Before making progress on this problem, we take a few moments to set up our notation and outline the approach that we will use throughout the text. The goal of employing mathematical notation is conciseness, though sometimes an unintended consequence of its use is a sacrifice of clarity. We will try to keep this in mind and introduce no more notation than is truly necessary.

In general, we denote the amount of time that has elapsed from the start of a problem by t and the population in question by P. The time, t, is called the **independent variable** because it does not depend on anything else; since the population P varies with time, we call P the **dependent variable**. Making use of function notation, we write $P(t)$ to mean the population t years after the start of the problem. Thus, the notation $P(t-1)$ means the population 1 year earlier, or the previous year's population. For example, if $t = 5$, then $P(t) = P(5)$ represents the population after 5 years. Then $P(t-1) = P(5-1) = P(4)$, which gives the population in the previous year. Along the same lines, the notation $P(0)$ represents the **initial population**, or the population after 0 years have passed. The function notation $P(t)$ is easily confused with multiplication as in "P times t"; it is important to remember that $P(t)$ is just a shorthand way of writing, "the population after t years have passed." All of our notation should gradually seem more natural as we gain experience using it. When dealing with a population that is growing or declining by a set percentage each year, it will be our habit to use a lowercase r to denote this percentage, and we call r the **growth rate** parameter.

Returning to Example 1.1, we make our notation explicit:

- t = the number of years since 1993.
- P = the population of grizzly bears in Yellowstone.
- $P(t)$ = the population of grizzlies t years after 1993.
- $P(0)$ = the initial population of grizzlies (so using the information in Example 1.1, $P(0) = 197$).
- r = the annual growth rate (so in this problem, $r = 1\%$ or $r = 0.01$).

With our notation in place, we solve the problem by constructing a type of mathematical model known as a **discrete dynamical system** (**DDS**). A DDS is a mathematical model that relates a quantity at one point in time to the same quantity at a previous point. In our current example, this means our model should relate the population of grizzlies in 1 year to the population in the previous year.

The key to setting up any DDS is to develop a thorough understanding of how the dependent variable—in this case, P, our population of grizzlies—changes from one point in time to the next. A helpful first step in developing such an understanding is to visualize the situation by drawing a **flow diagram**. In a flow diagram:

1. We represent any dependent variable by drawing an oval with an appropriate label.
2. We indicate increases in the variable by drawing appropriately labeled arrows pointing into the oval.
3. We indicate decreases in the variable by drawing appropriately labeled arrows pointing out of the oval.

In the grizzly example, we have one dependent variable, the bear population, and each year it increases by 1% of its previous value. Thus, our flow diagram for the situation is given in Figure 1.1.

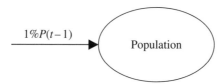

$$1\%P(t-1)$$

Population

FIGURE 1.1   Flow diagram for grizzly population growing by 1% each year.

The diagram tells us that from 1 year to the next, the population increases by 1% of its previous value. To find $P(t)$, we start with the population from the previous year, $P(t-1)$, and we add 1% of $P(t-1)$ to it. Translating this statement into a mathematical equation produces our first DDS:

$$P(t) = P(t-1) + 0.01P(t-1).$$

We have just constructed a mathematical model of the Yellowstone grizzly bear population by translating information about how the population is growing into a mathematical equation. Though future situations may be more complex, this is the process we mean when we say that we are going to model a situation mathematically: (i) learn about how the situation is changing, (ii) visualize the situation with a flow diagram, and (iii) translate the diagram into an equation.

One purpose of our model is to allow us to calculate values for the dependent variable over time. As an example of how to do this, we use the present model to calculate the grizzly population in 1994. Since 1 year has passed since 1993, we have $t=1$ (and $t-1=0$). According to our DDS, the population in 1994 will be

$$P(1) = P(0) + 0.01P(0)$$
$$= 197 + 0.01 \cdot 197$$
$$= 197 + 1.97 = 198.97$$
$$\approx 199.$$

We predict that—based on our assumptions and available data—there will be 199 grizzly bears in Yellowstone in 1994. Note that we substituted in the initial value $P(0) = 197$ and rounded to the nearest bear. Finding the population in 1994 is progress, but the original question asked us to predict the population in 2002. We get closer by calculating the population in 1995. As we continue to make predictions farther into the future, we save rounding for our final answer. We use unrounded numbers in intermediate steps. Since 2 years have passed since 1993, $t=2$ and the DDS tells us that

$$P(2) = P(1) + 0.01P(1)$$
$$= 198.97 + 0.01 \cdot 198.97$$
$$= 200.96 \approx 201.$$

We have found that $P(2) = 201$, that is, we predict there will be 201 Yellowstone grizzlies in 1995.

Hopefully now we see how to proceed: we calculate the population for *any* year past 1993 by repeatedly applying the DDS. This may require some patience, but if need be we can do it. We summarize the rest of the necessary steps for Example 1.1 in Table 1.1.

TABLE 1.1   Model Predictions for Grizzly Population to 2002

Year	t	Apply DDS	Substitute Previous Value	Result
1995	2	See previous	See previous	$P(2) = 200.96$
1996	3	$P(3) = P(2) + 0.01P(2)$	$P(3) = 200.96 + 0.01 \cdot 200.96$	$P(3) = 202.97$
1997	4	$P(4) = P(3) + 0.01P(3)$	$P(4) = 202.97 + 0.01 \cdot 202.97$	$P(4) = 205.00$
1998	5	$P(5) = P(4) + 0.01P(4)$	$P(5) = 205.00 + 0.01 \cdot 205.00$	$P(5) = 207.05$
1999	6	$P(6) = P(5) + 0.01P(5)$	$P(6) = 207.05 + 0.01 \cdot 207.05$	$P(6) = 209.12$
2000	7	$P(7) = P(6) + 0.01P(6)$	$P(7) = 209.12 + 0.01 \cdot 209.12$	$P(7) = 211.21$
2001	8	$P(8) = P(7) + 0.01P(7)$	$P(8) = 211.21 + 0.01 \cdot 211.21$	$P(8) = 213.23$
2002	9	$P(9) = P(8) + 0.01P(8)$	$P(9) = 213.23 + 0.01 \cdot 213.23$	$P(9) = 215.36$

Finally we arrive at an answer to the original question—we predict there will be 215 grizzly bears in Yellowstone in the year 2002. □

Two important features to remember about the DDS method are:

1. In order to calculate the population in one year, we need to know the population in the previous year.
2. By applying the DDS enough times, we can compute the predicted population in *any* future year.

It does not take many hand calculations with a DDS to realize that the aid of a computer would be welcome. Next we see how to arrange for Excel to help in carrying out the tedious calculations.

E.1   Introduction to Excel

In this Excel section we introduce the layout of an Excel worksheet and discuss entering text, entering a formula, copying a formula down, toggling between formulas and values, highlighting cells for formatting, and rounding.

Start Microsoft Excel and open a blank **workbook**. What we should see is pictured in Figure 1.2, though depending on the particular computer or software version, there may be minor differences in appearance.

The main part of the worksheet is a large grid with columns labeled alphabetically and rows labeled numerically. Excel refers to cells by first citing the column letter and then the row number; it also indicates the **active**, or current, cell by (i) highlighting the border of the chosen cell, (ii) displaying the cell reference

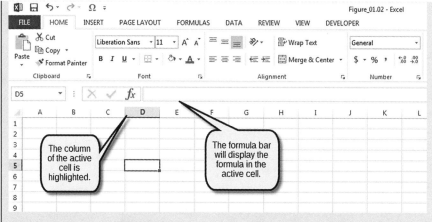

FIGURE 1.2   A blank Excel workbook.

in the top left corner, and (iii) highlighting the column and row headings. In Figure 1.2, we can see that the active cell is cell D5. To select a cell, we may either point and click on the cell or use the arrow keys to move around the grid. Once a cell is selected, we may type text or a formula into it. Note that we will see the text or formula appear in the cell itself and the **formula bar** (see Fig. 1.2) as we type.

Next we see how to use our spreadsheet to work Example 1.1. We begin by giving our worksheet an appropriate title. Select cell A1 by placing the mouse pointer over the cell and (left) clicking. Once the cell is highlighted, type "Yellowstone Grizzly Population," and press Enter (or Return). The worksheet should now look like Figure 1.3.

	A	B	C	D	E	F
1	Yellowstone Grizzly Population					
2						
3						
4						
5						
6						
7						

FIGURE 1.3   Excel workbook for grizzly population with title.

Next we enter column headings for t (the number of years past 1993) and the population. We are free to use any location we wish, but for the sake of consistency, let us agree to use cells A3 and B3. After typing in the headings and the initial values for t and the population, the spreadsheet should look like Figure 1.4.

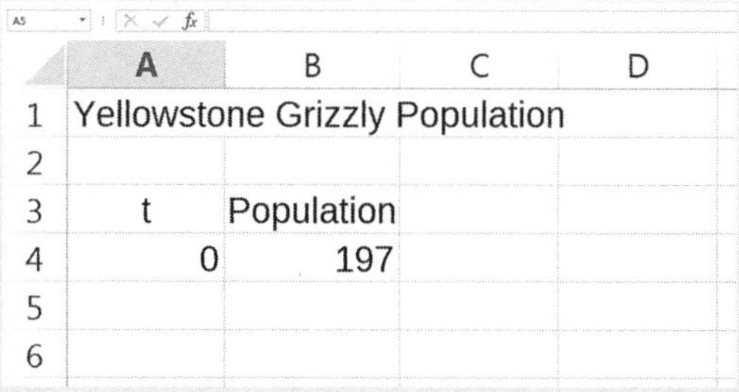

FIGURE 1.4 Excel workbook for grizzly population with time and population columns.

In column A we could manually type the numbers up to $t = 9$, but for later applications this would become tiresome. Instead, we enter a formula into cell A5 that will tell Excel how to produce the values we want for t. Since going from 1 year to the next amounts to adding 1 to t, in cell A5 we tell Excel to do just that by typing the formula "= A4 + 1" as pictured in Figure 1.5. The equals sign tells Excel that we are typing a formula and not just text. *We must first type "=" every time we want Excel to compute a formula!* The formula itself tells Excel to take the number that is stored in cell A4 and add 1 to it. Once we hit Enter, Excel will calculate the formula and display the result, in this case "1."

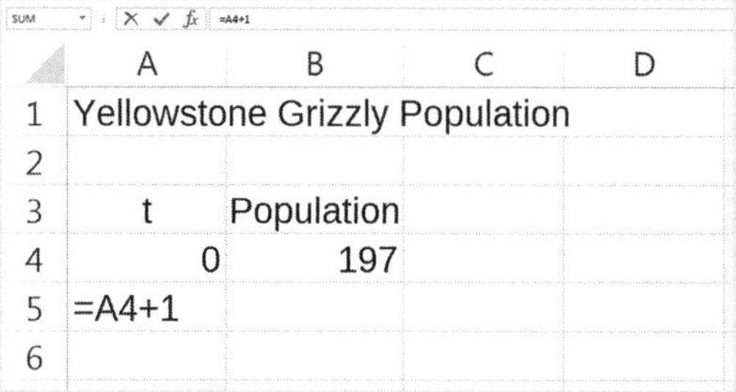

FIGURE 1.5 Excel workbook for grizzly population with formula for time entered.

Notice how Excel helps us keep track of what is going on in the formula by highlighting any referenced cell with a color that matches the reference. After we hit Enter, we should only see the resulting number "1" and not the formula. At this point it is natural to ask, "Why would I bother typing a formula when I could just type in a 1?" We would do just that if we were going to stop at $t = 1$. Here, however, we want to go all the way to $t = 9$, and inputting the formula into Excel will make that easier.

Click on cell A5 so that it is highlighted. Next, position the pointer over the dot in the bottom right-hand corner of the cell. This dot is called the **fill handle**. When you do this, the thick cross should turn into a thin cross. While the pointer is a thin cross, left-click on the fill handle, and without releasing the mouse button, drag the pointer down a few rows. Now release the button. The screen should appear as in Figure 1.6.

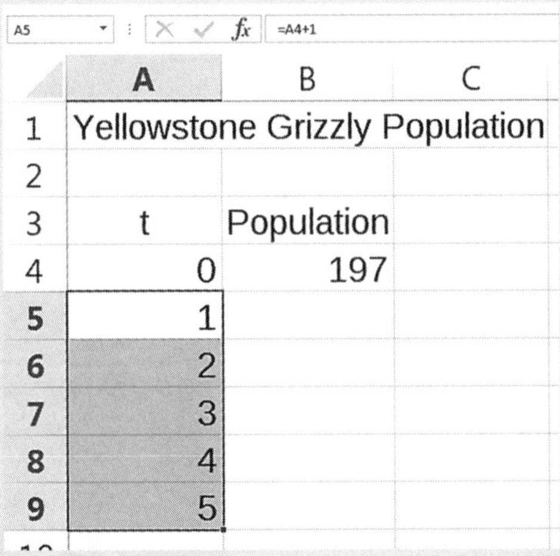

	A5	:	X ✓	f_x	=A4+1	

	A	B	C
1	Yellowstone	Grizzly	Population
2			
3	t	Population	
4	0	197	
5	1		
6	2		
7	3		
8	4		
9	5		

FIGURE 1.6   Excel workbook for grizzly population with time formula copied down.

If we were to display the formulas in each cell instead of the numerical results, we would see what appears in Figure 1.7. What Excel has done is copy the original formula to all of the highlighted cells while at the same time updating the formula for each cell to preserve our original intent: add one to the value in the cell above.

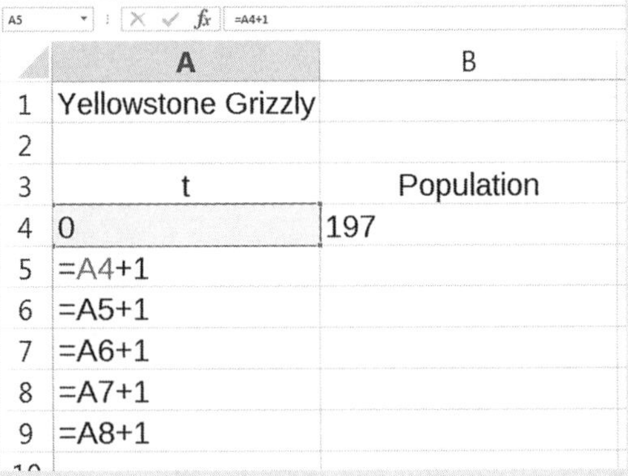

	A5	:	X ✓	f_x	=A4+1	

	A	B
1	Yellowstone Grizzly	
2		
3	t	Population
4	0	197
5	=A4+1	
6	=A5+1	
7	=A6+1	
8	=A7+1	
9	=A8+1	

FIGURE 1.7   Grizzly population Excel workbook with formulas displayed.

Formulas can be displayed by using the keyboard shortcut "CTRL+`" (hold down the control key and hit the single *left* quote key, located in the upper left corner of the keyboard). Repeating the "CTRL+`" shortcut takes you back to displaying numerical results instead of formulas. Again, Excel has automatically changed the formula so that our original intent—"take the value from the cell above and add 1 to it"—is preserved in every cell.

Now we turn to the population itself. As we did in the discrete dynamical system, we need Excel to take the previous year's population and add 1% of its value.

The initial population is stored in cell B4, so we enter the appropriate formula in cell B5 (see Fig. 1.8).

It is important to understand the formula in cell B5: it tells Excel to take the value in the cell above and add 1% of that value to it. Now press Enter and view the result. We should get exactly what we first did by hand. Finally we get a glimpse of the power of Excel in handling a DDS when we copy the formula down as before: select cell B5, grab the fill handle using the thin cross, and drag it down to cell B13. In Figure 1.9 we see the results of Excel having automatically calculated the population for all 9 years.

| SUM | ▾ | ⁞ | ✕ ✓ f_x | =B4+0.01*B4 | |
| | A | B | C |

	A	B	C
1	Yellowstone Grizzly Population		
2			
3	t	Population	
4	0	197	
5	1	=B4+0.01*B4	
6	2		
7	3		
8	4		
9	5		

FIGURE 1.8 Excel workbook for grizzly population.

	A	B	C
1	Yellowstone Grizzly Population		
2			
3	t	Population	
4	0	197	
5	1	198.97	
6	2	200.96	
7	3	202.97	
8	4	205.00	
9	5	207.05	
10	6	209.12	
11	7	211.21	
12	8	213.32	
13	9	215.46	

FIGURE 1.9 Grizzly population Excel model over 9 years.

What was formerly laborious to do by hand, we have just accomplished in seconds with Excel. To reinforce our understanding of what Excel has done, we once again press "CTRL+`" and examine the copied formulas (see Fig. 1.10).

	A	B
1	Yellowstone Grizzly	
2		
3	t	Population
4	0	197
5	=A4+1	=B4+0.01*B4
6	=A5+1	=B5+0.01*B5
7	=A6+1	=B6+0.01*B6
8	=A7+1	=B7+0.01*B7
9	=A8+1	=B8+0.01*B8
10	=A9+1	=B9+0.01*B9
11	=A10+1	=B10+0.01*B10
12	=A11+1	=B11+0.01*B11
13	=A12+1	=B12+0.01*B12

FIGURE 1.10   Grizzly population Excel model with formulas displayed.

It is critical before going on to understand both the relationship of the Excel formula to our original DDS and how Excel automatically updates formulas when they are copied down.

All that is left to do now is to have Excel round to the nearest bear. We round all values simultaneously by first selecting all cells that contain an unrounded number, in this case cells B5–B13. To do this we use the *thick* cross pointer to click on cell B5, drag down to cell B13, and then release the mouse button. The numbers should all remain as they are, but the cells will be highlighted. Under the Home tab on the toolbar, the Number group includes controls for formatting numbers. On the bottom right of that group are two buttons that control rounding by either increasing or decreasing the number of decimal places shown (see Fig. 1.11).

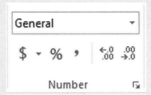

FIGURE 1.11   Excel Number group under Home tab.

Keep clicking the button until only whole numbers appear. *Note that even when Excel displays whole numbers, it uses the unrounded versions in all of its computations*. It is worth taking a moment to verify that all of the population values agree with those we produced by hand in Table 1.1.

We end this Excel section by noting a couple of useful Excel shortcuts:

- When entering a formula, we can just left-click on the cell whose address we want to enter rather than having to type the cell address in ourselves. Excel automatically inserts the address of the cell we click into the formula.
- If we have already copied a formula down in one column, then we can save time when copying formulas down in adjacent columns. Instead of left-clicking the fill handle and dragging down, we can just double-click on the fill handle. Excel will automatically copy the formula down to the same length as the adjacent column.

In 2002 the National Park Service put together a new estimate of the grizzly population and found that there were actually around 416 bears (Gunther, 2003). This indicates that the bear population fared much better than our initial model predictions, which in turn means that something happened between 1993 and 2002 that we did not account for in our simplifying assumptions.

It turns out that the difference between our model's 2002 predicted value and the actual 2002 estimate can be attributed to the successful implementation of the 1993 *Grizzly Bear Recovery Plan*, prepared by the US Fish and Wildlife Service, which outlined three recovery goals for the US grizzly population (Servheen, 1993). By 2002 all plan goals were achieved (Gunther, 2003), and in a somewhat controversial move, the Yellowstone grizzly population was removed from the Endangered Species List in late 2005 (*USA Today*, 2005). In the next example we investigate the effect that implementation of the recovery plan had on the growth rate of the population.

Example 1.2: Recognizing that the grizzly population must have grown at a faster rate than 1% per year, estimate the actual annual growth rate from 1993 to 2002.

The basic setup for the problem is the same as before—the difference now is that instead of a growth rate of 1%, the growth rate is unknown. If we call this unknown growth rate r, then our flow diagram will look as it does in Figure 1.12.

The diagram tells us that in order to find the population of grizzlies in year t, we take the population in the previous year and increase that value by r. The DDS is then

$$P(t) = P(t-1) + rP(t-1).$$

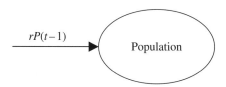

FIGURE 1.12   General exponential growth model.

We still know that in 1993 (when $t = 0$) there were 197 bears, and now we also know that in 2002 (when $t = 9$) there were 416 bears. Translating this information into our notation, we write $P(0) = 197$ and $P(9) = 416$. Our task is to find the value for the growth rate, r, that results in $P(9) = 416$.

E.2 Absolute Addressing

In this Excel section we show how to insert rows, how to work with parameters using absolute cell addressing, and how to use Excel for trial-and-error estimates.

We need to make some modifications to our original grizzly bear spreadsheet to make it easier for us to find r. To this end, instead of typing the growth rate 0.01 into the formula for population, we store the growth rate as a parameter in its own cell and then refer to that cell in the population formula. Excel then uses whatever value is in the referenced cell when it computes the population formula, allowing us to vary the growth rate with ease. Each time we type in a new value for the growth rate, Excel will automatically recalculate all of the population values without us having to change the actual formulas.

In order to make room for the cell where we have stored the growth rate parameter, we must insert rows into the spreadsheet between the title and our original work. This is accomplished by clicking on the Insert drop-down menu that is located in the Cells group of the Home tab. We then select "Insert Sheet Rows." The new setup with 1% entered for the growth rate parameter is shown in Figure 1.13. (To get the growth rate to appear as a %, use number formatting under the Home tab.)

	A	B	C
1	Yellowstone Grizzly Population		
2			
3	Growth rate, $r =$		1%
4			
5	t	Population	
6	0	197	
7	1	199	
8	2	201	
9	3	203	

FIGURE 1.13 Growth rate parameter stored in its own cell.

Now instead of typing in 0.01 when we create the formula for the population, we refer to the location of the parameter, that is, cell C3. There is, however, a catch. Since we want Excel to always refer to C3 to get the growth rate, we are not going to want Excel to automatically update that address when we copy our formula down. The remedy for this is to type dollar signs in front of values we do not want Excel to change. Thus, instead of typing "C3," we type "$C$3." This is referred to as **absolute addressing** or **absolute referencing**. Our finished formula appears in Figure 1.14.

Copying the formula down and displaying all of the formulas yields Figure 1.15. Notice that the absolute reference to cell C3 remains unchanged in all of the formulas while all other references are automatically updated by Excel.

	A	B	C
1	Yellowstone Grizzly Population		
2			
3	Growth rate, $r =$		1%
4			
5	t	Population	
6	0	197	
7	1	=B6+$C$3*B6	
8	2		
9	3		

FIGURE 1.14 Growth rate cell referenced with absolute addressing.

	A	B
1	Yellowstone Grizzly	
2		
3	Growth rate, $r =$	
4		
5	t	Population
6	0	197
7	=A6+1	=B6+$C$3*B6
8	=A7+1	=B7+$C$3*B7
9	=A8+1	=B8+$C$3*B8
10	=A9+1	=B9+$C$3*B9

FIGURE 1.15 Absolute addressing prevents cell address from changing.

Even though this spreadsheet is set up differently from our original, it is computing the same values, and in fact, the numerical answers at the moment are identical to the ones in Figure 1.9. The crucial difference in the new setup is that each time we type a new growth rate into cell C3, Excel automatically recalculates all of our formulas without us having to type in a new DDS.

What remains to do is to experiment with different values of the growth rate until we find one that leads to 416 bears after 9 years. Because we have already made the effort to set the spreadsheet up properly, all we need to do is keep typing different values into cell C3 until we get what we want.

First we try $r = 5\%$ by typing the value 5% into cell C3. Note in Figure 1.16 that once we press Enter, Excel instantly recalculates all of the populations to reflect the change in the growth rate.

	A	B	C
1	Yellowstone Grizzly Population		
2			
3	Growth rate, $r =$		5%
4			
5	t	Population	
6	0	197	
7	1	207	
8	2	217	
9	3	228	
10	4	239	
11	5	251	
12	6	264	
13	7	277	
14	8	291	
15	9	306	

FIGURE 1.16   Excel output automatically updates with change in growth rate.

Cell B15 tells us that a 5% growth rate would lead to 306 bears in 2002, and so we must have guessed too low. Next we try $r = 10\%$. This growth rate would produce 465 bears in 2002, so it is too high. After a few more tries, we arrive at the correct growth rate, about $r = 8.65\%$. Depending on rounding there may be some slight variation in the value for r. To contextualize our result, what we have learned from our model is that due to conservation efforts, the Yellowstone grizzly bear population grew by an average of about 8.65% per year from 1993 to 2002 rather than by 1% as was estimated in 1993. □

For future reference we mention another Excel shortcut for use with absolute addressing. Recall that instead of typing the address of a cell into a formula, we can just click on the cell itself and its address appears in the formula. To get Excel to enter an *absolute* address into the formula, press "F4" after clicking on the cell. This causes Excel to automatically add the "$'s" without us having to type them.

The next example asks us to use our model to project the effects of the conservation efforts into the future and compare the result with more recent estimates for the grizzly population.

Example 1.3: Assuming that the grizzly population continued to grow by 8.65% per year, what would the bear population be in 2005? How does that compare with the estimate of over 600 bears given in a November 15, 2005, *USA Today* article (*USA Today*, 2005)?

E.3 Multiple Formulas

In this Excel section we discuss how to work efficiently with multiple formulas.

To solve this problem we need to drag our formulas for the year and population down a few more rows to get to the year 2005, or $t = 12$. A convenient shortcut for doing so is to first use the thick cross pointer to select both cells that contain formulas, in this case A15 and B15. (Be careful not to use the thin cross for this since that will copy the formula from A15 to B15.) Next, we use the thin cross to grab the fill handle in the bottom right corner of the highlighted area and drag it down three more rows. Excel should copy and automatically update both formulas simultaneously, and what we should see is shown in Figure 1.17.

	A	B	C
1	Yellowstone Grizzly Population		
2			
3	Growth rate, $r =$		8.65%
4			
5	t	Population	
6	0	197	
7	1	214	
8	2	233	
9	3	253	
10	4	275	
11	5	298	
12	6	324	
13	7	352	
14	8	383	
15	9	416	
16	10	452	
17	11	491	
18	12	533	

FIGURE 1.17   Grizzly population predictions 1993–2005.

Our model with an estimated growth rate of 8.65% predicts approximately 533 bears for 2005. Since the *USA Today* article reported the population to be over 600, we see that the growth rate improved even more and that conservation efforts continued to be successful. □

1.1.2   Counting Yellowstone Grizzlies

The population estimates that we have used in this section are not easy to obtain. It might seem like they ought to be, but consider what counting an entire population of bears entails. First of all, to count bears directly, one must be able to spot all of the grizzlies over difficult, forested terrain; second, one must be careful not to count bears more than once.

The estimates given by Mattson and Gunther were produced through a combination of direct counting, field observations about the structure of grizzly populations, and some basic mathematics. Field scientists have found that the easiest bears to count are adult females who have newborn cubs. As groups they are easier to sight than single bears, and repeat counting is more easily avoided because litters are unique in the number and coloring of the cubs present. Since it is also known that adult female grizzlies breed every 3 years, the number of females with cubs is counted for a 3-year period, thus ensuring with some reliability that all adult females are counted exactly once. Next, the number of known adult female deaths is subtracted. Finally, adult females are known to comprise roughly 27.4% of grizzly populations. Consequently, the net total of adult females is divided by 0.274 to get the overall population number (Gunther, 2003).

As an example we look at the specifics of how the 2002 estimate of 416 was computed.

Example 1.4: In the year 2000, there were 35 adult females with newborn cubs sighted; in 2001, there were 42; and in 2002, there were 50, bringing the 3-year total to 127. During this same time period, there were 13 known deaths of adult females, so the adjusted total is $127 - 13 = 114$. Knowing that adult females account for about 27.4% of the total population means that $0.274 \times (\text{total population}) = 114$ adult females. Dividing both sides of the equation by 0.274 gives us our total: total population $= \frac{114}{0.274} = 416.05$. This estimate is considered a minimum population estimate because of the difficulties in sighting bears mentioned above. □

To reinforce what we have discussed so far, we now consider the case of another endangered species, the California condor.

1.1.3   California Condors

In April 1996, the US Fish and Wildlife Service, Pacific Region, prepared a document entitled *Recovery Plan for the California Condor* (Kiff, Mesta, & Wallace, 1996).

Section G of that document briefly reviews some of the historical estimates for the size of the condor population, and it indicates that the population was declining since at least as far back as the 1930s or 1940s. Fred Sibley (Sibley, 1969) determined that 50–60 condors were in existence in the late 1960s, while Sanford Wilbur estimated that by 1978 the number had dropped to 25–30 (Wilbur, 1980). Since we would like to deal in specifics, we assume conservatively that the population was 50 in 1968 and 25 in 1978.

Example 1.5: What was the average annual rate of decline for the condor population from 1968 to 1978?

To solve this problem we develop a mathematical model for the situation like we did for the grizzly population. We set up our notation, develop a flow diagram, translate the diagram into an equation, and finally implement the model with Excel. The notation for this problem is the same as we have been using, but we set it up explicitly for emphasis:

- t = years since 1968 (the independent variable).
- P = the population of condors in California (the dependent variable).
- $P(t)$ = function notation for the population of condors t years after 1968.
- $P(0)$ = the initial population of condors, so $P(0) = 50$.
- r = the annual rate of decline (presently an unknown parameter).

To create the flow diagram, recall that we represent any dependent variable by an oval. (In this case, there is only one—the condor population.) We then represent any additions or subtractions to the population by arrows leading in or out of the oval as appropriate. In this situation, the condor population is decreasing, so we draw an arrow leaving the population oval and label it with the (unknown) amount of decrease. The result is Figure 1.18.

FIGURE 1.18   Flow diagram for condor population with unknown rate of decrease.

Once we have a carefully constructed flow diagram, finding the DDS is relatively straightforward. In this case, the diagram says that to find the condor population in any year, we take the previous year's population and subtract r times that value. Thus, our DDS is given by

$$P(t) = P(t-1) - rP(t-1).$$

Until we use Excel to find a value for r, this is as far as we can go.

The setup we use for our spreadsheet is the same as for Example 1.2 where we stored the parameter r in its own cell. Referencing the cell where r is located with absolute addressing allows us to easily experiment with different values for r. The screenshot in Figure 1.19 was taken just after the formula for the DDS was entered but before copying it down. The rate of decline is entered as 10% as a temporary stand-in value.

	A	B	C
1	California Condor Population		
2			
3	Rate of decline, $r=$		10.00%
4			
5	t	Population	
6	0	50	
7	1	=B6-$C$3*B6	
8	2		

FIGURE 1.19   Condor population Excel model.

Time in this problem starts in 1968, and we need the population to turn out to be 25 in 1978. In other words, we need $P(10) = 25$. After copying the year and population formulas down to year 10, we experiment with different values for r until we find the value that produces $P(10) = 25$. As we see in Figure 1.20, the correct value for r is about 6.6%.

Our result tells us that under our model assumptions the population of condors decreased by about 6.6% each year from 1968 to 1978. □

	A	B	C
1	California Condor Population		
2			
3	Rate of decline, $r=$		6.6%
4			
5	t	Population	
6	0	50	
7	1	47	
8	2	44	
9	3	41	
10	4	38	
11	5	36	
12	6	33	
13	7	31	
14	8	29	
15	9	27	
16	10	25	

FIGURE 1.20   Condor rate of decline found via trial and error.

In this section we have studied two populations of endangered, or formerly endangered, species. The grizzly bear population was growing and the condor population declining, but under our assumptions each was changing by a fixed percentage every year. The models for both populations can be captured with the single formula:

$$P(t) = P(t-1) + rP(t-1).$$

If r is positive, we have exponential growth as we did in the case of the grizzlies, and if r is negative, we have exponential decline as exhibited by the condors.

1.1.4 Section Exercises

1 Consider the flow diagram in Figure 1.21.
   a. Find the corresponding DDS.
   b. Use a calculator to predict the population after 2 years if $P(0) = 50$.
   c. Use Excel to project the population in year 10.

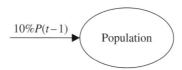

FIGURE 1.21 Flow diagram for Exercise 1.

2 Consider the flow diagram in Figure 1.22.
   a. Find the corresponding DDS.
   b. Use a calculator to predict the population after 2 years if $P(0) = 300$.
   c. Use Excel to project the population in year 10.

FIGURE 1.22 Flow diagram for Exercise 2.

3 Consider the flow diagram in Figure 1.23.
   a. Find the corresponding DDS.
   b. Use a calculator to predict the population after 2 years if $P(0) = 100$.
   c. Use Excel to project the population in year 10.

FIGURE 1.23 Flow diagram for Exercise 3.

4   Consider the flow diagram in Figure 1.24.

   a. Find the corresponding DDS.

   b. Use a calculator to predict the population after 2 years if $P(0) = 500$.

   c. Use Excel to project the population in year 10.

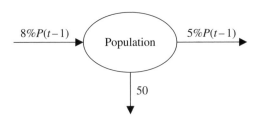

FIGURE 1.24   Flow diagram for Exercise 4.

5   Draw a flow diagram that corresponds to the following DDS:

$$P(t) = P(t-1) + 4\%P(t-1).$$

6   Draw a flow diagram that corresponds to the following DDS:

$$P(t) = P(t-1) - 0.05P(t-1).$$

7   Draw a flow diagram that corresponds to the following DDS:

$$P(t) = P(t-1) - 0.30P(t-1).$$

8   Draw a flow diagram that corresponds to the following DDS:

$$P(t) = 1.05 \cdot P(t-1).$$

9   Give the flow diagram and corresponding DDS for a grizzly population that is growing by 8% per year and has five bears illegally poached annually.

10   Give the flow diagram and corresponding DDS for a population that has a birth rate of 5% per year and a death rate of 2% per year.

11   Suppose you know that the DDS for a population is given by

$$P(t) = P(t-1) + 3\%P(t-1) - 50.$$

a. Draw a flow diagram that would lead to this DDS.

b. Explain in a complete sentence how the population is changing from year to year.

12 In Example 1.2 based on an initial population estimate of 197 Yellowstone grizzlies in 1993 and a later estimate of 416 Yellowstone grizzlies in 2002, we found that the population grew by about 8.65% per year.

a. Using the 8.65% growth rate, what would the exponential model predict for the grizzly population in the year 2193?

b. What does your answer in part a say about the long-term validity of the exponential growth model for the grizzly population?

13 Suppose that the *1993 Grizzly Bear Recovery Plan* had never been implemented and that the 1993 estimate of a 1% growth rate continued to hold. How long would it have taken for the population to reach 416 bears?

14 Based on the 1993 estimate of 197 for the total population of Yellowstone grizzlies, how many adult females were there in 1993?

15 Suppose that the numbers of adult females with cubs sighted in Yellowstone were 52 in 2003, 60 in 2004, and 65 in 2005. Estimate the total grizzly population in 2005.

16 Suppose that the goal of the National Park Service is for the Yellowstone grizzly population to reach 1000 bears by the year 2020. Are the current conservation efforts sufficient to reach this goal? Explain how you arrived at your conclusion.

17 Table 1.2 contains more population data for the wild California condor population from the 1996 *Recovery Plan for the California Condor* (US Fish and Wildlife Service, 1996).

a. Compare the population values in the table to what our model would predict using the rate of decline found in Example 1.5 and an initial population of 50 condors. In general, how well did our model do?

b. Can you think of possible reasons for any discrepancies?

TABLE 1.2 The Number of California Condors Remaining in the Wild between 1982 and 1985 (US Fish and Wildlife Service, 1996)

Year	Number Wild California Condors
1982	21
1983	19
1984	15
1985	9

18 Based on the Table 1.2, what value for *r* would give the best predictions for the condor population? Explain precisely how you made your determination.

19 Recall that our estimate for the California condor's rate of decline was based on the lower population estimates given by Sibley, Mailed, and Wilbur. Reestimate the rate of decline from 1968 to 1978 using three other combinations from the population estimates:

  a. The lower value from 1960s and the higher value from 1978.

  b. The higher value from 1960s and the lower value from 1978.

  c. The higher value from 1960s and the higher value from 1978.

  d. How much difference do you see in r?

20 *Extension:* The models in this chapter used a single growth rate, r, to describe the change in a population from 1 year to the next. This growth rate represents a combination of all factors influencing the population including births and deaths. In some situations it is more useful to include separate parameters for the population's birth rate and death rate.

  a. Give the flow diagram that represents the annual change in a population if b is the annual birth rate and d is the annual death rate.

  b. Find the DDS for the new model.

  c. Implement the new model in Excel.

  d. Find the population in year 30 for a population that starts at 100, has a birth rate of 5%, and has a death rate of 3%.

1.2 EXPONENTIAL GROWTH WITH STOCKING OR HARVESTING

Whether intentionally or unintentionally, humans often have an impact on wildlife populations. In this section we look at intentional influence, and we see how to incorporate the effects of such influence into our population models from the previous section. The two types of influence we investigate here are **harvesting**, the systematic removal of members from a population, and **stocking**, the systematic addition of members to a population.

1.2.1 Stocking Mississippi Sandhill Cranes

Based on data presented in *Mississippi Sandhill Cranes* (Gee & Hereford, 1993) and *Recovery Plan: Mississippi Sandhill Crane* (Valentine & Lohoefener, 1991), the Mississippi sandhill crane population was 50 in 1980 and was declining at an average rate of approximately 6% per year.

Example 1.6: Based on these estimates, when would the crane population become extinct without some kind of intervention?

At the moment there is nothing new for us in this problem; we just have a population that is declining exponentially. First we create a flow diagram, shown in Figure 1.25, and from the diagram we formulate our DDS.

FIGURE 1.25   Flow diagram for sandhill crane population.

The arrow leaving the population indicates a subtraction so our DDS is given by

$$P(t) = P(t-1) - 6\% P(t-1).$$

Now that we have the DDS, we set up an Excel spreadsheet to handle the calculations. As is our custom, we store all parameters in their own cells and refer to them using absolute addressing. This takes a little longer to set up in the beginning, but if we need to change any of our values later, it will keep us from having to redo the entire worksheet. Figure 1.26 shows a screenshot of our sandhill crane spreadsheet with the first population formula displayed.

	A	B	C
1	Sandhill Crane Population		
2			
3	Rate of decline, $r =$		6.0%
4			
5	t	Population	
6	0	50	
7	1	=B6-$C$3*B6	
8	2	44.2	
9	3	41.5	
10	4	39.0	

FIGURE 1.26   Sandhill crane population Excel model.

Before proceeding we need to decide when to consider the population extinct. Of course we know that if the population falls to zero, then we have an extinct population; however, an exponential model will never actually produce a population of exactly zero. A reasonable way around this is to declare the population extinct once it drops below 1 crane. Now the solution to our problem involves copying the formula for our model down while looking for the first year where the population falls below one. Doing so reveals the prediction that it would take about 65 years for the population to become extinct. □

In 1973 the Mississippi sandhill crane was added to the *US List of Endangered Fish and Wildlife*, and in 1975 the Mississippi Sandhill Crane National Wildlife Refuge was established (US Fish and Wildlife Service, 2011). Stocking efforts began at the Patuxent Wildlife Research Center in 1965 and involved a program of **hacking**: removing eggs from the wild for chicks to be captive-reared and then released back into the population (US Fish and Wildlife Service, 2011). By 1980 there were enough captive-reared cranes to begin stocking, and in 1981, the first cohort of captive-reared cranes was released into the wild population (Gee & Hereford, 1993). Stocking efforts have continued ever since in what is now the largest crane release program in the world. The history of the development of the stocking effort is an interesting one; the brief description below is from Gee and Hereford (Gee & Hereford, 1993).

> The first releases of hand-reared birds failed. Thus, releases of Mississippi sandhills on the refuge during the 1980s were birds raised by their parents or surrogate parents. These parent-reared birds proved wilder than the hand-reared birds and adapted well to the pine savanna. Unfortunately, the parent-rearing technique reduced production and increased expenses.
>
> The PWRC developed a new hand-rearing technique that visually isolated chicks from humans and imprinted them on adult sandhill cranes in the chick-rearing area. Caretakers dressed in sheets to hide their human form when handling birds, and encounters with cranes were limited. Juveniles were placed in socialization pens in the fall to form three cohorts (parent-reared, hand-reared, and a mixed group). A gentle release on the refuge allowed the birds to leave the release pen when ready and to return for food for a period after release. Surprisingly, a greater percentage of hand-reared birds has survived than the parent-reared birds, although both groups have paired and produced fertile eggs.

There were 9 cranes in the original release of captive-reared birds. Our next problem asks us to determine the effect that continuing to add 9 cranes every year would have on the population. For now we assume the same annual rate of decline of 6%.

Example 1.7: Predict what the population of cranes would be in 1993 if 9 cranes were introduced into the population every year.

We must first realize that this situation calls for a fundamentally different kind of model, one that incorporates a constant influx of cranes each year. We assume the original 6% rate of decline. We also have 9 birds being added each year, and we account for this in the flow diagram as an arrow pointing into the population oval. The result is the flow diagram given in Figure 1.27.

FIGURE 1.27   Flow diagram for sandhill crane population with hacking.

To translate the diagram into a DDS, we must keep in mind what it is telling us. The diagram says that to calculate the population in 1 year, we must start with the previous year's population, subtract 6% from it, and add 9 to it. Thus, our new DDS is

$$P(t) = P(t-1) - 0.06P(t-1) + 9.$$

Now we must implement our model in Excel. Looking ahead to the possibility of wanting to change the number of cranes introduced each year, we store the number 9 in its own cell and refer to it with absolute referencing. The new model, with the formula already typed in, is shown in Figure 1.28.

	A	B	C
1	Sandhill Crane Population		
2			
3	Rate of decline, r =		6.0%
4	Stocking num., a =		9
5			
6	t	Population	
7	0	50	
8	1	=B7-$C$3*B7+$C$4	
9	2		
10	3		

FIGURE 1.28 Sandhill crane population Excel model with hacking.

To complete the problem, we now have to copy our formula down to year 13 and report the result. After doing so we see that the model predicts about 105 cranes in 1993. □

Our initial model predicts a population of 105 cranes in 1993, but the reintroduction program actually raised the total population to 135 (Gee & Hereford, 1993). This better-than-predicted increase could be due to several factors, including better survivorship of cranes (both captive-reared and wild), or the addition of more than an average of 9 cranes per year. We assume for now that the only relevant factor is the average addition of more than 9 cranes every year. Our next problem is to determine how many cranes per year were necessary to achieve 135 in 1993.

Example 1.8: Determine approximately how many cranes on average were captive-reared and released annually between 1981 and 1993.

We do not yet know how many birds were introduced, so we will use the letter "a" to denote this unknown parameter. Figure 1.29 is the modified flow diagram for this scenario.

Translating the diagram into a DDS yields

$$P(t) = P(t-1) - 6\%P(t-1) + a.$$

We now take advantage of our foresight in storing the number of captive-released birds as a parameter. Allowing Excel to automatically recalculate the worksheet

FIGURE 1.29 Flow diagram for sandhill crane population with unknown hacking number.

each time, we try different values for a until the 1993 population is at least 135. It may take a few tries to find the right number, but we should end up with needing to introduce an average of about 12.23 cranes per year. Since we are looking for an *average* value, keeping the decimal number is fine. However, if we insist on introducing the same number of birds each year, we should round up to 13 to make sure we do not fall short of the goal of 135. □

The previous example examined a population in decline that was assisted by systematic stocking. Next we introduce an example of what can be considered the opposite problem—a population that is growing exponentially and requires systematic harvesting to keep it from growing too large.

1.2.2 Harvesting White-Tailed Deer in the Northeast

In the early 1900s, the population of white-tailed deer had nearly vanished in the United States. Since then their numbers have increased to the point of overpopulation, and they have become a serious problem for foresters and farmers (Palmer & Storm, 1995). This overabundance has been caused in part by the extirpation of natural predators such as wolves and mountain lions (McCullough, 1997). Also contributing to deer overpopulation is the artificial confinement of populations due to barrier fences and development (McCullough, 1997). An increase in mixed-use land combined with the increase in the deer population has led to increases in crop damage, forest damage, and deer–vehicle collisions (Palmer & Storm, 1995). Deer management, including the harvesting, or culling, of deer, is seen by many as an essential part of stabilizing deer populations and minimizing harmful interactions between deer and humans. One of the important questions, then, is, how many deer should be harvested in a particular area every year to maintain the population at a level that is healthy for all concerned?

It should be noted that not all agree with either the moral or scientific basis for the use of hunting as a method for controlling deer populations. Allen Rutberg writes (Rutberg, 1997):

> The scientific arguments in favor of deer management are commonly founded more on dogma than on data and more on intuition than on logic.
>
> In spite of repeated assertions of the fundamental dogmas and their corollary—sport hunting is necessary to prevent deer overpopulation—scientific tests are rare enough and counterexamples are common enough to raise doubts in the minds of both scientists

and thoughtful laypersons. The exponential rise in white-tailed deer populations in the United States during the last two decades makes a strong case that sport hunting has not controlled deer populations.

Hunting, as commonly practiced, may have profound effects on the age and sex structure of deer populations. Sport hunter preferences for shooting bucks skew sex ratios toward females, sometimes dramatically. This problem has been widely recognized in the management community, though principally as a concern for population productivity and hunter satisfaction, and some state wildlife agencies are acting to reduce the more extreme biases. However, less concern is shown about the population, behavioral, and genetic effects of heavy, early adult mortality.

Still, according to the National Biological Service, harvesting has been an effective tool in managing deer populations in states such as Pennsylvania that harvest considerably more female than male deer. In states such as Massachusetts, however, where many more male than female deer are harvested, the deer population has continued to increase (Palmer & Storm, 1995). The fact that the male/female harvesting ratio plays such an important role indicates that, once we understand the basic harvesting model, it would be worthwhile to consider a more sophisticated model that accounts for sex differences. For now, though, we will not make such a distinction.

Reliable estimates of growth rates are sometimes difficult to find. Occasionally we have to deduce the growth rate from the data that is available. According to Curtis and Sullivan, given a favorable environment, deer populations can double in as little as 2 or 3 years (Curtis & Sullivan, 2001). This is a common way of describing growth rates called the **doubling time**, and it is generally considered a more intuitive way of understanding growth. We will be conservative and assume that the deer population doubles every 3 years, and in the next problem we use Excel to help us translate this information into an estimate for the annual growth rate of white-tailed deer.

Example 1.9: Assuming that a white-tailed deer population without harvesting will double every 3 years, find the equivalent annual growth rate.

We are assuming that the population grows by a constant percentage each year; we just do not yet know what the percentage is. This puts us in a situation we have seen before—exponential growth with an unknown growth rate parameter, r. We know the DDS for the situation is given by $P(t) = P(t-1) + rP(t-1)$. As before we estimate the unknown growth rate parameter using trial and error in Excel. The difference is that this time we do not have an initial population to start with. The good news is that the growth rate we find will not depend on the starting population so we can use any value we like.

If we start with a million deer, for example, we have $P(0) = 1,000,000$. If the population doubles every 3 years, then we must have $P(3) = 2,000,000$. Our job is to find a value for r that produces $P(3) = 2,000,000$. Storing the growth rate in its own cell and referring to it with absolute addressing, we experiment with different rates until we find the one we need. Figure 1.30 shows a spreadsheet setup for this purpose.

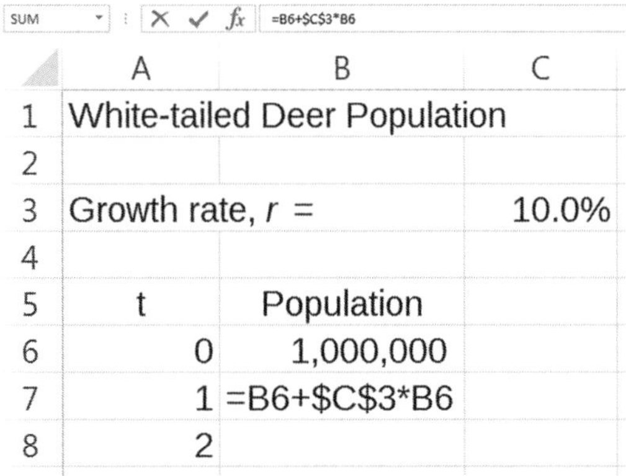

		SUM	▾	:	✕ ✓ f_x	=B6+$C$3*B6	

◢	A	B	C
1	White-tailed Deer Population		
2			
3	Growth rate, r =		10.0%
4			
5	t	Population	
6	0	1,000,000	
7	1	=B6+$C$3*B6	
8	2		

FIGURE 1.30 White-tailed deer population Excel model with unknown growth rate.

We type different growth rates into cell C3 until cell B9 equals 2,000,000. If we do not want to go out to three or four decimal places, the closest we can get is $r = 0.26$ or $r = 26\%$. So based on the Curtis and Sullivan estimate, a reasonable and fairly conservative growth rate for an unmanaged deer population is around 26% per year. It is worth taking a moment or two to try different values in cell B6 to verify that the initial population does not affect our answer. □

With a growth rate at 26% or even higher, an unmanaged deer population can quickly become a serious problem. A natural question, then, is, how many deer should be harvested each year to keep the population from becoming too large? At a more practical level, this question leads to issues such as how many hunting licenses should be issued and how long the hunting season should be. The definition of "too large" will vary depending on the point of view of the person being asked and on the particular needs of a state or region. For now we concentrate on harvesting enough deer so that the population stops increasing.

Our harvesting strategy is to harvest the same number of deer every year, and we investigate this scenario in the next example. For our initial population, we use the 1992 estimate of 3,000,000 deer in the Northeast given by Palmer and Storm (Palmer & Storm, 1995). Assuming a growth rate of $r = 26\%$, we answer the following.

Example 1.10: Determine the minimum number of deer that should be harvested in the Northeast each year in order to prevent the population from growing.

We need to build a mathematical model for the situation, and the first step in the process is to construct our flow diagram. We have an annual increase of 26% and an annual decrease due to harvesting that is constant but unknown. The diagram is similar to the one for the sandhill crane stocking example, and it appears in Figure 1.31.

FIGURE 1.31 Flow diagram for deer population with harvesting.

We once again use the letter "a" to stand for the unknown harvest number. Once we have the flow diagram constructed, we can immediately write down the DDS:

$$P(t) = P(t-1) + 26\%P(t-1) - a.$$

To implement the model in Excel, we once again take the trouble to store all of our parameters in their own cells so that the spreadsheet will be as flexible as possible. We show the setup of our spreadsheet in Figure 1.32 with an arbitrary number initially chosen for a.

	A	B	C
1	White-tailed Deer Population		
2			
3	Growth rate, r =		26.0%
4	Harvesting num., a =		1,000
5			
6	t	Population	
7	0	1,000,000	
8	1	=B7+$C$3*B7-$C$4	
9	2		

FIGURE 1.32 Deer population Excel model with harvesting.

Once the setup is complete, we drag our formula down a few years and experiment with different harvest numbers until the population no longer increases. We find that we must harvest at least $a = 780,000$ deer each year in order to prevent the population from increasing.

The actual number of deer harvested in the Northeast in 1992 is estimated to be 900,000 (Palmer & Storm, 1995). Thus, even this initial, relatively simple model has produced a result that compares reasonably well with reality. Unfortunately, the constant harvest approach that we have outlined above does have serious drawbacks. Some of these are investigated in the exercises. □

In this section we have taken our basic exponential model and added a component that takes into account possible human management of a population by harvesting or stocking. This additional component adds some measure of realism to our model, but it also adds some complexity. This is, unfortunately, the usual state of affairs in modeling: additional realism nearly always carries the price of extra complication.

As was the case for our exponential models, we can describe both of the cases presented in this section with a single DDS, namely,

$$P(t) = P(t-1) + rP(t-1) + a.$$

Models that have this general form are sometimes described as **affine** models. We describe populations that on their own would grow or decline by a fixed percentage each year by choosing r to be positive or negative, respectively. Similarly, we account for either stocking or harvesting by letting a be positive or negative, respectively. Note also that this DDS includes the basic exponential model as the special case where $a = 0$.

1.2.3 Section Exercises

1 Consider the flow diagram in Figure 1.33.

    a. Find the corresponding DDS.

    b. Use a calculator to predict the population after 2 years if $P(0) = 650$.

    c. Use Excel to project the population after 15 years.

FIGURE 1.33   Flow diagram for Exercise 1.

2 Consider the flow diagram in Figure 1.34.

    a. Find the corresponding DDS.

    b. Use a calculator to predict the population after 2 years if $P(0) = 30$.

    c. Use Excel to project the population after 15 years.

FIGURE 1.34   Flow diagram for Exercise 2.

3 Draw a flow diagram that corresponds to the following DDS:

$$P(t) = P(t-1) + 4\%P(t-1) - 40.$$

4 Draw a flow diagram that corresponds to the following DDS:

$$P(t) = P(t-1) - 0.05P(t-1) + 23.$$

5 *Extension*: In Example 1.8 we determined that an average of 12.23 Mississippi sandhill cranes were captive-reared and released annually between 1981 and 1993. Table 1.3 gives the actual numbers, taken from table 2 in Valentine and Lohoefener (1991), for the years 1981–1990.

a. How does 12.23 compare to the average number of cranes that were actually released?

b. What factor(s) might account for the difference?

c. Using a rate of decline of 6%, and the actual release values from 1981 to 1990, estimate how many cranes there were in 1990. (Note: this exercise will require a significant modification of the crane Excel spreadsheet.)

d. Determine, on average, how many cranes must have been released in 1991, 1992, and 1993 in order to end up with 135 cranes in 1993.

TABLE 1.3   The Number of Mississippi Sandhill Cranes Captive-Reared and Released between 1981 and 1990 (Valentine & Lohoefener, 1991)

Year of Release	Number Captive-Released
1981	9
1982	4
1983	8
1984	4
1985	10
1986	7
1987	2
1988	10
1989	13
1990	29

6 Regarding white-tailed deer, recall the Curtis and Sullivan (2001) estimate that deer populations can double every 2–3 years. In Examples 1.9 and 1.10, we based our Excel work on a doubling time of 3 years.

a. Rework Examples 1.9 and 1.10 in the text, this time assuming a doubling time of 2 years.

b. How does the new harvesting number compare to the previous estimate of 780,000?

c. Explain why the answer for b makes sense in the context of the problem.

7 When deciding on a real harvesting strategy, it is not just the total number harvested that is important. Rather, the sex ratio of the harvest number is also very important. In fact, Palmer and Storm write, "During the past decade, deer populations in the Northeast have continued to increase except in states that harvested markedly more antlerless than antlered deer" (Palmer & Storm, 1995). In a sentence or two, discuss why the sex ratio of deer harvests should play such an important role in controlling deer populations.

8 *Extension*: Consider a deer population model with harvesting where we would like to project the population of males and the population of females over time. What kinds of new information would we need for such a model? What would a flow diagram look like?

1.3   TWO FUNDAMENTAL EXCEL TECHNIQUES

In this section we take a look at some basic and powerful tools for visualizing our models and for finding unknown parameters. We begin by learning how to produce basic graphs in Excel.

E.4   Graphing in Excel

We can often glean valuable information from a model by visualizing it using a graph, and Excel makes creating graphs relatively easy. It takes some time and practice to learn how to use all of the many options available to us, but to create a basic, no-frills graph, the process only takes a few moments. The steps are:

1. Use the thick cross pointer to select all of the output we want to graph, including the column headings.
2. From the Charts group under the Insert tab, select "Insert Scatter (X,Y) or Bubble Chart" (the bottom right drop-down).
3. Select the particular type of graph desired, often "Scatter with Straight Lines and Markers."

Excel will automatically choose the title and scale for each axis based on the information selected. When the graph is selected, the Chart Tools group becomes active, with tabs for Design and Format. Just about any formatting change can be made from these two tabs. In particular, the Add Chart Element group under the Design tab provides many of the most common options, including customizing the axes, axis labels, and legend. We illustrate the creation of a basic graph with the following example.

Example 1.11: Graph the grizzly bear population for the years 1993–2002. Recall that the initial population was 197 in 1993 with an annual estimated growth rate of 1%.

Step one is to use the thick cross pointer to select all of the output that should appear on the graph, including the column headings. We are interested in seeing how the population changes over time, so our selection should include both the column for time and the column for population, as well as the column headings for each. Figure 1.35 shows the result of our selection just before clicking on the type of scatter graph we want, in this case "Straight Lines with Markers."

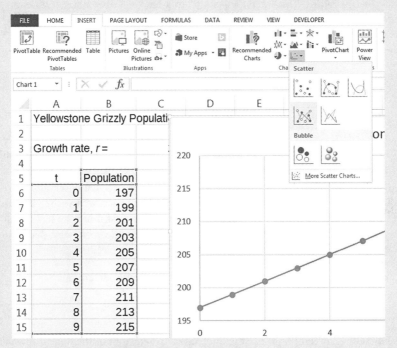

FIGURE 1.35   Creating graphs in Excel.

After clicking the graph type the graph will appear in the worksheet. The graph by itself appears in Figure 1.36.

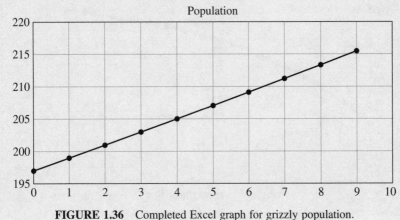

FIGURE 1.36   Completed Excel graph for grizzly population.

Note what Excel has done. The time values have automatically been placed on the horizontal, or X-axis, and the values for population have been placed on the vertical, or Y-axis. The title has been automatically created using our column heading "Population." Excel creates a graph just as we would by hand—by plotting points and then connecting the dots—but Excel is, of course, much better at it. $\square$

We do a second example to reinforce our understanding.

Example 1.12: Assuming that the original conditions in Example 1.8 hold all the way to 2015, graph the Mississippi sandhill crane population for the years 1980–2015. What can you say about the population trend based on the graph?

We open our sandhill crane spreadsheet, which should already be set up correctly. We still assume an annual rate of decline of 6% per year and an initial population of 50 in 1980. For the number being added each year, we use the value from Example 1.8, namely, $a = 12.23$ cranes per year. The first thing we have to do is drag our formulas down a few more rows until we get to the year 2015, that is, until $t = 35$. Once we do that, we follow the basic steps for graphing:

1. Select the output to be graphed including column headings.
2. Select the Chart group from the Insert tab.
3. Select the desired type of graph.

For step 1, remember to use the thick cross to select all of the data you want to graph including the column headings. We should end up with the graph given in Figure 1.37.

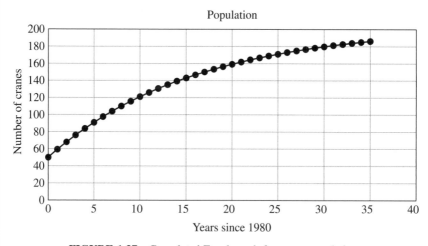

FIGURE 1.37 Completed Excel graph for crane population.

Here we have taken advantage of the features in the Add Chart Element group under the Design tab and added descriptive axis labels to the graph. Labeling axes clearly is a good habit to get into. Doing so allows someone unfamiliar with the problem to still understand the graph, and it can serve as a refresher if one has not looked at the graph in a while.

The graph shows that the crane population continues to increase over the years, but the downward bend in the graph indicates that this increase is slowing down over time. We could have discovered this feature of the population by carefully examining the numerical output in the population column, but this is an example of something that a graph allows us to see easily that numerical investigations would not. □

A nice feature of Excel graphs is that they automatically update when we change the value of a relevant parameter. Try changing the number of cranes added each year and watch what happens to the graph.

E.5 Goal Seek

Up until now we have concentrated on using Excel to solve a variety of population problems numerically. When faced with finding an unknown parameter such as a growth rate, our method has been to store the parameter in its own cell, refer to it with absolute referencing, and try different parameter values until we find the right one. It would be nice if we could find the correct parameter value without the tedium of the guess and check method, and fortunately Excel has the capability to do this for us in a couple of different ways. The method we introduce now is to use an Excel feature called **Goal Seek**. We show how Goal Seek works by revisiting Example 1.2.

Example 1.13: Recall that the population of grizzlies was 197 in 1993 and 416 in 2002. Use Goal Seek to determine an estimate for the growth rate of the Yellowstone grizzly population from 1993 to 2002.

We start by opening the spreadsheet we used to solve the original problem. Recall that we have stored the unknown growth rate as a parameter in its own cell and referred to it using absolute referencing so that Excel will not automatically update the cell address. Instead of using trial and error to find the growth rate that produces $P(9) = 416$, we use the Goal Seek feature of Excel to find the rate.

First we select the Data tab. From the Data Tools group, we select the What-If Analysis pull-down menu and select Goal Seek as shown in Figure 1.38.

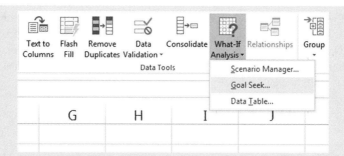

FIGURE 1.38   Locating Goal Seek in Excel.

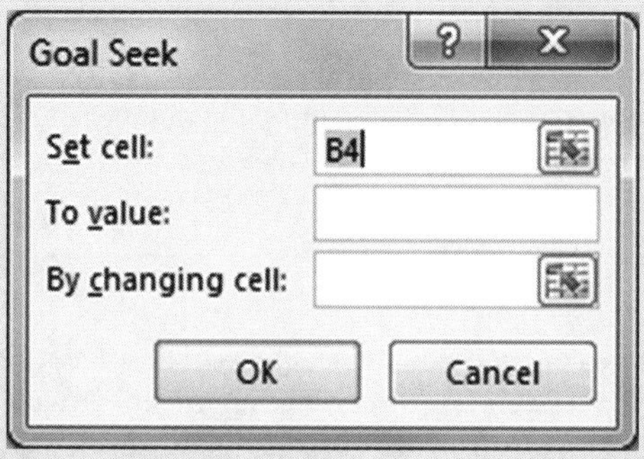

FIGURE 1.39   The Goal Seek dialog box.

Once we click on Goal Seek, the dialog box pictured in Figure 1.39 appears. The default address in the "Set cell" box is the cell that happens to be selected in the worksheet when the Goal Seek tool is opened. We have three values to enter in the dialog box:

1. In the "Set cell" box, we type the cell address where our target value resides. For this example we enter B15 since that is where the goal population for 2002 resides.
2. In the "To value" box, we type in our goal for cell B15. In this case, we want it to be 416, so we type in 416.
3. In the "By changing cell" box, we enter the cell address of the parameter whose value we need to change in order to achieve our goal. In this case, the parameter is the growth rate, and it lives in cell C3.

After we enter this information, the Goal Seek box should appear as it does in Figure 1.40.

It may help to interpret the Goal Seek box as a complete sentence. What we are saying in the Goal Seek box is "Set cell B15 to value 416 by changing cell C3," or equivalently, "Make cell B15 equal 416 by varying cell C3." A third

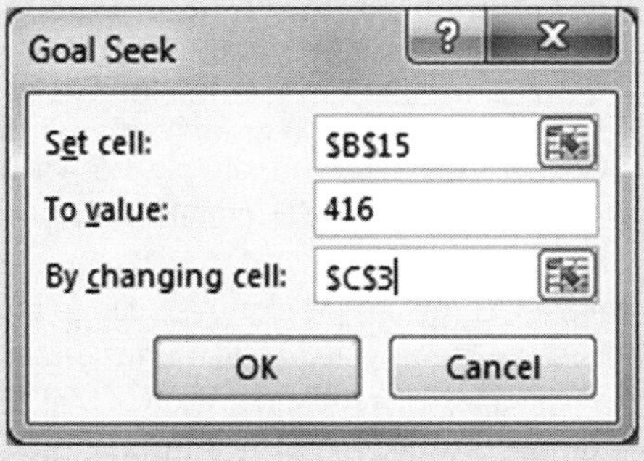

FIGURE 1.40 Filled-in Goal Seek dialog box for Example 1.13.

and more concrete version would be, "Arrange for the grizzly population to be 416 in 2002 by finding the right growth rate."

Once we click OK, Excel instantly finds the growth rate that causes our model to produce 416 grizzlies in 2002. The result is seen in Figure 1.41.

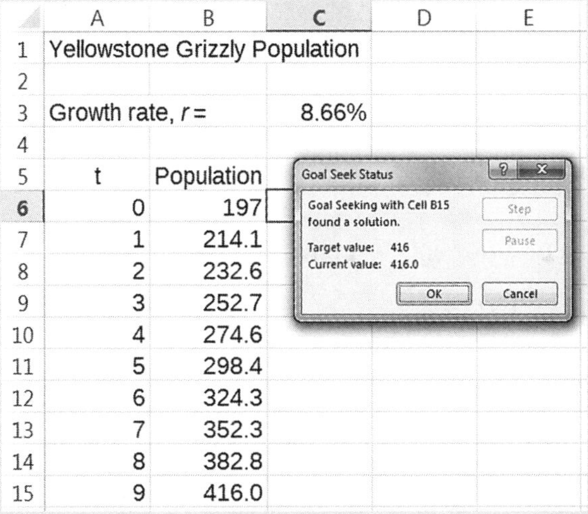

	A	B	C	D	E
1	Yellowstone Grizzly Population				
2					
3	Growth rate, *r* =		8.66%		
4					
5	t	Population			
6	0	197			
7	1	214.1			
8	2	232.6			
9	3	252.7			
10	4	274.6			
11	5	298.4			
12	6	324.3			
13	7	352.3			
14	8	382.8			
15	9	416.0			

FIGURE 1.41 The result of a successful Goal Seek for Example 1.13.

The "Goal Seek Status" box that appears verifies that Goal Seek was successful. Just click OK to go on. Note that cells C3 and B15 have automatically been changed to reflect the assigned population goal and the growth rate required to produce it. Also note that the value $r = 8.66\%$ agrees with the value $r = 8.65\%$ that we originally found using trial and error. The slight difference is due to Excel keeping more decimal places in the value for the goal. □

Next we rework another problem to make sure we know how to use the Goal Seek command.

Example 1.14: Use Goal Seek to rework Example 1.5 where we were asked to find the annual rate of decline for a condor population that was 50 in 1968 and 25 in 1978.

Since we are returning to a previous problem, there is no need to go through the setup of the flow diagram and DDS again. All that is necessary is to use Goal Seek on the spreadsheet we created for the problem instead of trial and error. Figure 1.42 shows the Goal Seek dialog box correctly filled in.

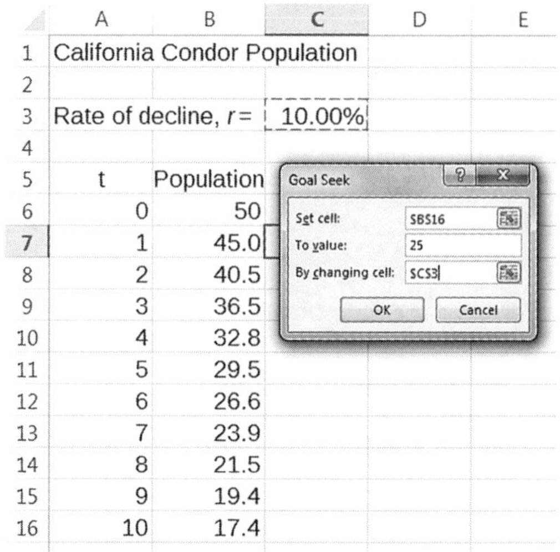

	A	B	C	D	E
1	California Condor Population				
2					
3	Rate of decline, $r=$		10.00%		
4					
5	t	Population			
6	0	50			
7	1	45.0			
8	2	40.5			
9	3	36.5			
10	4	32.8			
11	5	29.5			
12	6	26.6			
13	7	23.9			
14	8	21.5			
15	9	19.4			
16	10	17.4			

Goal Seek dialog box:
Set cell: $B$16
To value: 25
By changing cell: $C$3
OK   Cancel

FIGURE 1.42   Filled-in Goal Seek dialog box for Example 1.14.

Once we click okay, we should see both the goal and the parameter value automatically update to the desired values. Minor rounding differences notwithstanding, Goal Seek has produced the same value, $r = 6.7\%$, that we found using the somewhat clumsier method of trial and error. □

Goal Seek can be a very powerful tool when working with DDS models, and we will make extensive use of it throughout the text.

In this section we introduced two basic Excel techniques: graphing and Goal Seek. Graphs provide a way of quickly understanding some basic behavior of a model, such as whether a quantity is increasing or decreasing and whether the increase or decrease is speeding up or slowing down. We also saw how to label a graph and observed that a nice property of Excel graphs is that they automatically update to reflect changes we make to relevant parts of the spreadsheet.

This section was also devoted to introducing Goal Seek, a powerful tool for finding unknown parameter values. Goal Seek will save us a lot of time and tedium as we will no longer have to rely on the guess and check method for finding parameters.

1.3.1 Section Exercises

1 Consider the DDS below:

$$P(t) = P(t-1) + 0.10P(t-1) - 50.$$

    a. Graph the population over a period of 10 years if $P(0) = 550$.

    b. Graph the population over a period of 10 years if $P(0) = 450$.

    c. Describe the difference in the behavior of the population in the two cases.

2 Consider the DDS below:

$$P(t) = P(t-1) - 0.10P(t-1) + 50.$$

    a. Graph the population over a period of 10 years if $P(0) = 550$.

    b. Graph the population over a period of 10 years if $P(0) = 450$.

    c. Describe the difference in the behavior of the population in the two cases.

3 Describe the difference in population behavior between Exercises 1 and 2.

4 For each of the graphs in Exercises 1 and 2, describe the population behavior as "increasing at an increasing rate," "increasing at a decreasing rate," "decreasing at an increasing rate," or "decreasing at a decreasing rate."

5 Sketch a graph by hand for each of the following situations.

    a. A population where $P(0) = 100$ and the population is increasing over time at an increasing rate.

    b. A population where $P(0) = 200$ and the population is increasing over time at a decreasing rate.

    c. A population where $P(0) = 50$ and the population is decreasing over time at a decreasing rate.

    d. A population where $P(0) = 100$ and the population is decreasing over time at an increasing rate.

6 Consider the DDS given by $P(t) = P(t-1) + 0.10P(t-1) - a$. If the initial population is 500, use Goal Seek to determine the value for a that results in a population of 600 12 years later.

7 Consider the DDS given by $P(t) = P(t-1) + rP(t-1) - 50$. If the initial population is 400, use Goal Seek to determine the value for r that results in a population of 800 10 years later.

8 *Extension*: In Example 1.8 we determined that an average of 12.23 Mississippi sandhill cranes were captive-reared and released annually between 1981 and

1993. Table 1.3 gives the actual numbers, taken from table 2 in Valentine and Lohoefener (1991), for the years 1981–1990.

a. How does 12.23 compare to the average number of cranes that were actually released?

b. What factor(s) might account for the difference?

c. Using a rate of decline of 6%, and the actual release values from 1981 to 1990, estimate how many cranes there were in 1990. (Note: this exercise will require a significant modification of the crane Excel spreadsheet.)

d. Use Goal Seek to determine, on average, how many cranes must have been released in 1991, 1992, and 1993 in order to end up with 135 cranes in 1993.

9 Regarding white-tailed deer, recall the Curtis and Sullivan (2001) estimate that deer populations can double every 2–3 years. In Examples 1.9 and 1.10, we based our Excel work on a doubling time of 3 years.

a. Use Goal Seek to rework Examples 1.9 and 1.10 in the text, this time assuming a doubling time of 2 years.

b. How does the new harvesting number compare to the previous estimate of 780,000?

c. Explain why the answer for b makes sense in the context of the problem.

1.4   EXPLICIT FORMULAS

So far our study of population models has been largely computational. In this chapter we take a more analytic approach and try to learn as much as possible about how the models will behave without having to resort to Excel.

Our goal is to see how much we can learn just by a careful examination of our DDS formulas. The main idea is to repeatedly apply the DDS formula until a pattern emerges; we then use the pattern to write down an algebraic formula for our model known as an **explicit formula**. While the algebra in this section can be somewhat involved, the payoff will be a powerful new tool for studying our models.

1.4.1   Explicit Formula for Exponential Growth

We return to the first type of DDS we encountered—one that changed exponentially, either by growing or declining by a fixed percentage each year. Recall that the general version of the DDS for such a situation was found to be

$$P(t) = P(t-1) + rP(t-1),$$

where r is the annual rate of change so that a declining population will have a negative value for r. Originally we did not simplify the DDS. Instead, we concentrated on the setup of the model and its implementation in Excel. Now we do simplify the expression slightly because doing so will make our subsequent work easier. The simplification is a matter of factoring out the common term $P(t-1)$. Beginning with the DDS, we get

$$P(t) = P(t-1) + rP(t-1)$$
$$= 1 \cdot P(t-1) + rP(t-1)$$
$$= (1+r)P(t-1).$$

We now have a more compact version of our DDS: $P(t) = (1+r)P(t-1)$. This is often the form that is presented in discussions of exponential models.

Next, starting in year 1, we apply this simplified version of the DDS over and over, looking for a pattern to emerge. Our simplified DDS tells us that to find the population after 1 year, we compute

$$P(1) = (1+r)P(0).$$

To find the population after 2 years, we calculate

$$P(2) = (1+r)P(1).$$

Here is where we do something different. From the first step, we have an alternate expression for $P(1)$, namely, $P(1) = (1+r)P(0)$. Thus, we can substitute for $P(1)$ on the right-hand side of the expression for $P(2)$. The details of the substitution are

$$P(2) = (1+r)P(1)$$
$$= (1+r)[(1+r)P(0)]$$
$$= (1+r)^2 P(0).$$

Thus, we now know $P(2) = (1+r)^2 P(0)$ and we repeat the process.

From the DDS we know that to compute the population after 3 years, we must compute

$$P(3) = (1+r)P(2).$$

But from our recent work, we know that $P(2) = (1+r)^2 P(0)$. As before we substitute this expression for $P(2)$ on the right-hand side to get

$$P(3) = (1+r)P(2)$$
$$= (1+r)\left[(1+r)^2 P(0)\right]$$
$$= (1+r)^3 P(0).$$

So now we know that $P(3) = (1+r)^3 P(0)$. We may see a pattern emerging, but we do one more example to be sure.

We know that $P(4) = (1+r)P(3)$ so we make a substitution and write

$$P(4) = (1+r)P(3)$$
$$= (1+r)\left[(1+r)^3 P(0)\right]$$
$$= (1+r)^4 P(0).$$

To summarize our findings so far, we have found that

$$P(1) = (1+r)P(0);$$
$$P(2) = (1+r)^2 P(0);$$
$$P(3) = (1+r)^3 P(0); \text{and}$$
$$P(4) = (1+r)^4 P(0).$$

Based on these observations, we can with some confidence write down how to find $P(t)$ for *any* value of t, namely,

$$P(t) = (1+r)^t P(0).$$

This is our first example of an explicit formula for a DDS, and it differs from the DDS in important ways. Perhaps the most important difference is that using the DDS to calculate the population in 1 year requires that we first compute the population in the previous year; with the explicit formula there is no such limitation. We may calculate the population in any year directly by plugging the correct value for t into the formula. The next example shows us how the explicit formula provides us with an alternate way to answer a previous problem.

Example 1.15: Use the explicit formula to rework Example 1.1 without resorting to Excel.

For the grizzly bear population in Example 1.1, we had $r = 1\%$ and $P(0) = 197$ in 1993. We need to find the population 9 years later in 2002, so $t = 9$. Since we have the explicit formula $P(t) = (1+r)^t P(0)$ at our disposal, we plug in all of the necessary information on the right-hand side to get

$$P(9) = (1+0.01)^9 \cdot 197;$$
$$P(9) = 1.0937 \cdot 197 \approx 215.$$

It is worth looking back at our first solution to verify that this is the same result we achieved by applying the DDS nine times. $\square$

Oftentimes in working problems dealing with exponential models, it will be up to us whether to work with the DDS or the explicit formula. This will largely be a matter of personal preference, and we will see below that each approach has its pros and cons.

1.4.2   The Geometric Series Formula

Before we proceed to finding the explicit formula for exponential growth with stocking or harvesting, we need to lay some groundwork. In particular, we will need the **geometric series formula**, a way of quickly computing long sums of numbers that follow the pattern

$$c + cx + cx^2 + cx^3 + \cdots + cx^{n-1}.$$

In other words, the sum is made up of powers of some number, x, where each term is multiplied by the same constant, c. The notation "$+ \cdots +$" means that we include all terms in between. When a sum is in the form of a geometric series, the geometric series formula gives us an efficient way of summing the terms. This formula is given as Theorem 1.1.

Theorem 1.1   Given a sum that is in the form of a geometric series, we have

$$c + cx + cx^2 + cx^3 + \cdots + cx^{n-1} = c\frac{x^n - 1}{x - 1}.$$

The message contained in the formula is that instead of finding the sum by adding term by term, we can just compute the right-hand side. This can be a great time saver as we show in the examples below.

Example 1.16: Use the geometric series formula to compute the sum

$$3 + 3 \cdot 2 + 3 \cdot 2^2 + 3 \cdot 2^3 + \cdots + 3 \cdot 2^{12}.$$

The hard part is recognizing the pattern and seeing that $c = 3$, $x = 2$, and $n - 1 = 12$ (so $n = 13$). The rest is a matter of plugging these values into the right-hand side of the formula:

$$3 + 3 \cdot 2 + 3 \cdot 2^2 + 3 \cdot 2^3 + \cdots + 3 \cdot 2^{12} = 3\frac{2^{13} - 1}{2 - 1}$$
$$= 3\frac{8192 - 1}{1}$$
$$= 3 \cdot 8191 = 24{,}573.$$

Instead of having to add up 13 individual numbers, we get away with only a few computations. It is worth verifying such a result at least once by adding up all of the individual terms. $\square$

Even if we are patient enough not to mind adding the 13 numbers in the previous example, surely we can agree that in the next example the geometric series formula is the way to go.

Example 1.17: Compute the sum $2 + 2\left(\frac{8}{9}\right) + 2\left(\frac{8}{9}\right)^2 + 2\left(\frac{8}{9}\right)^3 + \cdots + 2\left(\frac{8}{9}\right)^{99}$.

Here we recognize that $c = 2$, $x = \frac{8}{9}$, and $n - 1 = 99$, so $n = 100$. Thus, the entire sum can be found by computing $2\frac{\left(\frac{8}{9}\right)^{100} - 1}{\frac{8}{9} - 1} \approx 17.99986$. $\square$

Though we have used the geometric series formula productively, we should note that we have not explained why it is true. It is not difficult to find an explanation with a quick search of the web, and we also provide a brief account in Appendix A. We should note that it is not always straightforward to apply. Often expressions are not written in a form that is easy to recognize as a geometric series, and the real difficulty is knowing *when* to apply the formula, not *how* to apply the formula.

In the next section we see how to use the geometric series formula to develop an explicit formula for our affine models.

1.4.3 Explicit Formula for Harvesting and Stocking

The algebra involved in finding an explicit formula for exponential growth with harvesting or stocking is somewhat more complicated than in the basic exponential case. However, the method is the same: apply the DDS repeatedly and look for a pattern.

Recall that the DDS for exponential growth with stocking or harvesting has the general form

$$P(t) = P(t-1) + rP(t-1) + a.$$

where r is the growth rate and a is the number that is being added to or subtracted from the population each year. If a is positive we are stocking, while if a is negative we are harvesting.

Our first step is to simplify the DDS just as we did for the exponential model to get

$$P(t) = (1+r)P(t-1) + a.$$

For $t = 1$ we have $P(1) = (1+r)P(0) + a$, and for $t = 2$ we have $P(2) = (1+r)P(1) + a$. Note that just like we did for the exponential model, we can make a substitution for $P(1)$ on the right-hand side of the second equation. Doing so yields

$$P(2) = (1+r)P(1) + a$$
$$= (1+r)[(1+r)P(0) + a] + a.$$

If we multiply through by $(1+r)$ on the right, we get

$$P(2) = (1+r)^2 P(0) + (1+r)a + a.$$

Similarly, consider the DDS for $t = 3$:

$$P(3) = (1+r)P(2) + a.$$

Though it becomes slightly messy, we substitute for $P(2)$ on the right-hand side, multiply through by $(1+r)$, and arrive at

$$P(3) = (1+r)P(2) + a$$
$$= (1+r)\left[(1+r)^2 P(0) + (1+r)a + a\right] + a$$
$$= (1+r)^3 P(0) + (1+r)^2 a + (1+r)a + a.$$

There is the hint of a pattern emerging, which we highlight by summarizing the results in Table 1.4.

TABLE 1.4   Looking for a Pattern for the Affine Model

t	$P(t)$
1	$P(1) = (1+r)P(0) + \{a\}$
2	$P(2) = (1+r)^2 P(0) + \{(1+r)a + a\}$
3	$P(3) = (1+r)^3 P(0) + \left\{(1+r)^2 a + (1+r)a + a\right\}$

We do one more iteration, $t = 4$. The DDS tells us that

$$P(4) = (1+r)P(3) + a.$$

Substituting in for $P(3)$ and multiplying through by $(1+r)$ gives

$$P(4) = (1+r)P(3) + a$$
$$= (1+r)\left[(1+r)^3 P(0) + (1+r)^2 a + (1+r)a + a\right] + a$$
$$= (1+r)^4 P(0) + (1+r)^3 a + (1+r)^2 a + (1+r)a + a.$$

We now have one more row to add to Table 1.5.

TABLE 1.5   Continuing to Look for a Pattern for the Affine Model

t	$P(t)$
1	$P(1) = (1+r)P(0) + \{a\}$
2	$P(2) = (1+r)^2 P(0) + \{(1+r)a + a\}$
3	$P(3) = (1+r)^3 P(0) + \left\{(1+r)^2 a + (1+r)a + a\right\}$
4	$P(4) = (1+r)^4 P(0) + \left\{(1+r)^3 a + (1+r)^2 a + (1+r)a + a\right\}$

The pattern seems to be that we get an exponential term (identical to the one from our first explicit formula) followed by a geometric series inside the curly brackets. If this pattern holds for all t, we should have that

$$P(t) = (1+r)^t P(0) + \left\{(1+r)^{t-1} a + \cdots + (1+r)^2 a + (1+r)a + a\right\}.$$

To simplify the geometric series part, we must recognize that the constant, "a," plays the role of "c," that "$t-1$" plays the role of "$n-1$," and that "$1+r$" plays the role of "x."

We apply the geometric series formula and finally arrive at the explicit formula for our affine DDS:

$$P(t) = (1+r)^t P(0) + a\frac{(1+r)^t - 1}{(1+r) - 1}.$$

A slight simplification gives the form we will most often use:

$$P(t) = (1+r)^t P(0) + a\frac{(1+r)^t - 1}{r}.$$

In the following section we show how our explicit formulas provide us with an alternate way of solving some of our previous problems.

1.4.4 Applying New Tools to Old Problems

In this section we do not tackle any new problems, but we solve some of our previous problems in a new way—this time relying on our explicit formulas rather than Excel.

Example 1.2 asked us to determine the annual growth rate for the Yellowstone grizzly population between 1993 and 2002. The way we first solved the problem was to implement our DDS in Excel and experiment with different growth rates until we achieved the desired 2002 population. As a second method, we also used Excel's Goal Seek tool. Now we present a third option, this time solving the problem without Excel. We will need a calculator—any scientific calculator will do, including the one built in to most computers or smart phones.

Example 1.18: Rework Example 1.2 by applying the appropriate explicit formula.

Recall that there were an estimated 197 bears in 1993 and 416 bears in 2002. Thus, $P(0) = 197$, and we need to find a value for r that will produce $P(9) = 416$. The appropriate explicit formula to use is the one for an exponential DDS: $P(t) = (1+r)^t P(0)$. In this example we know that $t = 9$ so we have $P(9) = (1+r)^9 P(0)$.

We also know that $P(0) = 197$, so we have

$$P(9) = (1+r)^9 \cdot 197.$$

Since $P(9) = 416$, we plug in 416 for $P(9)$ and solve for r, the only remaining unknown. This will involve taking ninth roots of both sides of an equation. We get

$$416 = (1+r)^9 \cdot 197$$
$$\frac{416}{197} = (1+r)^9$$
$$2.112^{1/9} \approx \left[(1+r)^9\right]^{1/9}$$
$$1.0866 \approx 1 + r$$
$$0.0866 \approx r.$$

We have found that $r = 8.66\%$, a result that agrees both with our original trial-and-error approach and with the slicker Goal Seek method. $\square$

Using the explicit formula to solve a harvesting or stocking problem is more computationally complex, but the methods are exactly the same as in the previous example.

Example 1.19: Rework Example 1.7 from Section 1.2.1 using the appropriate explicit formula.

Recall that the situation in Example 1.7 was a sandhill crane population declining at a rate of 6% per year, so $r = -0.06$. The population in 1980 was $P(0) = 50$, and in an attempt to preserve the population, $a = 9$ captive-reared cranes were being added to the population each year. We need to find $P(13)$, the population in 1993. Noting that $t = 13$, we plug everything we know into the explicit formula for stocking and use a calculator to compute the population. The initial setup will be

$$P(t) = (1+r)^t P(0) + a \frac{(1+r)^t - 1}{r}$$

$$P(13) = (1-0.06)^{13} \cdot 50 + 9 \frac{(1-0.06)^{13} - 1}{-0.06}.$$

From this point it is a matter of being careful with our calculator and the order of operations. We get

$$P(t) = (0.94)^{13} \cdot 50 + 9 \frac{(0.94)^{13} - 1}{-0.06}$$

$$= 0.447 \cdot 50 + 9 \cdot \frac{-0.553}{-0.06}$$

$$= 22.35 + 9 \cdot 9.217$$

$$= 105.303 \approx 105.$$

Once again we have about 105 cranes in 1993. $\square$

We do one more example.

Example 1.20: Rework Example 1.8 using the appropriate explicit formula.

The task in Example 1.8 was to find how many cranes should be added to the population each year in order to end up with 135 birds in 1993. This amounts to solving for the unknown a in the explicit formula, where $r = -0.06$, $P(0) = 50$, $t = 13$, and $P(13) = 135$. We plug in everything we know and solve for a. The initial setup is

$$P(t) = (1+r)^t P(0) + a\frac{(1+r)^t - 1}{r}$$

$$135 = (1-0.06)^{13} \cdot 50 + a\frac{(1-0.06)^{13} - 1}{-0.06}.$$

Next we carefully work through the numerical calculations until we are able to solve for the unknown a:

$$135 = (0.94)^{13} \cdot 50 + a\frac{(0.94)^{13} - 1}{-0.06}$$

$$135 = 22.35 + 9.21 \cdot a$$

$$112.65 = 9.21 \cdot a$$

$$12.23 \approx a.$$

This is the same answer that Excel gave us earlier, and once again if forced to round, we would round up to $a = 13$ to ensure the goal of 135 cranes is met. $\square$

This section was dedicated to finding explicit formulas for the DDS models we have used so far. The algebra became somewhat involved for the harvesting and stocking models, but the result of our hard work is that we now have formulas that allow us to compute populations for any year without first having to calculate all previous years' populations. Table 1.6 summarizes our results.

TABLE 1.6 A Comparison of the DDS and Explicit Formulas for the Exponential and Affine Models

	DDS	Corresponding Explicit Formula
Exponential model	$P(t) = P(t-1) + rP(t-1)$	$P(t) = (1+r)^t P(0)$
Affine model	$P(t) = P(t-1) + rP(t-1) + a$	$P(t) = (1+r)^t P(0) + a\dfrac{(1+r)^t - 1}{r}$

Even after only a handful of examples, we can begin to see some of the strengths and weaknesses of the different approaches. For someone comfortable with algebra and perhaps less so with Excel, the explicit formula approach would seem the clear leader. For someone who enjoys working with Excel or who likes to avoid algebra whenever possible, the trial-and-error or Goal Seek method makes more sense. Either method is fine as long as it is appropriate for the problem that needs to be solved.

While it is true that there are times when an explicit formula will allow us to answer questions that Excel cannot easily answer, we should issue a very important caution against growing too fond of explicit formulas. We noted, particularly in the affine derivation, that the algebra involved in finding an explicit formula in the first place can quickly become complicated. In fact for most of the DDS models that we will encounter, it is impossible (or at the very least unreasonable) to find an explicit formula at all. This is why Excel is such a natural choice as our primary method of dealing with DDS models: Excel allows us to get results that would otherwise be intractable.

1.4.5 Section Exercises

For all of the exercises below, use the appropriate explicit formula to find the solution.

1. Consider the DDS given by $P(t) = P(t-1) + 0.10P(t-1)$. Determine the population in year 5 if the initial population is 400.

2. Consider the DDS given by $P(t) = P(t-1) - 0.10P(t-1)$. Determine the population in year 8 if the initial population is 300.

3. Consider the DDS given by $P(t) = P(t-1) + 0.10P(t-1) - 5$. Determine the population in year 5 if the initial population is 400.

4. Consider the DDS given by $P(t) = P(t-1) - 0.20P(t-1) + 30$. Determine the population in year 8 if the initial population is 1000.

5. Consider the DDS given by $P(t) = P(t-1) + 0.10P(t-1) - a$. If the initial population is 500, determine the value for a that results in a population of 600 12 years later.

6. Consider the DDS given by $P(t) = P(t-1) + 0.20P(t-1) - 50$. Determine the initial population if the population in year 10 is 400.

7. Given the initial population estimate of 197 Yellowstone grizzlies in 1993 and the later estimate of 416 Yellowstone grizzlies in 2002, we found that the population grew by about 8.65% per year.
   a. Using the 8.65% growth rate, what would your model predict for the population in the year 2193?
   b. Does your answer in part a seem reasonable? Why or why not?
   c. Suppose that the 1993 Grizzly Bear Recovery Plan had never been implemented and that the 1993 estimate of a 1% growth rate continued to hold. How long would it have taken for the population to reach 416 bears?

8. Regarding white-tailed deer, recall the Curtis and Sullivan estimate that deer populations can double every 2–3 years. We based our Excel work on a doubling time of 3 years.
   a. Rework Examples 1.9 and 1.10 in the text, this time assuming a doubling time of 2 years.
   b. How does the new harvesting number compare to the previous estimate of 780,000?
   c. Explain why the answer in b makes sense in the context of the problem.

9. Consider the explicit formula for our harvesting/stocking model. Show that if there is no stocking or harvesting, then the formula is the same as the explicit formula for plain exponential growth.

10. According to the Population Reference Bureau (PRB), the world population was 7.2 billion in 2014. The PRB projects a population of 9.7 billion in the year 2050 (PRB, 2014). Most of the world's population growth is expected to occur in developing nations, particularly in Africa.

    a. Assuming an exponential model for the world's population growth through 2050, determine the growth rate given the PRB estimates.

    b. Using the growth rate you found in a, what does your model predict for the world population in 2055?

11 The US Census in 2000 (see www.census.gov) estimated the population of the United States to be 281.4 million. Without immigration, the population would grow by approximately 0.6% each year. Data available at www.census.gov indicates that approximately 1,000,000 immigrants enter the United States each year.

    a. Create a flow diagram for the US population.

    b. From the flow diagram, give the DDS.

    c. Use the explicit formula to predict the US population in the year 2050.

    d. How does your projection compare to the 419.9 million projected for 2050 by the US Census?

12 *Extension*: Suppose a population of cranes declines by 6% per year without human intervention. A hacking program is begun with 100 cranes added the first year. Each subsequent year the number of cranes added increases by 10%. Thus, in year 2, 110 cranes are added, and in year 3, 121 are added.

    a. Give a flow diagram for the crane population.

    b. Give the corresponding DDS.

    c. Implement the model in Excel where the growth rate, r, the initial hacking number, a_0, and the annual percentage increase in the hacking number, s, are all stored as parameters.

    d. Project the crane population in year 10 if initially there are 400 cranes.

13 *Extension*: Following the spirit of the derivation for the explicit formula for an affine model, find the explicit formula for the general model described in Exercise 12. Confirm your result in 12(d) by using the explicit formula.

1.5 EQUILIBRIUM VALUES AND STABILITY

In this section we take an analytic approach to studying the global behavior of our models. In particular, we are interested in understanding their long-term behavior without having to resort to Excel calculations.

1.5.1 Equilibrium Values

Fundamental to the study of a DDS is finding its **equilibrium values**. An equilibrium value is a number, which we denote by P^* in the context of population, at which the DDS does not change. In other words, P^* is an equilibrium value if setting $P(t-1) = P^*$ results in $P(t) = P^*$ also.

For the models we have seen so far, we can find the equilibrium values numerically by appealing to Excel or algebraically by using the DDS. Below we look at examples of each approach.

We begin by finding the equilibrium value for the Mississippi sandhill crane population modeled in Example 1.7. In that problem we assumed that the crane population was naturally declining at a rate of 6% per year and that on average 9 cranes were added to the population each year.

Example 1.21: Using the information in Example 1.7, find the equilibrium value for the crane population.

First we find the required value via trial and error in Excel. We use the spreadsheet that we developed for the crane population, and we experiment with different values for the initial population until we find a value for P^*. This amounts to searching for an initial population that causes all subsequent populations to stay at that value. A few moments spent trying different values for $P(0)$ yield the result that $P^* \approx 150$. The resulting spreadsheet is shown in Figure 1.43. Notice that the population is no longer changing.

	A	B	C
1	Sandhill Crane Population		
2			
3	Rate of decline, $r =$		6.0%
4	Stocking num., $a =$		9
5			
6	t	Population	
7	0	150	
8	1	150.0	
9	2	150.0	
10	3	150.0	
11	4	150.0	
12	5	150.0	
13	6	150.0	
14	7	150.0	

FIGURE 1.43 Sandhill crane population at equilibrium.

An easy mistake to make at this point is to see that the population values are not changing and think that this must mean no cranes are entering or leaving the population. This is not the case. Cranes are still dying and still being added; it is just that the additions and subtractions are exactly balancing each other out.

Finding equilibrium values by trial and error can be tedious, especially since Goal Seek is no help in these situations. A more efficient, and more exact, approach is to use the DDS for the problem and a little bit of algebra. Recall that the DDS for the crane population is given by

$$P(t) = P(t-1) - 0.06P(t-1) + 9.$$

The defining property for an equilibrium value, P^*, is that plugging P^* in for $P(t-1)$ results in $P(t) = P^*$. This leads us to substitute P^* into both sides of the DDS to get the equation

$$P^* = P^* - 0.06P^* + 9.$$

All that remains is to solve for P^*:

$$P^* = P^* - 0.06P^* + 9$$
$$0 = -0.06P^* + 9$$
$$0.06P^* = 9$$
$$P^* = \frac{9}{0.06} = 150.$$

This is the same value we found using Excel. We can always check that our equilibrium value is correct by plugging it into Excel as the initial population. If we have really found an equilibrium value, the population will stay at that value. □

Another nice property of finding the equilibrium value algebraically is that it is no more difficult to find in the general case than it is in particular examples. We supply the details below for the general affine model.

Recall that the general DDS for exponential growth with stocking or harvesting is given by

$$P(t) = P(t-1) + rP(t-1) + a.$$

To find the equilibrium value for such a system, we proceed as before. We know that if we plug in P^* on the right, then we must also get P^* on the left. Thus, we have that

$$P^* = P^* + rP^* + a.$$

Solving for P^*, we get

$$P^* = P^* + rP^* + a$$
$$0 = rP^* + a$$
$$-rP^* = a$$
$$P^* = \frac{a}{-r}$$
$$P^* = -\frac{a}{r}.$$

We see that finding an equilibrium value for such a model turns out to be a relatively straightforward calculation—just divide the harvesting or stocking number by the growth rate and be careful with the minus signs.

1.5.2 Stability

Finding the equilibrium values for a DDS is not the end of the story. In fact there are many different types of equilibrium values, and we begin looking at some of these different types now.

We start by considering two main types of equilibrium values: those that are **stable** and those that are **unstable**. It is possible for an equilibrium value to be a bit of both, but for now we stick with just the two types. A **stable equilibrium** is an equilibrium value where the DDS tends toward that value even if we start away from it. An **unstable equilibrium** is an equilibrium value where the DDS tends away from that value if we start away from it. We illustrate the different types using two familiar examples: the sandhill cranes and the white-tailed deer.

Example 1.22: Consider the sandhill crane population that we assume is naturally declining by 6% each year and which we stock with 9 cranes per year. We have already found that $P^* = 150$. Determine the stability of this equilibrium value.

The question is whether this equilibrium is stable or unstable. In other words, if the population of cranes begins at a value other than 150, will the population tend toward 150 (stable) or further away from it (unstable)? We answer the question with Excel by graphing the population over time with several different initial populations.

E.6 Working with Multiple Columns

In this Excel section we show how to copy formulas across columns and how to graph multiple columns simultaneously.

First we open the crane spreadsheet and create four additional columns for population as in Figure 1.44. Note in Figure 1.44 that we have entered a different initial population in each column, including the equilibrium value for one of them.

	A	B	C	D	E	F
1	Sandhill Crane Population					
2						
3	Rate of decline, r =	6.0%				
4	Stocking num., a =	9				
5						
6	t	Pop. 1	Pop. 2	Pop. 3	Pop. 4	Pop. 5
7	0	130	140	150	160	170
8	1	131.2				
9	2	132.3				
10	3	133.4				

FIGURE 1.44 Sandhill crane Excel model with multiple population columns.

Next we use the thin cross and fill handle to copy our formulas from Population 1 over to the other columns. Excel automatically updates the formulas just as it does when we copy them down; only here it automatically updates columns rather than rows. Once we copy the formulas over, we copy them all down at once. In this example we copy them down to year $t = 50$. Figure 1.45 shows our setup with the formulas displayed for the first two initial populations.

	A	B	C
1	Sandhill Crane Pop		
2			
3	Rate of decline, r =		0.06
4	Stocking num., a =		9
5			
6	t	Pop. 1	Pop. 2
7	0	70	110
8	=A7+1	=B7-$C$3*B7+$C$4	=C7-$C$3*C7+$C$4
9	=A8+1	=B8-$C$3*B8+$C$4	=C8-$C$3*C8+$C$4
10	=A9+1	=B9-$C$3*B9+$C$4	=C9-$C$3*C9+$C$4
11	=A10+1	=B10-$C$3*B10+$C$4	=C10-$C$3*C10+$C$4
12	=A11+1	=B11-$C$3*B11+$C$4	=C11-$C$3*C11+$C$4

FIGURE 1.45   Sandhill crane Excel model with population DDS copied across.

Next we graph all five populations on the same axes by selecting all columns (including the column for time) as well as the column headings. Using a scatter graph with straight lines but no markers, we produce the graph given in Figure 1.46.

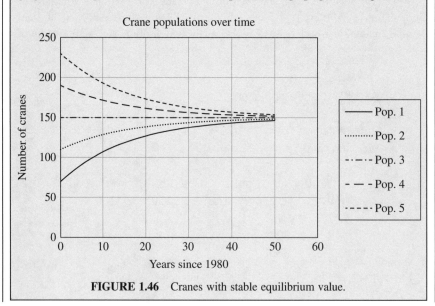

FIGURE 1.46   Cranes with stable equilibrium value.

Note what is happening in Figure 1.46. No matter what value we select for the initial population of cranes, all populations tend *toward* the equilibrium value of 150. This is an example of a stable equilibrium—even if we begin off of the equilibrium value, the population still tends toward it. ☐

In the next example we highlight the difference between a stable and an unstable equilibrium.

Example 1.23: Consider the white-tailed deer population that we assume is naturally growing by 26% each year and from which we harvest 780,000 deer per year. Find the equilibrium value for the deer population and determine its stability.

Our previous work has shown us that finding P^* is relatively painless. We compute $P^* = -\frac{a}{r} = -\frac{-780,000}{0.26} = \frac{780,000}{0.26} = 3,000,000$. Thus, the equilibrium value in this example is 3,000,000 deer.

To see if it is stable or unstable, we create a graph as we did for the crane population that shows the graphs for several different initial populations of deer. In the graph below we use initial populations of 2,500,000, 2,750,000, 3,000,000, 3,250,000, and 3,500,000. We copy all of our formulas down to year 6 and give the completed graph in Figure 1.47.

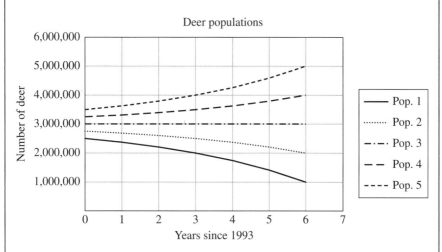

FIGURE 1.47   Deer with unstable equilibrium value.

Note that unlike the crane populations, the deer populations tend to move *away* from the equilibrium value, except for the one that started right on it. Because populations that start off of the equilibrium tend to move away from it, we call the equilibrium for the deer population unstable. ☐

In this section we introduced the notion of an equilibrium value (where a DDS stays constant), and we discussed two main types of equilibrium values: stable and unstable. Stability is determined by what happens when a population starts away from the equilibrium value. Stable equilibrium values are marked by having populations that start off the equilibrium and tend toward them, while unstable equilibrium values have populations that start off the equilibrium and move further away from them. Note that what happens at the equilibrium value is the same in both cases: if you plug an equilibrium value into your Excel model, the population will stay at that value. Staying at the equilibrium value does not make it stable; it just confirms that it is an equilibrium.

1.5.3   Section Exercises

1   Consider the DDS given by $P(t) = P(t-1) + 0.05P(t-1) - 10$.

    a. Find all equilibrium values for the DDS.

    b. Use Excel to confirm that the values found in a are in fact equilibrium values.

    c. Determine the stability of any equilibrium values found in a by producing an appropriate Excel graph.

2   Consider the DDS given by $P(t) = P(t-1) - 0.05P(t-1) + 10$.

    a. Find all equilibrium values for the DDS.

    b. Use Excel to confirm that the values found in a are in fact equilibrium values.

    c. Determine the stability of any equilibrium values found in a by producing an appropriate Excel graph.

3   *Extension*: Consider the DDS given by $P(t) = P(t-1) + 0.004(5 - P(t-1)) \cdot P(t-1)$.

    a. Find all equilibrium values for the DDS.

    b. Use Excel to confirm that the values found in a are in fact equilibrium values.

    c. Determine the stability of any equilibrium values found in a by producing an appropriate Excel graph.

4   *Extension*: Consider the DDS given by
$P(t) = P(t-1) + 0.0004(50 - P(t-1)) \ (P(t-1) - 10) \cdot P(t-1)$.

    a. Find all equilibrium values for the DDS.

    b. Use Excel to confirm that the values found in a are in fact equilibrium values.

    c. Determine the stability of any equilibrium values found in a by producing an appropriate Excel graph.

5   *Extension*: Consider the DDS given by
$P(t) = P(t-1) + 0.05 \left(1 - \frac{P(t-1)}{10,000}\right) P(t-1) - 125$.

    a. Find all equilibrium values for the DDS.

    b. Use Excel to confirm that the values found in a are in fact equilibrium values.

    c. Determine the stability of any equilibrium values found in a by producing an appropriate Excel graph.

6 Consider the Yellowstone grizzly population where the growth rate is assumed to be 8.65% and there is no stocking or harvesting. Show algebraically that the only equilibrium value for the population is zero.

7 Show that for *any* exponential model where there is no harvesting or stocking, the only equilibrium value is 0.

8 Consider the sandhill crane population where we assume that the rate of decline is $r = -0.06$. Find the number of cranes we would have to stock each year in order to make the long-term crane population turn out to be 400 cranes.

9 The US Census in 2000 (see www.census.gov) estimated the population of the United States to be 281.4 million. Without immigration, the population would grow by approximately 0.6% each year. Data available at www.census.gov indicates that approximately 1,000,000 immigrants enter the United States each year.

    a. Suppose that instead of growing by 0.6% per year, the US population was declining by 0.6% each year. Give the DDS for this situation.

    b. At what value would the US population stabilize in the long run?

    c. Produce a graph that indicates the US population would stabilize at this value no matter where it started.

    d. If the US government wanted to stabilize the population at 400,000,000, how many people should it allow to immigrate each year?

2

PERSONAL FINANCE

From calculating car payments to projecting retirement income to pricing stock options, mathematical modeling plays a vital role in the financial world. For many people the time during or soon after college is the time when they begin to assume financial independence. Bank accounts, credit cards, cars, houses, and retirement accounts become realities that must be addressed, and the more one understands about the underlying mathematics, the better prepared one will be able to address them.

In this chapter we begin to get a sense of not only the power of mathematical modeling but also the power of mathematical abstraction: the mathematics we developed in Chapter 1 in the context of population growth is the very mathematics that we will need to model many financial concepts. It is this wide applicability that makes the hard work of mathematical abstraction worth it.

In the sections that follow, we develop spreadsheet models for a variety of financial situations that students have already faced or likely soon will. By the end of the chapter, we will have created a set of financial tools that we can actually use to make decisions about our financial lives. While the goal of the present text is to introduce material that is as realistic and relevant as possible, some choices and simplifying assumptions have to be made to begin making progress. These assumptions will be made explicit in the text.

On the topic of relevance, we should note that no example will be equally relevant to every student's financial situation. Some students rely on student loans, some do not. Some have credit card debt, some do not. Some work long hours to pay for school,

Models for Life: An Introduction to Discrete Mathematical Modeling with Microsoft® Office Excel®,
First Edition. Jeffrey T. Barton.
© 2016 John Wiley & Sons, Inc. Published 2016 by John Wiley & Sons, Inc.

while some are fortunate enough not to have to. Even within broad categories of similarity, every financial situation is unique, and the static cases and examples presented here cannot hope to account for this diversity completely. There are, however, numerous exercises that invite the reader to personalize a situation by using his or her own salary, debt situation, taste in cars, or dream home.

We begin with a discussion of compound interest and savings accounts.

2.1   COMPOUND INTEREST AND SAVINGS

Money that is deposited into a bank savings account does not remain locked away in a vault. Rather the bank uses that money to try to make money, for example, by lending it out to other customers. Because the bank is using the money and not the account owner, the bank pays the account owner a small fee for allowing it the use of that money. This fee is called **interest** and is the reason depositing money into a bank savings account is better than stashing it under a mattress. We call the amount deposited into an account the **principal** and the amount that is in the account at any given time the **balance**.

We begin with an example of how interest is calculated.

Example 2.1: Suppose $10,000 is deposited into an account that earns 6% interest every year, credited to the account at the end of the year. Find the amount of interest earned and the balance in the account at the end of the year.

In this example the principal is the original deposit of $10,000. The interest rate is 6%, so at the end of the year, the account will be credited with

$$\$10,000 \times 6\% = \$10,000 \times 0.06 = \$600.00$$

in interest. The balance in the account at the end of the year is the principal plus the interest, or $10,600.00. □

When interest is left in an account so it can earn yet more interest, the interest is said to be **compounding**; this idea is fundamental to a great number of financial calculations. As we shall gradually come to see, **compound interest** is a powerful force that can dramatically influence the course of our finances in a good way when saving and in a bad way when borrowing. We examine how compounding works in the next example.

Example 2.2: Suppose the balance from the previous example is left in the account for another year. Find the balance at the end of the second year.

We start with $10,600. The account pays 6%, so at the end of the year, we will earn

$$\$10,600 \times 0.06 = \$636.00$$

in interest. The balance will be

$$\$10,600.00 + \$636.00 = \$11,236.00.$$

Notice that in the second year we earned more interest ($636.00) than we did in the first ($600.00). This difference is due to the effect of compounding: the $600.00 in interest we earned the first year was left in the account to earn interest during the second year. The $36.00 difference may seem insignificant, but over long periods of time, compounding can make a huge difference in our bottom line. ☐

Before doing any further examples, we take a moment to set up a discrete dynamical system (DDS) for our interest-bearing savings account. The balance in the account is our dependent variable, and the only factor influencing the balance is the annual interest deposit. When interest is paid once a year, we say that the interest is **compounded annually**. We let r stand for the interest rate, so for our previous examples, we have $r = 6\%$. We denote the balance in the account after t years by $B(t)$.

The flow diagram is straightforward to create. Because interest serves to increase our balance, we represent it as an arrow pointing into the oval for account balance. The resulting flow diagram is shown in Figure 2.1.

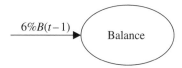

FIGURE 2.1   Flow diagram for a savings account.

Translating the information in the diagram into an equation gives us the corresponding DDS: $B(t) = B(t-1) + 6\%B(t-1)$.

At this point we should recognize something that will prove to be very useful. The context is different, but what we have here is another example of exponential growth—exactly the kind of model we encountered in our beginning population models. Instead of a population we have an account balance, and instead of a growth rate, we have an interest rate, but all of the mathematics is the same. A benefit of recognizing this fact is that we can immediately use the explicit formula we developed in Chapter 1 and we can also use the Excel models we created with only minor modifications.

Example 2.3: Use the explicit formula for exponential growth to rework Examples 2.1 and 2.2.

Recall that the explicit formula for exponential growth without harvesting or stocking is given by

$$P(t) = (1 + r)^t P(0).$$

In our financial context we instead use $B(t) = (1+r)^t B(0)$, where $B(t)$ is the account balance after t years, $B(0)$ is the initial deposit, and r is the annual interest rate. For the balance after 1 year, we have $t = 1$, so we write

$$B(1) = 1.06^1 \cdot 10,000$$
$$= 10,600.$$

For the balance after 2 years, we have $t = 2$ and thus

$$B(2) = 1.06^2 \cdot 10,000$$
$$= 1.1236 \cdot 10,000 = 11,236.00.$$

Both answers agree with the ones we found before. $\square$

A complication in financial computations is that interest is often not just paid once a year but instead can be paid at a variety of different frequencies. Sometimes interest is paid annually, sometimes semiannually, sometimes monthly, and sometimes even daily. Most of the financial instruments that we will study use monthly compounding, and we will assume monthly compounding from now on unless explicitly stated otherwise.

The idea behind **monthly compounding** is that instead of receiving all of the interest once a year, the account is credited with one twelfth of the interest each month. This arrangement is better for the saver because interest deposited, say, in February, will itself earn interest for the rest of the year. In the next example we assume that our 6% rate is compounded monthly instead of annually, and we verify that monthly compounding is to our advantage.

Example 2.4: Assuming an initial deposit of $10,000 and an interest rate of 6% compounded monthly, find the account balance after 1 year and after 2 years. Compare these results to the balances in Examples 2.1 and 2.2.

The fact that we are using monthly compounding means we have to be careful both with our interest rate and with our time units. This is our first example of using time units other than years, but this will not affect the problem-solving process. Also, an easy mistake to make is to read the statement "6% compounded monthly" and assume that we earn 6% *every* month. Unfortunately this is not the case, and we will need to exercise care in remembering that "6% compounded monthly" means that we earn *one twelfth* of 6% each month. Thus for the current example, we earn a monthly rate of $\frac{6\%}{12} = 0.5\%$.

Now that we know the monthly rate, we set up our flow diagram just as before. Since time is now in months, the diagram in Figure 2.2 shows the *monthly* change in the account balance, not the annual change.

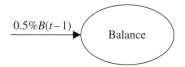

FIGURE 2.2   Flow diagram for account with monthly compounding.

From the diagram we write down the DDS:

$$B(t) = B(t-1) + 0.5\%B(t-1).$$

Our task is to find the balance after 1 year and after 2 years. Because the time units for the DDS are months instead of years, this amounts to computing the balance for $t = 12$ and $t = 24$ months. We can use either Excel or the explicit formula to do so, but here we use Excel. We need to keep in mind that because the balance changes monthly, we will have a column for t in months instead of years, and we must drag our formulas down to $t = 24$ to get to 2 years. Figure 2.3 shows the setup of the Excel spreadsheet including the formula for the balance.

	A	B	C
1	Savings Account		
2			
3	Interest rate, r =		6.00%
4			
5	t (months)	Balance	
6	0	10000	
7	1	=B6+($C$3/12)*B6	

FIGURE 2.3   Excel savings account model setup.

Note that we store the interest rate in its own cell, refer to it with absolute addressing, and divide it by 12 in the formula. Including the division by 12 in the formula saves us a step and also has the benefit of not introducing a rounding error.

After copying the formula down to month 24, we see in Figure 2.4 that Excel gives us a balance of $10,616.78 after 1 year. If we look further down, we will see a balance of $11,271.60 after two.

If we compare these results to the examples using annual compounding, we see that monthly compounding nets us an additional $16.78 over 1 year and an additional $35.60 over two. □

	A	B	C
1	Savings Account		
2			
3	Interest rate, r =		6.00%
4			
5	t (months)	Balance	
6	0	10,000	
7	1	10,050.00	
8	2	10,100.25	
9	3	10,150.75	
10	4	10,201.51	
11	5	10,252.51	
12	6	10,303.78	
13	7	10,355.29	
14	8	10,407.07	
15	9	10,459.11	
16	10	10,511.40	
17	11	10,563.96	
18	12	10,616.78	

FIGURE 2.4   Excel results for Example 2.4.

The preceding example points to the following general rule: *the more frequently interest is compounded, the better it is for the saver.* This rule will be explored more in the exercises.

The interest rate we have been using, 6%, is higher than what is typically paid on an ordinary savings account. At this writing a good interest rate for a savings account is about 1% (Bankrate, 2015). Six percent is, however, a reasonable **rate of return** for an **investment account**, say, one that is invested in stocks and bonds. The phrase "rate of return" differs from "interest rate" in that it includes many factors that influence the growth of an investment, such as dividends, interest, and appreciation. For now we do not really care exactly why the account is growing; all we care about is the rate of this growth, represented as a single number. We use this rate of return, or growth rate, exactly the same way we use interest rates when calculating balances. Our habit will be to express the rate of return in terms of its equivalent interest rate.

For many savings or investment accounts, the owner does not just make an initial deposit and leave it to grow. Often after opening an account, the owner makes regular deposits as a way of saving for some future goal such as a car, tuition, a house, or retirement. In the next example we examine such a regular savings program, assuming monthly deposits and monthly compounding.

Example 2.5: Find the balance in an account after 10 years if initially $5000 is deposited and every month thereafter a deposit of $100 is made. The account earns the equivalent of 7% annual interest compounded monthly.

First we create a flow diagram to represent the situation. We should have an arrow each for the monthly interest rate and the monthly deposit. As usual we must remember that the monthly interest rate is found by dividing the stated rate by 12: $\frac{7\%}{12} = 0.58333\%$. Note in the flow diagram in Figure 2.5 that since interest and deposits both serve to increase our account balance, both are represented by inward-pointing arrows.

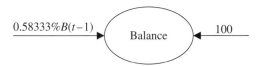

FIGURE 2.5   Flow diagram for savings account with monthly deposits.

Both arrows represent additions to the account balance, so our DDS is

$$B(t) = B(t-1) + 0.58333\%B(t-1) + 100.$$

Recall that this is the same form that our DDS would take if we were considering a growing population that included stocking: $P(t) = P(t-1) + rP(t-1) + a$. Once we recognize that fact, we can immediately write down the explicit formula for the account balance in any month, namely,

$$B(t) = \left(1 + \frac{r}{12}\right)^t B(0) + a\frac{(1+(r/12))^t - 1}{r/12},$$

where t is the number of months, r is the interest rate compounded monthly (as a decimal), $B(0)$ is the initial deposit, and a is the monthly deposit. In the current example the formula becomes

$$B(t) = (1.0058333)^t 5000 + 100\frac{(1.0058333)^t - 1}{0.0058333}.$$

Using the explicit formula to find the balance after 10 years is now a matter of plugging in the appropriate value for t, in this case 120 (10 years = 120 months). Plugging in $t = 120$ yields

$$B(120) = (1.0058333)^{120} 5000 + 100\frac{(1.0058333)^{120} - 1}{0.0058333}$$
$$= 2.0096533 \cdot 5000 + 100\frac{2.0096533 - 1}{0.0058333}$$
$$= 10,048.267 + 100 \cdot 173.084$$
$$= 27,356.67.$$

So after 10 years our savings account will have grown to \$27,356.67. $\square$

E.7   Improving Spreadsheet Readability

In this Excel section we show how to make some basic changes to our spreadsheets to make them easier to read and use including hiding/unhiding rows and making use of number formats.

In this section we set up an Excel spreadsheet that will allow us to easily investigate the effects of all of the different parameters on a savings account. The relevant quantities are the interest rate, the initial deposit, the monthly deposit, and the length of time we maintain the account. We allow for easy changes to the first three by storing them in their own cells, and we may change the length of time by dragging our formulas down to the appropriate time. A screenshot of the resulting spreadsheet is shown in Figure 2.6. Note that in the formula for the DDS, we divide the interest rate by 12 to account for monthly compounding.

	A	B	C
1	Savings Account		
2			
3	Interest rate, r =		7.00%
4	Monthly deposit, a =		100
5			
6	t (months)	Balance	
7	0	5,000	
8	1	=B7+($C$3/12)*B7+$C$4	
9	2	5,259.09	

FIGURE 2.6   Excel savings account model with monthly payments.

Excel allows us to choose the format for numbers in the spreadsheet. To change numbers to dollars-and-cents, or currency, format, we select the cell(s) we wish to change, and then from the Home tab, select Currency from the drop-down menu in the Number group. The spreadsheet as it appears with updated formatting is shown in Figure 2.7.

	A	B	C
1	Savings Account		
2			
3	Interest rate, r =		7.00%
4	Monthly deposit, a =		$100.00
5			
6	t (months)	Balance	
7	0	$5,000.00	
8	1	$5,129.17	
9	2	$5,259.09	
10	3	$5,389.76	

FIGURE 2.7   Excel savings account model with updated formatting.

By dragging the formulas down to the 120th month, we can confirm the result we obtained via the explicit formula. Excel produces a balance of $27,356.79 after 10 years, or 120 months. The slight discrepancy between this value and the one obtained earlier is due to the fact that Excel keeps many more decimal places in its calculations. Generally the Excel result will be more accurate than those we produce with the explicit formula and a calculator unless we are willing to keep many more decimal places in our calculations.

When we have to copy formulas down so far to arrive at an answer, it can be awkward having to scroll up and down in order to see the parameters involved and the final answer simultaneously. Excel provides a way around this by allowing us to hide rows that we do not want to see.

First, we point to the gray row number on the left side of the spreadsheet. The pointer should become a small, black, right-pointing arrow. Once the arrow appears, (left) click on the number of the first row to be hidden, and drag down to the last row to be hidden. Release the mouse button. Once all of the desired rows are selected, there are two ways to hide them. One way is to select the Format drop-down out of the Cells group from the Home tab. Then select down to Hide and Unhide and click Hide Rows. Alternately, we can right- (ctrl) click on the selected area, and choose Hide. After successfully hiding all but the first and last rows, we should see the screen pictured in Figure 2.8.

	A	B	C
1	Savings Account		
2			
3	Interest rate, $r =$		7.00%
4	Monthly deposit, $a =$		$100.00
5			
6	t (months)	Balance	
7	0	$5,000.00	
127	120	$27,356.79	
128			

FIGURE 2.8 Excel results from Example 2.5 with rows hidden.

Now if we want to investigate the effect of changing a parameter on the final balance, we can type in a new value for the parameter and not have to scroll down pages and pages to see the result. We can unhide the rows by first selecting across the break line using the right-pointing arrow pointer then choosing Unhide Rows from the Format drop-down or via right-clicking on the relevant area.

2.1.1 Saving for a Car

Oftentimes people save with a particular goal in mind whether it is college tuition, a car, or a down payment on a house. The spreadsheet we have just created is an excellent tool for deciding whether or not a given savings goal is attainable. Perhaps we know how much we can deposit initially, and we need to figure out if we can afford the required monthly deposit to reach our goal in a certain amount of time. Or maybe we know how much we can deposit a month and need to know how long it will take to accomplish the goal. We typically will not have a lot of control over the interest rate available to us, but we can still investigate the effects of a change in the rate.

To fix things we suppose that a first-year college student wants to save for a car for after graduation. The car is a 4-year-old Volkswagen Jetta SEL sedan in good or better condition with 48,000 miles on it, which as of this writing has a fair market value of approximately $14,500 according to the Kelley Blue Book website (Kelley Blue Book, 2015). In the following problems we use our spreadsheet and the Goal Seek command to answer questions which should be of interest to our student.

Example 2.6: Suppose that the student has already saved $5000.00 to put toward the purchase of the Jetta and has an account that earns 5% annual interest compounded monthly. Find the monthly contribution required for the student to be able to afford the car in 5 years.

Since we have stored all of our parameters in their own cells, they are easy to change. We type 5% for our interest rate and $5000 for our initial deposit. We have also copied our formulas down to $t = 60$ months and hidden most of the rows we do not need to see. The spreadsheet with a monthly deposit stand-in value of $100 should appear as in Figure 2.9.

	A	B	C
1	Savings Account		
2			
3	Interest rate, $r =$		5.00%
4	Monthly deposit, $a =$		$100.00
5			
6	t (months)	Balance	
7	0	$5,000.00	
8	1	$5,120.83	
66	59	$13,062.97	
67	60	$13,217.40	

FIGURE 2.9 Excel setup for Example 2.6.

By glancing at cell B67, we can see that a $100.00 monthly deposit is not enough to enable the student to purchase the Jetta. To find a sufficient monthly deposit, we use Goal Seek as shown in Figure 2.10.

Remember that we are telling Goal Seek to "make cell B67 (the balance after 60 months) $14,500.00 by changing cell C4 (the monthly deposit)." Once we click okay, Goal Seek tells us that the student needs to save at least $118.86 per month. □

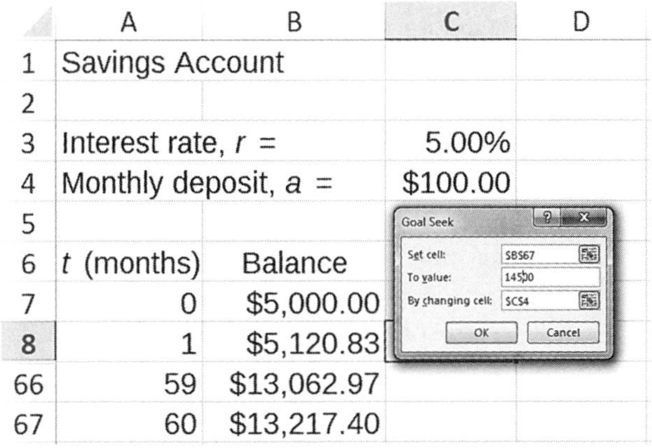

	A	B	C	D
1	Savings Account			
2				
3	Interest rate, r =		5.00%	
4	Monthly deposit, a =		$100.00	
5				
6	t (months)	Balance		
7	0	$5,000.00		
8	1	$5,120.83		
66	59	$13,062.97		
67	60	$13,217.40		

Goal Seek

Set cell: $B$67
To value: 14500
By changing cell: $C$4

OK   Cancel

FIGURE 2.10 Setting up Goal Seek for Example 2.6.

Sometimes we know how much we can afford to save every month, and so the question is not "How much do I need to save?," but rather "How much car can I afford based on what I can save?" Our spreadsheet handles this kind of question as well.

Example 2.7: Suppose we still have an account that earns 5% annual interest compounded monthly and that we start the account with an initial deposit of $5000. If we can manage to save $50 per month over the 5 years until graduation, how much can we spend on a car?

All we need to do here is type in $50.00 for our monthly deposit. The spreadsheet then reports that the balance in 60 months (still located in cell B67) will be $ $9817.10. The website www.cars.com allows a potential used car buyer to specify a desired make and model and the maximum price she or he is willing to pay (Cars.com, 2015). Selecting Volkswagen Jetta for under $10,000.00 yields a long list of options, all of which are either (i) more than 4 years old, (ii) have relatively high mileage, or (iii) have fewer options than the $14,500 car. Still, if we are fortunate enough to have $5000 to use as our initial deposit, our $50.00 a month could put us into a good vehicle in 5 years' time. □

Still another relevant way of looking at a car purchase is to set a savings goal and then figure out how long it will take to reach it if we know how much per month we can save.

Example 2.8: Given a $5,000.00 initial deposit, an interest rate of 5% compounded monthly, and the ability to save $75 per month, how long will it take to save $14,500 for the 4-year-old Jetta?

Once we type in $75.00 for the monthly deposit, we only need to drag our formulas down until the balance reaches at least $14,500. (We may do this with or without unhiding the hidden rows.) After doing so, we see in Figure 2.11 that it would take 84 months, or 7 years, to reach our savings goal. □

	A	B	C
1	Savings Account		
2			
3	Interest rate, r =		5.00%
4	Monthly deposit, a =		$75.00
5			
6	t (months)	Balance	
7	0	$5,000.00	
8	1	$5,095.83	
90	83	$14,479.50	
91	84	$14,614.83	

FIGURE 2.11   Excel results for Example 2.8.

The previous series of examples shows just how flexible our Excel model can be. Furthermore, even though all of our examples can be done via the explicit formula, once we have the spreadsheet set up, the Excel version will be faster—especially for questions like Example 2.8 that ask us to find an unknown time.

It can be eye-opening to convert a required monthly deposit into the time it would take to earn that amount at a typical wage. A useful website for wage information is the US Department of Labor, Bureau of Labor Statistics: www.bls.gov (BLS, 2015). In particular, the *Occupational Outlook Handbook* contains a variety of information about hundreds of jobs and careers, including their required training and expected earnings (BLS, 2012). While working exercises that ask for a current salary, consider answering the same questions using potential jobs for after graduation using information found in the *Occupational Outlook Handbook*.

Example 2.9: Suppose that a student's current job is waiting tables at a full service restaurant. How many extra hours per week would the student need to work in order to accomplish the savings goal from Example 2.6?

From Example 2.6 we know the student needs to save $118.86 every month. According to the Bureau of Labor Statistics, as of May 2012, the median hourly wage for waiters and waitresses was $8.92 (including tips). Thus we compute that our student will need to work an additional $\dfrac{\$118.86}{\$8.92} = 13.33$ h per month, or approximately $\dfrac{13.33}{4} = 3.33$ h per week. This is the equivalent of about one extra lunch shift per week. Depending on the other demands on the student's time and finances, this may or may not be feasible, but it puts the attainability of the goal into a useful context. □

In the next section we examine a savings goal with a much longer time horizon: saving for retirement. In such an example we really get to appreciate the power of compounding over the long term.

2.1.2 Saving for Retirement

With pensions becoming increasingly unavailable, saving for retirement is a necessity if we want our golden years to be comfortable. It is also something that takes careful planning, including the consideration of many issues beyond the scope of this text. However, most of the technical difficulties in retirement planning involve *how* to save, not *whether* to save. When saving for retirement, there are a few basic rules to go by:

1. *Start early.* Nowhere is the power of compound interest more evident or more important than in saving for retirement. The reason is that retirement is usually such a long-term goal that the effects of compounding build over decades instead of just a couple of years. However, the only way to take advantage of such long-term benefits is to start saving as early as possible, even if the amount does not seem like much now. Unfortunately the very fact that retirement *is* so far off makes it psychologically difficult to save during the years when it is most beneficial. And of course early in a person's career is also when salary is usually the lowest.

2. *Use automatic payroll deduction.* If we have our monthly retirement contribution deducted directly from our paycheck and deposited automatically into our retirement account, we will not be tempted to spend the money.

3. *Take advantage of any employer-sponsored retirement plans that offer matching funds.* Matching funds are contributions that an employer will make to an employee's retirement, provided the employee also contributes at a certain level.

4. *Take advantage of tax-advantaged accounts such as IRAs, Roth IRAs, 401(k)'s, etc.* It is not always the case that these instruments are advantageous, but they make sense for most people. Good web sources of information about these kinds of plans include www.fidelity.com, www.vanguard.com, and http://www.irs.gov/Retirement-Plans/Traditional-and-Roth-IRAs.

In the examples that follow, we examine the soundness of *rule #1: start early.* We assume that a person enters the workforce at age 25 and retires at age 65. By the time the person retires, he or she needs to have saved enough money to live on for the rest of their lives. Since different people will have different needs and different goals for their level of retirement income, there is no one savings goal that will be right for everyone. A fairly conservative goal for a retirement nest egg would be $1 million

(Franklin, 2006). In the next example we find the monthly retirement contribution necessary for someone who wants to retire with $1 million.

Example 2.10: Determine how much per month a person would need to save over a 40-year career in order to retire with a nest egg of $1 million. We assume no initial deposit and that we have a retirement account that earns the equivalent of 8% annual interest, compounded monthly.

The context of this problem is different than when we were saving for a car, but the mathematics is exactly the same. We can use the savings account spreadsheet that we already created and modify it to include an initial deposit of $0.00 and an interest rate of 8%.

We save for a total of 40 years, or 480 months. So the first thing we do is copy our formulas down to $t = 480$ and hide most of them. After making these modifications, we see the screenshot given in Figure 2.12.

	A	B	C
1	Savings Account		
2			
3	Interest rate, r =		8.00%
4	Monthly deposit, a =		$75.00
5			
6	t (months)	Balance	
7	0	$0.00	
8	1	$75.00	
486	479	$260,017.14	
487	480	$261,825.59	

FIGURE 2.12   Excel setup for Example 2.10.

Next we use Goal Seek to figure out the monthly deposit that would be required to end up with $1,000,000 in cell B487 by changing cell C4. By now we should be reasonably comfortable with using Goal Seek in this manner, so we omit the details. After a successful Goal Seek, we see that a monthly deposit of $286.45 is required in order to become a millionaire at age 65. This result shows clearly the power of compounding. Over our 40-year career, we made 480 deposits of $286.45 to the account. That means our total contribution to the account was $480 \times \$286.45 = \$137,496$, but we ended up with $1 million in the account. Thus we earned $\$1,000,000 - \$137,496 = \$862,504$ in interest! □

In the next example we emphasize the role that time plays in saving for retirement by reworking Example 2.10 for different start dates for our saving.

Example 2.11: Under the same assumptions as Example 2.10, determine the monthly retirement contribution that would be required to save $1 million by age 65 if we start saving at ages 30, 35, 40, 45, 50, 55, and 60.

In this example we need only copy down the formula for shorter amounts of time and then reapply Goal Seek. If we start saving at age 30, for example, we will save for 35 years rather than 40. Therefore we only need to copy the formula down to $t = 35 \times 12 = 420$ months. We summarize all of the calculations in Table 2.1.

TABLE 2.1   **Monthly Contributions Required to Save $1 Million by Age 65 for Different Starting Ages**

Age When Savings Begins	Years until Retirement	Monthly Contribution Required for $1 Million Goal ($)
25	40	286.45
30	35	435.94
35	30	670.98
40	25	1,051.50
45	20	1,697.73
50	15	2,889.85
55	10	5,466.09
60	5	13,609.73

Notice how much a delay of even 5 years can affect the required monthly deposit. If we consider that a delay in saving at the beginning of a career also puts many people into prime child-rearing years, we can see that saving the required amount would become even more difficult. The moral is that the longer we allow compounding to work for us, the easier it is to attain our goals. Our mathematical model makes this moral precise by showing us exactly how delaying the start of saving will affect the attainability of our goal. □

In this section we learned some of the language and mathematics required for dealing with compound interest. Though we focused on savings accounts, the results are all valid for more general investing accounts as well provided we know the interest rate equivalent to the account's overall rate of return. We developed a spreadsheet model for saving that included the interest rate, initial deposit, and constant monthly deposits, and we used this model to answer many relevant questions in the context of saving for a car or retirement.

In the next section we examine the effects of compounding in the context of borrowing for major purchases such as a car or home.

2.1.3   Section Exercises

1   Consider a savings account that earns 4% interest compounded monthly. You initially deposit $1000.00 into the account and make no further deposits or withdrawals.

a. Draw a flow diagram for the monthly balance in the account.

b. Find the DDS.

c. Determine the account balance 2 years from the initial deposit.

d. Determine how long it will take the deposit to double.

2   Consider a savings account that earns 6% interest compounded monthly. You make an initial deposit and then monthly deposits thereafter.

a. Give the flow diagram for the monthly balance.

b. Find the DDS.

c. Determine the balance after 10 months using Excel.

d. Determine the balance after 10 months using the appropriate explicit formula.

e. Suppose the initial deposit is $5000. Determine the monthly deposit required for the balance to grow to $10,000 in 2 years.

3   Consider a savings account that earns 5% interest compounded monthly. Suppose you will withdraw $500 per month from the account.

a. Find the DDS for the balance.

b. Find the equilibrium value for the balance.

c. Determine the stability of the equilibrium value.

d. Interpret the meaning of the equilibrium value in the context of a savings account balance.

4   Decide on something you would like to save up for, for example, a car, a down payment on a house, a charitable contribution, etc., and choose your time horizon, that is, how long you want it to take. Find an online savings or money market account and use the interest rate given. Assume monthly compounding. If you can afford to deposit 10% of your goal initially, determine the monthly deposit required for you to reach your goal in the time you choose.

5   The "Rule of 70" is a useful financial rule of thumb that estimates the doubling time for an investment. The Rule of 70 says that if r is the interest rate (given as a percent) for a savings account (or rate of return for an investment), then the time it takes for an initial deposit to double, assuming no further deposits or withdrawals, is approximately $\dfrac{70}{r}$ in years.

a. Compare what the rule estimates for doubling time with the actual doubling time for $r = 10\%$.

b. Compare what the rule estimates for doubling time with the actual doubling time for $r = 7\%$.

c. Compare what the rule estimates for doubling time with the actual doubling time for $r = 5\%$.

d. Compare what the rule estimates for doubling time with the actual doubling time for $r = 3\%$.

6 *Extension*: Use the fact that $\ln(1 + r) \approx r$ for r near 0 to derive the Rule of 70.

7 Consider a savings account that earns 3% interest compounded monthly. Suppose you initially deposit $200 and subsequently deposit $50.00 per month.

    a. Use Excel to find the balance in the account after 4 years.

    b. Use the appropriate explicit formula to find the same balance.

    c. Over the entire 4 years, how much did you actually deposit into the account?

    d. By comparing your answer in c. to the account balance after 4 years, determine how much total interest you earned over the 4 years.

8 Suppose you have an account that earns 6% interest. Initially you deposit $1000 into the account.

    a. Find the balance after 1 year if the 6% is paid annually.

    b. Find the balance after 1 year if the 6% is compounded monthly.

    c. When compounding monthly, the actual monthly rate is found by taking 6%/12. What is the analogous daily rate if interest is compounded daily?

    d. Find the balance after 1 year if the 6% is compounded daily.

9 *Extension*: Suppose you have found the world's greatest savings account, and it pays 100% interest. You deposit $1.00 into the account initially and make no further deposits or withdrawals.

    a. Find the balance after 1 year if interest is paid annually.

    b. Find the balance after 1 year if interest is compounded monthly.

    c. Find the balance after 1 year if interest is compounded daily.

    d. Find the balance after 1 year if interest is compounded hourly.

    e. Find the balance after 1 year if interest is compounded every second.

    f. Does there seem to be a limit on how high your balance can grow due to more frequent compounding? Do you recognize this limit?

10 Though individual needs of course vary, a rough estimate for the amount of savings someone needs to live on in retirement is $1,000,000. Determine how much per month a person would need to save in order to retire with a nest egg of $1 million. Assume that the person retires at age 65 and has an account that earns the equivalent of 9% compounded monthly.

    a. Carry out the calculations if the person starts saving at age 25, 30, 35, 40, 45, 50, 55, and 60. To organize your results, fill in a table like Table 2.1.

    b. What's the moral of the story?

11 The $1,000,000.00 goal we used for our retirement savings plan is based on someone who needs to generate $40,000 a year to live on in retirement. Some people may need less income and others more. Take a few moments to decide how much income you would like to have to live on during your retirement years.

    a. A reasonable rule of thumb is that you need to accumulate roughly 25 times your desired income in order to generate that income without dipping into your savings. Based on your income estimate and this rule of thumb, how much will you need to accumulate by the time you retire?

b. At what age would you like to retire?

c. At what age do you foresee being able to start saving for retirement?

d. Assuming that your retirement account will earn the equivalent of 9% interest compounded monthly, determine how much you will need to save each month during your career.

12 An estimate given by Franklin (Franklin, 2006) is that you should try to save approximately 15% of your income each month for retirement. If the contribution you found in 11(d) is 15% of your monthly income, what annual salary do you need to meet your goal?

13 *Extension*: As we advance in our careers, our income typically increases through periodic raises, and as our income rises, so should our retirement contributions. This is one good reason to base your retirement contribution on a percentage of your income (as in Exercise 12) instead of on a fixed amount over your entire career. Doing so allows you to increase your contribution over time without missing the extra money.

a. Set up an Excel spreadsheet for a retirement plan where the monthly contribution is a fixed percentage of your monthly salary. The following should be parameters stored in their own cells: monthly salary, monthly contribution %, and assumed interest rate.

b. Assume you start saving 15% of your income each month at age 25 with a salary of $36,000 a year. If your account earns the equivalent of 9% interest compounded monthly, use your spreadsheet to compute the amount of money you will have saved by age 65.

c. How much will you end up with if the situation is the same as in part b. except that now you get an annual 4% raise in salary?

14 *Extension*: Suppose a savings account earns 3% interest compounded monthly. After the first month $100.00 is deposited into the account. Each subsequent month the deposit increases by 1%. Thus in month 2, $101.00 is deposited, and in month 3, $102.01 is deposited.

a. Give a flow diagram for this situation.

b. Give the corresponding DDS.

c. Implement the model in Excel where the interest rate, r; the initial monthly deposit, a_0; and the monthly percentage increase in the monthly deposit, s, are all stored as parameters.

d. Find the account balance after 2 years if initially there is $400.00 in the account.

15 *Extension*: Following the spirit of the derivation for the explicit formula for an affine model, find the explicit formula for the general model described in Exercise 14. Confirm your result in 14(d) by using this explicit formula.

16 *Extension*: Suppose you have access to an account that earns the equivalent of 5% interest compounded monthly. You know that you will need to withdraw $250.00 from this account every month for the next 4 years. Find the amount you would need to deposit into the account today to exactly cover all of those future withdrawals.

2.2 BORROWING FOR MAJOR PURCHASES

The previous section has shown us what a powerful advantage compound interest can be when saving for a long-term goal. Unfortunately, the reverse is also true: compound interest can make the cost of long-term borrowing downright depressing. We examine this effect by looking at common types of consumer credit including car loans and home mortgages. We begin with car loans.

2.2.1 Car Loans

At some point in our lives, we may want or need to purchase an automobile. It is also the case that many of us will take out a **car loan** to finance at least part of this purchase. As we will see later, aside from some new terminology, we already have the mathematical tools necessary to begin to analyze car loans.

Generally speaking when taking out a loan, the lower the interest rate, the better it is for the borrower. It is not always quite that straightforward, however, since lenders can charge fees in connection with the loan in addition to interest. The Truth in Lending Act (TILA) is a federal law enacted in 1968 as a way of standardizing how costs of consumer loans are reported with the goal of making it easier for consumers to comparison shop among lenders. Included in the TILA was the introduction of the **annual percentage rate** (**APR**). The APR is a rate that is intended to represent the total cost of a loan, including interest as well as any fees associated with it, and hence it can be used to make a true comparison between lenders: the one offering the lower APR should be the best choice for the borrower. The APR still does not always tell the whole story, but it is a good place to start.

In the context of car loans, the APR is a fairly reliable way to make comparisons. Despite its name, for car loans the APR is actually a rate that is compounded monthly. Thus, an APR of 12% would mean a loan where 1% interest is charged each month.

When we borrow money to purchase a car, we have to make monthly payments over the **term** of the loan. The term of the loan is the length of time over which we will repay it, and most car loans are for terms of between 3 and 6 years, or between 36 and 72 months. The initial amount we borrow is called the loan **principal**, or the initial loan balance. We refer to the APR as the interest rate, denoted in the usual way by r, and we let a represent the monthly loan payment. The amount we borrow, the term, and the interest rate all affect how large our monthly payment will be.

Next we show how to use Excel and the mathematics we have already developed to calculate our monthly payment, and we investigate the effects of all of the relevant parameters on how much we pay for a loan. In the exercises the reader is asked to choose her or his own car complete with options and investigate different financing options.

Example 2.12: Karen decides to take advantage of a car dealer's no-money-down financing offer to buy a new car. This means that she will not make a down payment, opting instead to borrow the entire purchase price of the vehicle, in this case a 2010 Ford F-150 STX pickup truck with two-wheel drive and an automatic transmission. Karen can purchase such a vehicle with 70,000 miles on it for approximately $19,600 (Car Max, 2015). She decides on a 5-year loan term, and as of January 2015, the average APR for such a loan is approximately 3% (Bankrate, 2015). Use a DDS and Excel to compute Karen's monthly payment on the loan.

Following our usual notation, we let B represent the outstanding balance on the loan, that is, the amount Karen has left to repay. Since our time units are months, $B(t)$ denotes the outstanding balance after t months. We let r stand for the APR, so in this case $r = 3\%$, and we let a stand for the (as yet unknown) monthly payment. Also, because Karen is borrowing the entire purchase price of the car, the initial amount, or principal, of the loan is $B(0) = \$19,600.00$.

The only two factors that change the balance $B(t)$ are the APR, which increases the amount we owe, and the monthly payment, which decreases the amount we owe. It is important to remember that since the APR is compounded monthly, the monthly rate is found by taking $\frac{3\%}{12} = 0.25\%$. A flow diagram for the situation is shown in Figure 2.13.

FIGURE 2.13   Flow diagram for auto loan model.

This flow diagram should look familiar as it is the same diagram we needed for a population that was growing exponentially and being harvested. The DDS should also be familiar since it is an affine model:

$$B(t) = B(t-1) + 0.25\% \; B(t-1) - a.$$

One consequence of making such an observation is that we can use the explicit formula we already developed without having to start from scratch. For now we proceed using the DDS and Excel.

Once we set up our spreadsheet, the key to finishing the problem is being clear about exactly what it is we are looking for: the monthly payment that completely pays off the loan in 5 years. In mathematical language this means we need to find a

such that $B(60) = 0$. This is a familiar sort of exercise with Excel, and we use Goal Seek to find the value for a.

First we give the setup for our loan spreadsheet. We need to store the APR and monthly payment in their own cells and refer to them using absolute addressing. In Figure 2.14 we show the Excel loan spreadsheet with the formula for monthly balance displayed.

	A	B	C
1	Loan Calculator		
2			
3	APR, $r =$		3.00%
4	Monthly payment, $a=$		$150.00
5			
6	t (months)	Balance Owed	
7	0	$19,600.00	
8	1	=B7+($C$3/12)*B7-$C$4	

FIGURE 2.14   Excel auto loan model setup.

We have entered a monthly payment of $150 as a stand-in value, but we will use Goal Seek to find the actual required payment. To do so, we need to tell Goal Seek to make cell B67 (the outstanding balance after 60 months) equal to zero by changing cell C4 (the monthly payment). Goal Seek finds that a monthly payment of approximately $352.19 will completely pay off Karen's loan in 60 months. □

A sobering exercise can be to compute not only the monthly payment for the loan but also the total cost of the loan, that is, the amount of interest we pay over the life of the loan. A knee-jerk calculation would be to take 3% of the loan amount, $19,600, which would produce $588.00 in interest. Unfortunately such a calculation vastly underestimates the actual cost of the loan because it neglects the effects of compounding. Just as compounding can produce significant benefits when we are saving, it can produce significant negative effects when we borrow. The next example illustrates this point.

Example 2.13: Determine how much total interest Karen paid on the loan in Example 2.12.

This problem can be answered using Excel, but it is not necessary. What we need to do is compute the total amount Karen paid over the course of the loan and then subtract the initial loan amount: any amount paid above the initial loan amount must be due to interest. Since Karen made 60 payments of $353.19, she paid a total of $352.19 \times 60 = $21,131.40$ over the life of the loan. The amount she borrowed was $19,600.00; therefore, the amount of interest she paid is $21,131.40 - $19,600.00 = $1,531.40$. This amounts to Karen having paid about 8% of the truck's price in interest charges. □

One of the reasons Karen ended up paying so much for the F-150 is that she financed the entire cost of the truck. Typically buyers must make a substantial **down payment** on the car before financing the rest. A down payment is cash paid up front toward the cost of the car. For example, if we buy a $20,000.00 car and make a down payment of $5,000, we will only need to borrow the remaining $15,000.00. One effect of making a down payment is that by reducing the amount we borrow, we also reduce the monthly payment as well as the total cost of purchasing the car. While it is generally true that the larger the down payment the better, we will assume that the best we can do is 20% of the car's price. This is the minimum down payment level recommended by www.bankrate.com (Sizing Up Your Down Payment, 2005).

Example 2.14: Assuming the same term and APR as before, determine Karen's monthly payment and total cost of the F-150 if she makes a 20% down payment and finances the rest.

A down payment of 20% means that Karen pays $19,600 \times $20\% = $3,920$ cash up front and borrows the rest, in this case $19,600 - $3,920 = $15,680$. The inclusion of a down payment means that we have an additional parameter to include in our spreadsheet.

We follow our usual practice of making our spreadsheet as flexible as possible by inserting new rows for (i) the purchase price, (ii) the down payment percentage, and (iii) the down payment in dollars and cents. We get Excel to calculate the down payment from the purchase price and the down payment percentage, and we also get Excel to calculate the initial loan amount since it is no longer the entire price of the vehicle. Our modified Excel spreadsheet setup is shown in Figure 2.15.

	A	B	C
1	Loan Calculator		
2			
3	APR, r =		3.00%
4	Purchase price =		$19,600.00
5	Down payment % =		20.00%
6	Down payment =		$3,920.00
7	Monthly payment, a=		$352.19
8			
9	t (months)	Balance Owed	
10	0	$15,680.00	
11	1	$15,367.01	

FIGURE 2.15   Excel auto loan model with down payment.

The new spreadsheet with formulas displayed is given in Figure 2.16.

◢	A	B	C
1	Loan Calculator		
2			
3	APR, r =		0.03
4	Purchase price =		19600
5	Down payment % =		0.2
6	Down payment =		=C4*C5
7	Monthly payment, a=		352.186337015637
8			
9	t (months)	Balance Owed	
10	0	=C4-C6	
11	=A10+1	=B10+($C$3/12)*B10-$C$7	

FIGURE 2.16 Excel auto loan model with down payment and formulas displayed.

Note that the down payment and initial loan amount are being calculated automatically based on the car price and the down payment percentage.

Modifying the spreadsheet in this way takes some time, and it is not strictly necessary for solving this one problem. Keep in mind, however, that we may want to investigate different down payments or car prices later, and if we do, we will have already done most of the necessary work.

Now we use Goal Seek to find the monthly payment that results in a zero balance after 60 months. We have done this before, and so we only report the result of our Goal Seek here—Karen will need to pay \$281.75 per month. To find the total she pays for the truck, we once again multiply the monthly payment by the number of payments she makes and obtain \$281.75 × 60 = \$16,905.00. Together with her down payment of \$3,920, Karen ends up paying a total of \$20,825 for the truck, a savings of \$306.40 versus the total when borrowing the entire purchase amount. Thus Karen's 20% down payment saves her about \$300 in interest charges over the life of the loan. □

One way that consumers can lower their monthly payments on a car is to take out a loan with a longer term. A problem with such loans is that they can easily lead to what is called being "upside down." This means that because the loan is being paid back so slowly and because cars depreciate so quickly, the borrower can end up owing more on the loan than the car is worth. If the car were to be totaled in an accident or if the car was stolen, auto insurance would only pay out the value of the car, so in this situation the borrower could end up owing money on a car that they no longer own. In the exercises the reader is asked to investigate how taking out a longer loan affects both the monthly payment and the total amount paid for the vehicle.

2.2.2 Home Mortgages

For many people the largest purchase they will ever make is the purchase of a home, and most people will have to take out a **home mortgage loan** to do so. When shopping for a mortgage, borrowers have two main types to choose from: a fixed-rate mortgage where the interest rate is the same over the life of the loan or an adjustable-rate mortgage (ARM) where the interest rate can rise or fall over the course of the loan. Fixed-rate loans have the advantage of predictable, constant monthly payments that make budgeting easier, while ARMs typically offer lower initial payments. We will confine our focus to fixed-rate mortgages here.

In addition to deciding on the type of mortgage, borrowers also must select the term, which is how long the borrower has to pay back the loan. Typical terms in the United States for fixed-rate mortgages are 15 and 30 years. The longer the term, the lower the monthly payment, but a longer term also means a higher total cost for the loan. Since mortgage interest rates are also compounded monthly, our initial example will look very similar to examples from the car loan section. Letting r be the APR on the mortgage, a the monthly payment, and t time in months, we have the flow diagram in Figure 2.17.

FIGURE 2.17   Flow diagram for home mortgage loan model.

The total monthly mortgage payment typically includes more than just the cost of the loan. For example, it may also include additional payments for mortgage insurance and property taxes. The part of the monthly payment that is just for the mortgage loan is called the **principal and interest payment** (PIP), and it is this payment that we call a.

We begin with a typical example.

Example 2.15: Find the principal and interest payment for a 30-year fixed-rate mortgage loan of $150,000.00 that has an APR of 4%.

This is the same type of problem as before where we seek the unknown payment a that will result in a zero loan balance at the end of the term, in this case at month $t = 360$. We have a choice in this example whether to use Excel or the explicit formula, but here we will use Excel. First we make superficial changes to the car loan spreadsheet to reflect the new mortgage context. The result is shown in Figure 2.18 with a temporary stand-in value given for a.

Since we are told that we are borrowing $150,000, we assume that to be the purchase price with no down payment. After dragging our equations down to month 360, we Goal Seek for the monthly payment that would result in a zero balance at month 360. The result $a = \$716.12$ is shown in Figure 2.19. $\square$

	A	B	C
1	Home Mortgage Calculator		
2			
3	APR, r =		4.00%
4	Purchase price =		$150,000.00
5	Down payment % =		0.00%
6	Down payment =		$0.00
7	Monthly PIP, a=		$350.00
8			
9	t (months)	Balance Owed	
10	0	$150,000.00	
11	1	$150,150.00	

FIGURE 2.18   Excel home mortgage loan model.

	A	B	C
1	Home Mortgage Calculator		
2			
3	APR, r =		4.00%
4	Purchase price =		$150,000.00
5	Down payment % =		0.00%
6	Down payment =		$0.00
7	Monthly PIP, a=		$716.12
8			
9	t (months)	Balance Owed	
10	0	$150,000.00	
11	1	$149,783.88	
369	359	$713.74	
370	360	$0.00	

FIGURE 2.19   Excel results for Example 2.15.

In the next section we examine our PIP more closely and determine exactly how much of each payment is devoted to interest and how much to paying back the loan principal.

2.2.3   Amortization Schedules

Each PIP serves to pay off interest earned on the loan as well as paying back part of the original principal that was borrowed. At the beginning of a loan, nearly all of each payment goes toward paying interest, but gradually as the loan balance decreases, the interest owed also decreases and hence the amount that goes toward paying off the principal increases.

A table that shows exactly how much of each loan payment goes toward paying interest and how much goes toward paying off the principal is called an **amortization**

schedule. Amortization schedules are useful for keeping track of how much total interest has been paid and how much of the loan is still left to be paid off, which is necessary information if the borrower decides to pay the loan off early.

We show how to produce an amortization schedule in the next examples.

Example 2.16: Determine how much of the first principal and interest payment from Example 2.15 goes toward paying interest and how much goes toward paying off the principal.

We know that during the first month of our loan, the principal of $150,000 incurs an interest charge of $150,000 \times \dfrac{4\%}{12} = \$500.00$. Our first payment must pay all of the $500.00 in interest charges, and whatever is left over will go toward paying down the principal, in this case $716.12 - \$500.00 = \$216.12$. $\square$

In general the payment a that occurs at time t will first pay off the interest charged on the previous month's balance: $\dfrac{r}{12}B(t-1)$. The remainder of the payment, $a - \dfrac{r}{12}B(t-1)$, is devoted to paying off the principal. These observations and our experience with Excel will allow us to make fairly quick work of producing amortization schedules.

Example 2.17: Find the complete amortization schedule for the loan in Example 2.15.

An amortization schedule consists of columns to keep track of (i) the part of each monthly payment devoted to interest, (ii) the part of each payment devoted to paying down the principal, (iii) the total amount of interest paid to date, and (iv) the balance remaining on the loan. We show the general setup for the amortization spreadsheet in Figure 2.20.

	A	B	C	D	E
1	Amortization Schedule Calculator				
2					
3	APR, r =		4.00%		
4	Purchase price =		$150,000.00		
5	Down payment % =		0.00%		
6	Down payment =		$0.00		
7	Monthly PIP, a=		$716.12		
8					
9	t (months)	Balance Owed	Interest Payment	Principal Payment	Total Interest to Date
10	0	$150,000.00			
11	1	$149,783.88			

FIGURE 2.20   Excel amortization model setup.

We already know the monthly payment, and our spreadsheet already calculates the balance remaining for each month. We have also just discussed how to calculate the interest charge each month, noting the leftover payment amount that goes to pay off the principal. The only column left to consider is the column for total interest paid to date. To keep track of this, we just need to add the previous month's total interest figure to the interest charge for the current month. Figure 2.21 shows the formulas entered into our spreadsheet. Note that all new columns begin at month 1 since that is when the first payment is made.

	A	B	C	D	E
1	Amortization Schedule (				
2					
3	APR, r =		0.04		
4	Purchase price =		150000		
5	Down payment % =		0		
6	Down payment =		=C4*C5		
7	Monthly PIP, a=		716.12294319819		
8					
9	t (months)	Balance Owed	Interest Payment	Principal Payment	Total Interest to Date
10	0	=C4-C6			
11	=A10+1	=B10+($C$3/12)*B10-$C$7	=($C$3/12)*B10	=$C$7-C11	=C11
12	=A11+1	=B11+($C$3/12)*B11-$C$7	=($C$3/12)*B11	=$C$7-C12	=E11+C12

FIGURE 2.21   Excel amortization model with formulas displayed.

After dragging all of our formulas down to the 360th month, we have a complete amortization schedule, pictured in Figure 2.22 with most of the rows hidden.

	A	B	C	D	E
1	Amortization Schedule Calculator				
2					
3	APR, r =		4.00%		
4	Purchase price =		$150,000.00		
5	Down payment % =		0.00%		
6	Down payment =		$0.00		
7	Monthly PIP, a=		$716.12		
8					
9	t (months)	Balance Owed	Interest Payment	Principal Payment	Total Interest to Date
10	0	$150,000.00			
11	1	$149,783.88	$500.00	$216.12	$500.00
12	2	$149,567.03	$499.28	$216.84	$999.28
369	359	$713.74	$4.75	$711.37	$107,801.88
370	360	$0.00	$2.38	$713.74	$107,804.26

FIGURE 2.22   Excel amortization schedule for Example 2.17.

> Note that the row for the first month of the schedule agrees with the work we did on the first payment in Example 2.15. □

In the next section we examine one of the ways shopping for mortgages differs from shopping for an auto loan. As we will see, in the mortgage context, we cannot simply rely on the APR to make a useful comparison between loan offers.

2.2.4 Points

In the context of mortgage lending, a **point** is a fee paid by the borrower at closing (when all loan paperwork is signed and money changes hands) for the purpose of securing a lower interest rate than would otherwise be made available. One point is equal to 1% of the amount to be borrowed. So on a $200,000 loan, one point would cost the borrower $2,000. Not all loans require points, and the decision to take out a loan with or without points is an interesting one that we are now in a position to investigate.

Though it can vary with the lender, a good rule of thumb is that each point paid reduces the APR on the loan by 0.25–0.375%. A lower interest rate means a lower monthly loan payment, but paying points also means a higher up-front cost for the loan. This presents the borrower with a complicated decision: is it better to pay the points up front and enjoy lower monthly payments, or is it better to have no up-front costs but a higher rate? As we shall see in the following examples, the answer actually depends on the amount of time the borrower expects to keep the loan before selling the house or refinancing.

Example 2.18: Martin will take out a 30-year fixed-rate mortgage of $250,000.00. He can avoid paying points on the loan by accepting an interest rate of 4.25%, or he can pay 2 points to obtain a rate of 3.75%. Determine Martin's total cost for each loan.

First we must assume that Martin has enough cash to pay the 2 points, which comes to $2\% \times \$250,000 = \$5,000.00$. Next we note that the important issue here is the total amount he will pay for each loan including interest and points.

For the loan with no points and monthly payment a_1, the total amount paid on the loan will be the number of months $t = 360$ times the monthly payment: $360 \cdot a_1$. Then the total amount of interest paid will be $360 \cdot a_1 - \$250,000.00$. Using Goal Seek with an APR of 4.25% determines Martin's monthly payment in this case to be $a_1 = \$1229.85$, so his total loan cost will be $360 \cdot a_1 - \$250,000.00 = \$192,746.00$.

For the loan with points and monthly payment a_2, the total amount paid on the loan will be the number of months times the monthly payment plus the points paid: $\$5000 + 360 \cdot a_2$. This leads to a total loan cost of $\$5,000 + 360 \cdot a_2 - \$250,000$. All that is left is to find the monthly payment and total loan cost in this case and compare to the previous one. The monthly payment with an APR of 3.75% is of course lower, found by Goal Seek to be $a_2 = \$1157.79$. Thus the total cost of the loan with points is $\$5,000 + 360 \cdot a_2 - \$250,000 = \$171,804.40$.

Comparing the two total costs reveals that Martin could save nearly $21,000 in loan costs over the life of the loan by paying the 2 points up front. $\square$

Example 2.18 shows that if the borrower is going to keep the mortgage for the full 30 years, then paying the points at the outset can be a good idea if she has enough cash to pay them. However, most homeowners do not stay in the same house for 30 years, and even if they do, they may refinance or completely pay off their mortgage at some point before the 30 years are up. The next example shows that in such a case paying the points may not be a wise decision.

Example 2.19: Take an extreme case where Martin from our last example decides to completely pay off his mortgage after the first month. Compare total loan costs for the two loan options.

In the case of no points, Martin will only have to pay the first month's interest charge on the loan before he completely pays it off. This interest charge will equal $\$250,000 \times \dfrac{4.25\%}{12} = \$885.42$.

In the case of the second mortgage, Martin has unfortunately already paid $5000 in points for the lower rate of 3.75%. Thus after one month he will have paid the points plus one month's interest: $\$5,000.00 + \$250,000 \times \dfrac{3.75\%}{12} = \$5,781.25$. Here having paid the points costs Martin an additional $\$5781.25 - \$885.42 = \$4895.83$ on the loan. Thus paying the up-front cost of the points can work against the borrower who decides to exit the loan early. $\square$

In light of Examples 2.18 and 2.19, we see that the crucial factor in deciding whether or not to pay points is how long we plan to keep the loan. Our last example for this section shows how to determine the "break-even time," that is, the time at which the two options are equivalent.

Example 2.20: Determine how long Martin from Example 2.18 must keep his loan in order for paying the points to be a good idea.

We know the answer lies somewhere between 1 and 360 months. To fully answer this question, we have to compare the total amount paid for each loan side by side up to time t. For the first loan this is equivalent to the total interest paid; for the second loan it is the total interest plus the points. What we need is an amortization schedule for each loan, being careful to add points to the cost of the second one.

We begin by adding a second loan amortization schedule next to the original using all of the same formulas. This second loan will need a place for the user to enter points paid and a place for the second, lower APR. We are still assuming the purchase price and the loan amounts are the same since there is no mention of a down payment. Note that the final column for the second loan includes points paid along with total interest to date in order to reflect the total loan cost at each month. This modification of our original amortization spreadsheet is shown in Figure 2.23.

	A	B	C	D	E	F	G	H	I
1	Loan with no Points					Loan with Points			
2						Points Paid =		$5,000.00	
3	APR, r =		4.25%			APR, r =		3.75%	
4	Purchase price =		$250,000.00			Purchase price =		$250,000.00	
5	Down payment % =		0.00%			Down payment % =		0.00%	
6	Down payment =		$0.00			Down payment =		$0.00	
7	Monthly PIP, a=		$1,229.85			Monthly PIP, a=		$1,157.79	
8									
9	t (months)	Balance Owed	Interest Payment	Principal Payment	Total Interest to Date	Balance Owed	Interest Payment	Principal Payment	Total Interest to Date Plus Points
10	0	$250,000.00				$250,000.00			
11	1	$249,655.57	$885.42	$344.43	$885.42	$249,623.46	$781.25	$376.54	$5,781.25
369	359	$1,225.51	$8.67	$1,221.18	$192,741.56	$1,154.18	$7.20	$1,150.59	$171,800.43
370	360	$0.00	$4.34	$1,225.51	$192,745.90	$0.00	$3.61	$1,154.18	$171,804.03

FIGURE 2.23   Excel amortization model with points.

Once we have the amortization schedules for each loan that include total loan cost, we compare them side by side by hiding all columns we do not need. Figure 2.24 shows the result of hiding the columns.

	A	E	I
1	Loan with no Points		Loan with Points
2			
3	APR, r =		
4	Purchase price =		
5	Down payment % =		
6	Down payment =		
7	Monthly PIP, a=		
8			
9	t (months)	Total Interest to Date	Total Interest to Date Plus Points
10	0		
11	1	$885.42	$5,781.25
12	2	$1,769.61	$6,561.32

FIGURE 2.24   Excel amortization model comparison of points versus no points.

We see that for the first few months, the loan with no points has the lower total cost. This makes sense in light of the $5000 in up-front costs that the points require. As we scroll down, though, the difference in cost between the two loans shrinks until eventually the loan with points becomes cheaper. Once we pass this point, the loan with points increases its advantage each month for the remainder of the life of the loan. The first time at which the loan with points becomes the better option is in month 49 as shown in the results in Figure 2.25.

	A	E	I
1	Loan with no Points		Loan with Points
2			
3	APR, r =		
7	Monthly PIP, a=		
8			
9	t (months)	Total Interest to Date	Total Interest to Date Plus Points
10	0		
11	1	$885.42	$5,781.25
12	2	$1,769.61	$6,561.32
58	48	$41,046.19	$41,106.80
59	49	$41,867.91	$41,827.22
60	50	$42,688.18	$42,546.27

FIGURE 2.25   Excel total cost comparison results of points versus no points.

Thus if Martin believes that he will move or pay off the mortgage within the next 4 years, he should choose the loan with no points. However, if he believes he will stay in the house for more than 4 years, the loan with points is the better option. ☐

In the next section we consider a different kind of consumer credit: credit cards.

2.2.5   Section Exercises

1   Choose a car that you would like to buy, and use the Internet to find its cost and a source for current APRs for car loans. Assume that you have to finance the entire cost of the car.

Your car (year, make, and model): _____.
Price (give source): _____.
Source for APRs: _____.

Use Microsoft Excel to fill in Table 2.2. The table should show what your monthly payment would be for loans of each term, and it should also show the total amount you paid for the car.

TABLE 2.2   Investigating the Effect of Loan Term on the Monthly Payment and Total Amount Paid for the Car

Number of Years to Repay	APR	Monthly Payment Required	Total Amount Paid for Car
3			
3.5			
4			
4.5			
5			
5.5			
6			

2   Use an online auto loan calculator to check your results in Exercise 1.

3   Based on your table from Exercise 1, fill in the blanks:
   a. As the length of the loan increases, the monthly payment _____.
   b. As the length of the loan increases, the total amount paid _____.

4   For your car in Exercise 1, suppose that you are able to make a down payment of 25% of the cost of the car, and you finance the rest. In other words, you are now only borrowing 3/4 of what you did before. Using the same APRs as before, fill in a new copy of Table 2.2.

5   By making a down payment of 25%, how much do you save on the total amount paid for the car for a 5-year loan?

6   *Extension*: You are about to purchase a car that costs $20,000 by taking out a 5-year loan, and you are faced with a choice. The dealer is offering two special promotions: (i) you can get a rebate of $1000 cash back with an APR of 3.5%, or (ii) you can select a lower APR of 2% but no rebate.
   a. Which special is the better choice assuming no down payment in both cases? Explain.
   b. What would the regular APR need to be in order for the choice to be a tie? Explain.

7   The APRs you see advertised on television and the Internet are not always available to everyone. Typically these rates are reserved for customers who have excellent credit, and having a bad credit score can increase the APR you end up receiving. Suppose you borrow $20,000.00 with a 5-year term to purchase a new car. Determine how much additional total interest you pay on the loan for each additional percentage point in the APR. Consider APRs between 3 and 10%.

8 For the following mortgage loan problems, assume a term of 30 years and a fixed-rate mortgage with an APR of 4.15%.

    a. Estimate the gross annual income you expect to have by the time you are ready to purchase a home. You can base this on the average salary for your chosen profession. You can also base it on the assumption of a joint income if you envision having a partner and both of you will work.

    b. As a (very) general rule of thumb, you should be able to afford to purchase a home whose price is between 3 and 5 times your household's gross annual salary. For example, if you expect an income of $50,000 per year, you will likely be able to afford a house that costs between $150,000 and $250,000. Based on your own salary estimate and the range given above, set a realistic budget for your home purchase.
Home price = _____

    c. Using www.zillow.com, select a house in a location of your choosing that is within your budget. Include a screenshot of your house below.

    d. Assumin`g no down payment, use Excel to determine your monthly PIP for the loan and the total cost of buying the house.

    e. Supply a screenshot from Excel showing both the beginning and end of the full amortization schedule for the loan.

    f. Check your work by supplying a screenshot from an online mortgage calculator like the one found at www.bankrate.com: http://www.bankrate.com/calculators/mortgages/mortgage-calculator.aspx.

9 A standard down payment for a home mortgage is 20% of the price of the house. Repeat Exercises 8d–f assuming you make a down payment of 20%. How much interest do you save by making the down payment?

10 Suppose now that you do not make a down payment but instead pay 2 points to lower the APR to 3.65%. Repeat Exercises 8d–f under this scenario.

11 Based purely on total cost over the full 30 years, which is preferable: paying the 2 points or not? (Assume no down payment in either case.)

12 How much total do you pay for the loan with no down payment and no points if you completely pay the loan off after 6 months?

13 How much total do you pay for the loan with no down payment and 2 points if you completely pay the loan off after 6 months?

14 If you pay each loan off (with points or without points) after 6 months, which was the better choice?

15 At what point would paying each loan off completely result in equal costs between paying no points and paying 2 points?

2.3 CREDIT CARDS

Credit cards are a wonderfully convenient way to pay for goods and services, particularly when purchasing over the Internet. For some transactions such as renting a car, credit cards are virtually indispensable. If selected and used carefully, they can even provide real financial benefits in the form of cash back rewards or travel points. The catch is that credit cards can be fantastically expensive if even slightly mishandled, and a case can be made that credit card companies try to make it as easy as possible for the consumer to do so. If mishandled or abused habitually, they can also negatively impact one's credit report, which in turn can result in thousands of dollars in extra costs for transactions that would seem to be completely unrelated, for example, by having to pay a higher interest rate on a home mortgage loan.

Each time a credit card is used for a purchase, the user is taking out a loan from the credit card company or bank that provided the card. When signing up for the card, the consumer agrees to pay back these loans in a timely fashion, and the consequences of not doing so vary in important ways from card to card. The great benefit of these day-to-day loans is that they are a very safe, convenient way to pay for things and they usually carry what is called a **grace period**. The grace period is the amount of time the user has to pay back the charges before the company starts charging interest or fees. The grace period, however, applies only if the balance is paid in full every month before the grace period expires. It is when the balance is not completely paid off that things can turn ugly.

The TILA of 1968 was an important piece of legislation for the protection of consumers seeking credit. Since then the prevalence of credit cards and the abusive tactics employed by some credit card companies led to another important piece of legislation, the Credit CARD Act of 2009. This act places important limits on credit card issuers and the ways they can charge consumers, and it provides important protections for consumer rights. In the following we take a look at the ways credit card companies make money from consumers, and we examine the effects of the CARD Act on these practices.

2.3.1 The Schumer Box

The **Schumer Box** is named for New York Senator Charles Schumer who as a member of the House of Representatives was responsible for enacting an amendment to the TILA in 1988. This amendment is known as the Fair Credit and Charge Card Disclosure Act, and it required credit card issuers to spell out all fees and terms for any credit card solicitation in a straightforward, easy-to-read format. "Schumer Box" is the name given to the tabular format of these disclosures. Here we provide an example of a Schumer Box for a credit card, and we discuss what each part of the disclosure means. Table 2.3 shows an example of a Schumer Box.

The list below is a brief overview of some important terms and potential differences in the way credit cards work. These are all ways that credit card companies can legally separate users from their money, and the specific terms must be disclosed in the credit card agreement. We will discuss most of them in detail, but credit card companies

TABLE 2.3   Example of a Schumer Box

Interest Rates and Interest Charges	
Annual percentage rate (APR) for purchases	Introductory rate of 0% for 6 months Nonintroductory rate of 19.9% APR may vary with changes in the prime rate
APR for balance transfers	Introductory rate of 0% for 6 months Nonintroductory rate of 19.9% APR may vary with changes in the prime rate
APR for cash advances	25.9% APR may vary with changes in the prime rate
Penalty APR	29.9% Penalty APR may be applied if a payment is late and may apply indefinitely from that point forward. APR may vary with changes in the prime rate
Grace period for new purchases	25 days from end of billing cycle
Minimum interest charge	$0.50
Fees	
Annual fee	$0.00
Balance transfer fee	3% of the transfer amount
Cash advance fee	$10.00 or 3%, whichever is greater
Foreign purchase transaction fee	None
Late payment	$35.00
Returned payment	$35.00
Things You Should Know about the Card	
How balances are calculated	Average daily balance method (including new transactions)
Can the account terms be changed?	Yes, as permitted by law

think of new ways to charge consumers all the time so the list should not be considered exhaustive:

- APR for purchases: This line discloses the interest rate or **finance charge** assessed if the balance is not paid in full by the payment due date. Note that the disclosure includes an introductory rate (which will be prevalent in any advertising for the card) as well as what the real rate will be once the introductory rate expires. The real rate is given as 19.9% here but will sometimes depend on the user's creditworthiness. Note also that the APR "may vary with the market based on the Prime Rate." This means that if interest rates rise in general, so will the rate on the card.

- APR for transfers: This line discloses the interest rate the user will pay (after the introductory period) for credit card balances transferred from another card.

- APR for cash advances: This line discloses the interest rate the user will pay if the card is used to get cash. Note that the rate is extremely high—25.9% (and it can go higher if interest rates rise). This is typical of credit card cash advance rates. It is almost never a good idea to use a credit card to get cash.

- Penalty APR: This line discloses the maximum APR that the card issuer can charge the user and gives the conditions under which it can do so. Note that any late payment can trigger the penalty rate of 29.4% and that once triggered, the rate change can be permanent.

- Grace period: This line notes how long the user has to pay the balance in full before the user incurs finance charges. The 25 days referred to here is known as the grace period. Note that the grace period only applies to new purchases—balance transfers and cash advances incur finance charges starting on the day they are made.

- Minimum interest charge: This line states that if any interest at all is due, then the user will be charged at least $0.50 in interest.

- Annual fee: The annual fee is a fee charged simply for using the card. This agreement specifies that this card has no annual fee. It is generally not difficult to find credit cards with no annual fee, but rewards cards (e.g., cars that give you cash back or airline miles) often have them.

- Transaction fees: Certain types of uses for credit cards incur fees in addition to any finance charges. This line discloses those fees. We see that this card will charge the user 3% for any balance transfers and the larger of 3% and $10.00 for any cash advance. If a consumer used this card to take out a $20.00 cash advance, that user would be charged a $10.00 fee, and the total charge of $30.00 for the transaction would immediately incur the cash advance APR of 25.9%, compounded daily. On the plus side, this card does not charge a fee for foreign transactions, so it would be a good card to use for foreign travel. Foreign transaction fees are common and are typically 2–3%.

- Penalty fees: These are fees the user must pay for not abiding by the card agreement. This card charges $35.00 for any late payment (in addition to triggering the penalty APR) or any returned payment. There is no fee for going over the **credit limit**, which is the maximum balance the card allows.

- How we will calculate balances: There are four common ways for credit card issuers to calculate finance charges and balances, and this line specifies which one the card uses. Here it is the **average daily balance (ADB) (including new transactions)** method. We will discuss this method later in the section.

- Can the account terms be changed?: Like all credit cards, the terms for this card can change within the bounds set by the law. In particular the card issuer must give at least 45 days' written notice and inform the user of their right to opt out of the changes by closing the account.

Though the Schumer Box is a useful tool for quickly understanding the main costs associated with a credit card, there are things it does not include, and consumers should still read and understand the full agreement before signing up for a card. An important part of many card agreements that is not presented in the Schumer Box is specific information about any rewards program the card offers.

2.3.2   The Credit CARD Act

In 2009 an amendment to the TILA was passed. The Credit Card Accountability Responsibility and Disclosure Act, also known as the Credit CARD Act, provided many new and important consumer protections (Prater, 2012). For the full text of the act, the reader is referred to the Credit CARD Act (2009).

Below we highlight some of the changes brought about by the Credit CARD Act:

- The card issuer must notify the user at least 45 days in advance of any significant changes to terms of the agreement. The cardholder may opt out of the changes during this 45-day period by closing the account and paying off any remaining balance under the original agreement.
- Limits on changes in the APR: Unless the increase occurs due to the expiration of an introductory rate or the APR is tied to a variable rate, the APR for new transactions cannot be raised within the first year of issuance. Exceptions are increases due to at least two consecutive late payments, but the issuer must give 45 days' notice of such an increase.
- Notice of universal default: **Universal default** is a practice of some card issuers where they will raise the APR on a card for seemingly unrelated transactions by the user. For example, a card issuer might raise the APR on a card if the cardholder misses payments on her electric bill. Card issuers must now provide the cardholder 45 days' notice of this change. The use of universal default by a card issuer must be disclosed in the card agreement, and it is something to consider when shopping for a card.
- Fee limits: Overlimit fees are no longer standard practice. Unless the card user chooses to have overlimit fees in exchange for more spending flexibility, transactions in excess of the card limit will simply be rejected. If there are overlimit fees, they cannot exceed the amount that the user goes over their credit limit. For example, a user who exceeds their limit by $5.00 can no longer be hit with a $25.00 fee—that fee would now be capped at $5.00. Late fees for occasional late payments are now capped at $25.00.
- Minimum payment disclosures: The **minimum payment** is the least the user can pay on a card while still avoiding late fees. Rules for calculating the minimum payment will appear on the card's monthly statement and will vary depending on the card issuer. In general credit card companies would like nothing better than for the user to only make the minimum payment each month because they are then able to charge interest on the remaining balance. To ensure cardholders understand how detrimental only making the minimum payment can be, card issuers are now required to report how long it will take users to pay off their balance if they only make the minimum payment each month.
- Elimination of two-cycle billing: In the past if a cardholder failed to pay off a balance in full by the payment due date, some card issuers would not only charge interest on the current balance, but they would also charge interest retroactively on the previous month's balance even if that balance had been paid in full. This is now illegal.

• How payments are applied: Suppose a user has a balance of $1000.00 on a card, where $500.00 is from new purchases and $500.00 is from cash advances. The APR for purchases is lower than for cash advances, say, 15% for purchases and 25% for cash advances. In the past if this user sent in a payment of $500.00, that payment would be put toward the lower APR balance first so that he would still owe $500.00 but at 25%. Now any payments (above the minimum) by law must be applied to high APR balances first.

The first mandatory biennial review of the effects of the Credit CARD Act by the Consumer Financial Protection Bureau (CFPB) was completed in 2013 (CFPB, 2013). We conclude our discussion of the CARD Act by including an excerpt from the executive summary of this report below.

The CARD Act has impacted the way that consumers pay for credit in the credit card marketplace and has significantly enhanced transparency for consumers. Over-limit fees and repricing actions have been largely eliminated; those effects can be directly traced to the Act. The dollar amount of late fees is down as well, and the CARD Act directly caused this reduction.

The end result is a market in which shopping for a credit card and comparing costs is far more straightforward than it was prior to enactment of the Act. Many credit card agreements have become shorter and easier to understand, though it is not clear how much of these changes can be attributed directly to the CARD Act since it did not explicitly mandate changes to the length and form of credit card agreements. Limitations on "back-end" fees, along with restrictions on an issuer's ability to raise interest rates, have simplified a consumer's cost calculations. Credit card costs are now more closely related to the clearly disclosed annual fees and interest rates. This greater transparency means a consumer deciding whether to charge a purchase can now make that decision with far more confidence that costs will be a function of the current interest rate rather than some yet-to-be determined interest rate that could be reassessed at any time and for any reason by the issuer.

2.3.3   Calculating Finance Charges

An important difference between credit card interest and the interest on an auto or mortgage loan is that credit card interest is typically compounded daily, not monthly. Recall that if an interest rate is compounded monthly, it means that 1/12th of the APR is charged each month in interest. Similarly, if an interest rate is compounded daily, 1/365th of the APR is charged each day. We illustrate the change with an example.

Example 2.21: Suppose that the APR on a credit card is 15%; find the interest charge on a balance of $500.00 carried for one day.

The daily interest rate is $\frac{15\%}{365} = 0.0411\%$. If the account balance at the end of the day is $500, the card account would be charged $500.00 \times 0.0411\% = \$0.21$, or 21¢ in daily interest. That 21¢ would be then be added to the balance, and at the beginning of the following day, the balance would be $500.21. If no payments are made, the entire balance of $500.21 is subject to the daily rate at the end of the next day.

Before concluding that $0.21 is not so bad, we should remember that the charge is only for a single day and that the effects of compounding, especially daily compounding, can be dramatic over long time periods. □

Different cards compute finance charges in different ways, and these differences make it more difficult to compare credit card offers side by side even when they carry the same APR. The four most commonly cited methods are the **daily balance** method, the **ADB** method, the **previous balance** method, and the **adjusted balance** method. The daily balance and ADB methods can include or exclude new transactions, though most cards include them. When a credit card uses daily compounding, the daily balance and ADB methods are exactly equivalent, and there is only a very slight difference if the card using the ADB method uses monthly compounding. In general the adjusted balance method is best for card users, though cards that use it appear to be rare. Which method a card uses and how it works are disclosed in the user agreement.

By far the most common methods are the daily balance and the ADB methods (both with new transactions). Later we develop a spreadsheet model to calculate finance charges for these methods, and in the process we show that they are in fact equivalent.

Since purchases, balance transfers, and cash advances all typically carry different APRs, finance charges for each type of transaction are treated separately. We show how to calculate finance charges for purchases and note that the model we create will work for transfers and cash advances as well, we just need to update the APR to handle each case. Total finance charges on a credit card bill are found by adding together the finance charges for each class of transaction.

Suppose we fail to pay off our card balance in full by the due date. We made a payment, so we do not incur a late fee, just not a large enough payment to pay off the card. Our balance is now subject to daily finance charges, and our balance each day is determined as follows:

1. Our initial **daily balance** is our balance on the first day of the **billing cycle**, which is the time period (usually 30 or 31 days) covered by our billing statement. This balance is calculated as the unpaid balance from our previous bill, plus any new purchases minus any new payments that post on the first day.
2. On any other day the daily balance is found to be the previous daily balance, plus the finance charge on that balance $\left(\text{previous balance} \times \frac{\text{APR}}{365}\right)$, plus any new purchases on that day, minus any new payments on that day.

Since purchases and finance charges serve to increase our balance, we represent them as inward-pointing arrows on our flow diagram. Payments, which serve to decrease the amount we owe, are represented by an outward-pointing arrow. With the APR denoted by r, the flow diagram is given in Figure 2.26.

FIGURE 2.26 Flow diagram for credit card model using daily balance method for finance charges.

If we let $B(t)$ be the daily balance for day t, then our DDS can be written as

$$B(t) = B(t-1) + \frac{r}{365}B(t-1) + \text{purchases}(t) - \text{payments}(t).$$

In other words, "The new daily balance is the previous daily balance, plus the interest on that balance, plus new purchases, minus new payments." Both the new flow diagram and the DDS are slightly different than the other examples we have seen because new purchases and payments do not occur regularly. Instead we understand the diagram and DDS to mean that any day on which there is a new purchase or payment, we add or subtract it from the daily balance accordingly.

Next we give an example of how the daily balance method works using Excel.

Example 2.22: The credit card we examine is the Chase Bank Freedom Visa, which specifies in its user agreement (CFPB, 2015) that it uses the daily balance method for computing finance charges. We use an APR for new purchases of 15.99%, which is in the range specified in the agreement. Suppose that we were not able to pay off our balance completely and that we carried a balance of $500.98 from the last billing cycle to the current one. Suppose also that the current billing cycle covers the dates March 7 through April 6, inclusive. Our account activity during the current billing cycle is as follows:

- On March 10 we charged $57.63 for groceries.
- On March 15 we charged $35.06 for gas.
- On March 15 we charged $12.56 at a restaurant.
- On March 30 we charged $87.54 for car repairs.
- On March 30 we made a payment of $100.00.
- On April 6 we charged $47.88 for our electric bill.

Answer the following questions:

1. What will be the total amount of finance charges we will incur this month?
2. What will be the new balance owed on our next statement?

E.8   The SUM Command and Working with Dates

In this Excel section we show how to quickly find the sum of a large range of numbers using Excel's SUM command. We also take advantage of Excel's date format.

We set up an Excel credit card spreadsheet model that will compute finance charges for us using the daily balance method. The end result will be a spreadsheet a consumer can use to easily verify the information contained in his or her own credit card statement.

What makes implementing this DDS with a spreadsheet different and more complicated than our previous examples is that the purchases and payments will not be the same from day to day. Instead of treating them as single parameters, we need to set up our spreadsheet so that the user can enter new purchases or payments manually for each day of the billing cycle. (Note that this is similar to the crane problem in the exercises where we had a different stocking number each year.)

We store the new purchase APR and the balance carried over from the previous billing cycle as parameters in their own cells, and we create columns for the time, daily balance, new purchases, and payments. We also add a column for the daily finance charges so we can easily keep track of them. Figure 2.27 gives a screenshot of the setup.

	A	B	C	D	E
1	Credit Card Daily Balance Calculator				
2					
3	New purchase APR, r =		15.99%		
4	Previous balance =		$500.98		
5					
6	t (days)	Daily Balance	New Purchases	New Payments	Daily Finance Charges
7	0				
8	1				
9	2				

FIGURE 2.27   Excel credit card model setup for daily balance method.

Next we enter formulas for the daily balance as well as for each day's finance charge. There is no formula for purchases and payments because those occur at irregular intervals. The formula version of our spreadsheet is shown in Figure 2.28.

Note that we have to handle the initial daily balance in a slightly different way since it is based on the balance carried over from the last billing cycle.

Instead of numbers we want the time column to keep track of actual dates in our billing cycle. We can arrange this by typing the date "March 7, 2015," into the cell for $t=0$ and recopying the formula down to the date April 6. All that remains is to enter each purchase and payment on the correct date. A portion of the complete spreadsheet is given in Figure 2.29.

	A	B	C	D	E
1	Credit Card Daily B				
2					
3	New purchase APR		0.1599		
4	Previous balance =		500.98		
5					
6	t (days)	Daily Balance	New Purchases	New Payments	Daily Finance Charges
7	0	=C4+C7-D7			=($C$3/365)*B7
8	1	=B7+($C$3/365)*B7+C8-D8			=($C$3/365)*B8
9	2	=B8+($C$3/365)*B8+C9-D9			=($C$3/365)*B9
10	3	=B9+($C$3/365)*B9+C10-D1			=($C$3/365)*B10

FIGURE 2.28 Excel credit card daily balance model with formulas displayed.

	A	B	C	D	E
1	Credit Card Daily Balance Calculator				
2					
3	New purchase APR, r =		15.99%		
4	Previous balance =		$500.98		
5					
6	t (days)	Daily Balance	New Purchases	New Payments	Daily Finance Charges
7	7-Mar	$500.98			$0.22
8	8-Mar	$501.20			$0.22
9	9-Mar	$501.42			$0.22
10	10-Mar	$559.27	$57.63		$0.25
11	11-Mar	$559.51			$0.25
12	12-Mar	$559.76			$0.25
13	13-Mar	$560.00			$0.25
14	14-Mar	$560.25			$0.25
15	15-Mar	$608.11	$47.62		$0.27

FIGURE 2.29 Excel credit card model setup for Example 2.22.

In order to answer the first question in the example, we need to find the total finance charge for our next bill; thus, we need to add up all of the daily finance charges for the entire billing cycle. We could do this by entering a long formula, but fortunately Excel's SUM function will accomplish this more easily. First we choose a cell at the top of our spreadsheet where we will store the total finance charge. Click in that cell, type "=SUM(E7:E37)," and then press enter. This command tells Excel to add the contents of all cells between E7 and E37, inclusive. The result is the finance charge total we seek.

We can also access the SUM command from Excel's Home tab. Click the Autosum drop-down menu from the Editing group and select SUM. Excel will highlight the range it thinks is appropriate, but if this is not the desired range, we can select the correct one by using the thick cross pointer to highlight it. Figure 2.30 shows the updated Excel spreadsheet with the total finance charge formula displayed.

	A	B	C	D	E	F	G	H
1	Credit Card Daily Balance Calculator							
2								
3	New purchase APR, r =		15.99%		Finance charges =		=SUM(E7:E37)	
4	Previous balance =		$500.98				SUM(number1, [number2], ...)	
5								
6	t (days)	Daily Balance s	New Purchase	New Payments	Daily Finance Charges			
7	7-Mar	$500.98			$0.22			
8	8-Mar	$501.20			$0.22			

FIGURE 2.30   Excel daily balance model with total finance charge formula displayed.

The answer to question 1 is $8.02, and we answer question 2 by adding the last daily balance in the billing cycle, namely, $649.38, to the last finance charge, $0.28. The result will be the amount owed on the next monthly statement: $649.66. □

Next we show that for daily compounding, the ADB method of calculating finance charges will give the same result as the daily balance method. First we note the ADB method using daily compounding calculates the daily balance in the same way. Once we have all of the daily balances, we proceed as follows:

1. Compute the ADB over the billing cycle. This requires that we add up all of the daily balances and then divide by the number of days in the billing cycle.
2. Once we have the ADB, we multiply the average by the daily rate: $\text{ADB} \times \frac{\text{APR}}{365}$. Finally we multiply the result by the number of days in the billing cycle: $\text{ADB} \times \frac{\text{APR}}{365} \times \text{days in cycle}$.

Example 2.23: Show that in Example 2.22 the ADB method would yield the same finance charge as the daily balance method.

E.9   The AVERAGE and COUNT Commands

Finding the average of all of our daily balances is done most efficiently with Excel's AVERAGE function. This function computes the average of all values in a range specified by the user. Like the SUM command, it must be preceded by an equals sign, and the user can either type the initial and final values in the range separated by a colon or the user can select "Average" from the Autosum drop-down menu and select the range with the thick cross pointer. Figure 2.31 shows the formula required.

E	F	G
Average Daily Balance =		=AVERAGE(B7:B37)

FIGURE 2.31   Formula for average daily balance.

Once we have the ADB, we need to multiply it by our daily rate and the number of days in the billing cycle. Since the number of days in the billing cycle will vary with the month, we let Excel keep track by simply counting the number of days on our spreadsheet. The Excel COUNT function is also found in the Autosum drop-down menu, and it counts the number of numerical entries in a given range. Thus the total finance charge using the ADB method will be calculated by the formula shown in the screenshot in Figure 2.32. We note that the result is exactly the same as it was for the daily balance method. ☐

E	F	G
Average Daily Balance =		=AVERAGE(B7:B37)
Total finance charge =		=G3*(C3/365)*COUNT(B7:B37)

FIGURE 2.32   Total finance charge based on average daily balance.

Though the total finance charge can be computed without knowing the ADB, we note that some credit cards use the ADB to compute the required minimum payment each month. Thus for these cards, it is still necessary to compute the ADB.

Our last example for this section examines the calculation of the minimum payment.

Example 2.24: Find the minimum payment for the outstanding balance in Example 2.22 if the rule is 4% of the average daily balance or $25.00, whichever is larger.

E.10   The MAX Command

The Excel command for choosing the larger of two numbers is the MAX command, and the syntax for it is "=MAX(number1,number2)." We want to take the larger of $25.00 and 4% of $590.47; the formula is shown in Figure 2.33. Observe that the minimum payment on the next statement balance will be $25.00. ☐

	E	F	G	H
	Average Daily Balance =		$590.47	
	Total finance charge =		$8.02	
	Minimum payment =		=MAX(25,0.04*G3)	

FIGURE 2.33 Calculation of the minimum payment.

2.3.4 Section Exercises

For the following exercises, use the credit card with terms given in the abbreviated Schumer Box in Table 2.4.

TABLE 2.4 Schumer Box for Section 2.3 Exercises

Interest Rates and Interest Charges	
APR for purchases	18.9%
APR for balance transfers	21.9%
APR for cash advances	24.9%
Penalty APR	28.9%
Grace period	25 days from end of billing cycle
Minimum interest charge	$1.00
Fees	
Annual fee	$45.00
Balance transfer fee	$10.00 or 4%, whichever is greater
Cash advance fee	$10.00 or 5%, whichever is greater
Foreign purchase transaction fee	None
Late payment	$35.00
Returned payment	$35.00

1   Suppose you took out a $25,000 car loan for a term of 60 months using this credit card's cash advance APR as the APR for the car loan. How much in total would that car end up costing you?

2   Suppose you take out a cash advance of $40.00 using this credit card and do not pay it off for 2 years. How much in total would that $40.00 end up costing you? (For simplicity, assume no other charges or payments are made to your card over the 2-year period.)

3   Suppose you transfer a balance of $1200 from another credit card to this one. Assuming you make no other payments or transfers, how much would you owe at the end of one month?

4 Use the daily balance method for computing finance charges to answer the following. Suppose your balance from your last credit card statement was $1000.00. You charge $34.35 for groceries on March 12, $21.04 for gas on March 24, and $25.00 for an iTunes gift card on April 1. You also make a payment of $500.00 on April 2. Assume that you have not been paying your balance in full each month so that you do incur finance charges and that the billing cycle runs from March 7 to April 6.

   a. What will be the total finance charges on your next billing statement?

   b. What will be the total you owe on your next billing statement?

   c. What will be the minimum payment due on your next billing statement if the rule is "4% of your outstanding balance or $75.00, whichever is higher?"

2.4 THE TIME VALUE OF MONEY: PRESENT VALUE

Present value is a fundamental concept in finance, and it is defined to be the value today of a future payment or payments. Finding the present value allows us to make fair comparisons among different payment options and hence to make decisions about what option is best for us. For example, which is the better offer—a guaranteed payment of $200.00 one year from today or a guaranteed payment of $500.00 ten years from today? We should not automatically opt for the $500.00 just because it is the larger amount. Perhaps if we took the $200 one year from today and invested it, we would end up with more than $500 nine years later. If that were the case, then the $200 one year from today would be the better choice because we could put it to work earning interest for 9 years. What we need is a way of comparing future payment options that takes the availability of compound interest into account. Present value provides a way of doing this.

2.4.1 Present Value of Regular Payments

The idea of present value, that we can view future payments as having an equivalent, discounted value if they were made today, takes some getting used to. We start with considering an example of a single payment; then we move on to multiple payments of a constant amount.

Example 2.25: Assuming that we have access to a savings account that earns 4% interest compounded annually, compute the present value of a guaranteed payment of $500.00 to be made to us 3 years from today.

  Another way of stating this question is to ask: "How much money would I need to deposit into my account today in order for it to accumulate to $500.00 at the end of 3 years?" We know how savings accounts (without regular deposits) grow already; for reference the explicit formula is repeated below:

$$B(t) = (1+r)^t B(0).$$

Finding the present value of our promised \$500.00 is now a matter of finding the initial deposit required. In other words, find $B(0)$ such that $B(3) = 500$. We show the required algebra below:

$$B(3) = (1 + 0.04)^3 B(0)$$

$$500 = (1.04)^3 B(0)$$

$$\frac{500}{(1.04)^3} = B(0)$$

$$\$444.50 = B(0).$$

We have found that the present value of a \$500.00 payment that would occur 3 years from today is \$444.50, assuming we can earn 4% interest during that time. Another way to view our result is that mathematically speaking, there is no difference between an offer of \$444.50 today and an offer of \$500.00 three years from today. They are equivalent, provided our assumption about the interest rate is valid. $\square$

As the previous example demonstrates, there is not really a unique, definitive present value for a future payment—it always depends on the interest rate we assume. If our assumption about the interest rate changes, then so will the present value of any future payment.

Suppose that instead on one future payment, we want to know the present value of a series of regular future payments. In this case the concept and the computations are a little more difficult.

Example 2.26: Find the present value of a series of 10 annual payments of \$400.00 beginning 1 year from today. Assume that we could earn 5% interest compounded annually during this time.

We wish to find the amount of money we would need to deposit into an account today that would exactly generate our 10 annual payments of \$400.00. In other words, how much do we need to start with so that we could make 10 annual \$400 withdrawals and end up with a balance of zero after the 10th one? The presence of regular withdrawals from our account requires the use of the explicit formula given by

$$B(t) = (1 + r)^t B(0) + a \frac{(1 + r)^t - 1}{r}.$$

Here a is the annual withdrawal, so $a = -\$400.00$, r is the assumed interest rate of 5%, t is 10 years, and $B(0)$ is the initial balance that we seek. Since we require the account to be empty after the 10th payment, we must have that $B(10) = 0$. We plug in all of our known values and solve for $B(0)$:

$$B(10) = (1.05)^{10} B(0) - 400 \frac{(1.05)^{10} - 1}{0.05}$$

$$0 = 1.62889 B(0) - 400 \frac{0.62889}{0.05}$$

$$5031.157 = 1.62889 B(0)$$

$$3088.70 = B(0).$$

Thus the present value of 10 future payments of $400.00 is $3088.70, assuming we could earn 5% interest. Note that this number is quite a bit less than the face value of all of the payments: $400.00 \times 10 = 4000.00. This is again due to present value accounting for not just the face value of money but the time value of money as well: future payments are not worth as much to us as payments made today because of the availability of interest. $\square$

We note that we can also work the previous example with Excel. The work is mathematically equivalent to paying off a loan balance where we know the payment and the term but not the initial loan amount.

Example 2.27: Rework Example 2.26 using Excel.

After only superficial modifications to the loan spreadsheet, we have the setup shown in Figure 2.34.

	A	B	C
1	Present Value of Regular Payments		
2			
3	Assumed interest rate, $r =$		5.00%
4	Future payments, $a =$		$400.00
5	Present value =		$5,000.00
6			
7	t	Balance	
8	0	$5,000.00	
9	1	=B8+$C$3*B8-$C$4	
17	9	$3,346.02	
18	10	$3,113.32	

FIGURE 2.34   Excel model for present value of regular payments.

We have stored a stand-in value of $5000 for the present value. The initial balance in cell B8 refers to the cell where we store the present value, C5. Note that because we are using annual compounding, the interest rate is unmodified in the balance formula. Finally, we have hidden most of the rows that we do not need to see.

We Goal Seek on the present value that would make the account balance equal to 0 at time $t = 10$, and the result shown in Figure 2.35 confirms our earlier result: if $3088.70 were deposited into the account today, the growth due to interest would allow us to withdraw all 10 payments of $400.00 and be left with a zero balance at the end. In other words a sum of $3088.70 paid today would exactly generate the proposed stream of future payments if the interest rate for our hypothetical account is 5%. ☐

	A	B	C
1	Present Value of Regular Payments		
2			
3	Assumed interest rate, $r =$		5.00%
4	Future payments, $a =$		$400.00
5	Present value =		$3,088.69
6			
7	t	Balance	
8	0	$3,088.69	
9	1	$2,843.13	
17	9	$380.95	
18	10	$0.00	

FIGURE 2.35 Excel results for present value in Example 2.27.

Oftentimes future payments are not regular. Our next section shows how to modify our work in this section to handle irregular payments.

2.4.2 Present Value of Irregular Payments

In many situations the future payments we expect are not all the same amount, or they do not occur at regular intervals. Consider the following example.

Example 2.28: Consider the following stream of guaranteed future payments: $100 one year from today, $200 three years from today, and $500 five years from today. Assuming that we could earn an interest rate of 3% compounded annually, find the present value of the stream of payments.

We first approach this example by using the explicit formula. Note that we cannot use the formula $B(t) = (1+r)^t B(0) + a\dfrac{(1+r)^t - 1}{r}$ because it requires that the same payment a be made every year. Instead we find the present value of each payment separately using the formula $B(t) = (1+r)^t B(0)$. Then the present value of the entire stream will be the sum of the individual present values.

Let $B_1(0)$, $B_2(0)$, and $B_3(0)$ represent the present values of the first, second, and third payments, respectively. From our work in Example 2.25, we know that $B_1(0) = \frac{100}{1.03} = \$97.09$, $B_2(0) = \frac{200}{(1.03)^3} = \$183.03$, and $B_3(0) = \frac{500}{(1.03)^5} = \$431.30$. Adding up all of the individual present values gives us the present value for the series of payments: $\$97.09 + \$183.03 + \$431.30 = \$711.42$. Again we interpret this result as follows: if we were to deposit $\$711.42$ into an account earning 3% today, we would be able to withdraw each payment at the specified time and end up with a zero balance immediately following the last withdrawal. □

The approach we took in the previous example will always work. The problem with the approach is that it is very slow if the number of payments is large. For instances where we have a lot of payments, Excel is the clear choice. To demonstrate how to use Excel, we use it to rework Example 2.28.

Example 2.29: Create an Excel spreadsheet for finding the present value of the irregular stream of payments from Example 2.28.

The setup of this spreadsheet needs to be different than the one used previously because there is no longer a single payment parameter, a. Instead we may have different payment amounts at different times. The DDS for this situation is

$$B(t) = B(t-1) + rB(t-1) - a(t).$$

Here $B(t)$ represents the balance in our hypothetical account and r is the assumed interest rate. We understand $a(t)$ to be the future payment amount at time t and that amount can be zero if there is no payment scheduled for that time.

To use Excel to model this situation, we need to build in the flexibility to handle a payment of any amount at any time. We accomplish this by adding a column next to the balance column where the user can manually enter any future payments at the appropriate time. Then in our formula for the balance, we subtract the value in the payment column rather than subtracting the single parameter value from before. Figure 2.36 shows the new setup with a stand-in value of $5000 for the present value.

	A	B	C	D
1	Present Value of Irregular Payments			
2				
3	Assumed interest rate, r =		5.00%	
4	Present value =		$5,000.00	
5				
6	t	Balance	Future Payments	
7	0	$5,000.00		
8	1	$5,250.00		
9	2	$5,512.50		
10	3	$5,788.13		

FIGURE 2.36 Excel model setup for present value of irregular payments.

We also give the formula version in Figure 2.37. Note how the formula in column B refers to the possible payment in column C.

	A	B	C
1	Present Value of Irreg		
2			
3	Assumed interest rate		0.05
4	Present value =		5000
5			
6	*t*	Balance	Future Payments
7	0	=C4-C7	
8	=A7+1	=B7+$C$3*B7-C8	
9	=A8+1	=B8+$C$3*B8-C9	
10	=A9+1	=B9+$C$3*B9-C10	

FIGURE 2.37 Excel model for present value of irregular payments with formulas displayed.

The final step is for us to input the correct assumed interest rate and all of the future payments at the appropriate times and then use Goal Seek to find the present value in C4 that gives us a balance of zero at time $t = 5$. The results of our Goal Seek are shown below in Figure 2.38. Notice that the present value of $711.42 is the same as the result from Example 2.28. □

	A	B	C	D	E
1	Present Value of Irregular Payments				
2					
3	Assumed interest rate, *r* =		3.00%		
4	Present value =		$711.42		
5					
6	*t*	Balance	Future Payments		
7		0	$711.42		
8		1	$632.76	$100.00	
9		2	$651.75		
10		3	$471.30	$200.00	
11		4	$485.44		
12		5	$0.00	$500.00	

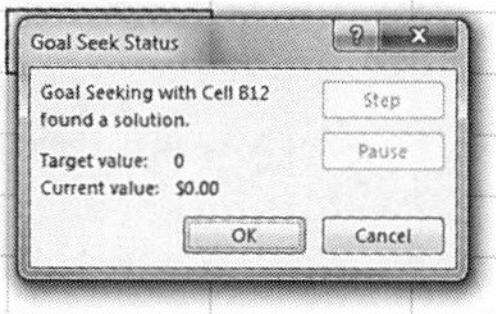

Goal Seek Status

Goal Seeking with Cell B12 found a solution.

Target value: 0
Current value: $0.00

Step   Pause   OK   Cancel

FIGURE 2.38 Excel present value result for Example 2.29.

The Excel spreadsheet model we created in Example 2.29 is a very useful one as we will see in the next sections. We begin with lotteries.

2.4.3 Lottery Payouts

For lotteries with large jackpots, the winner of the lottery has a choice to make: take a lump sum in cash today or take 30 guaranteed annual payments over the next 29 years starting today (called the **annuity** option). What occasionally surprises winners is that the lump-sum option is not the same as the jackpot. A jackpot of $20 million, for example, would pay a lump sum of about $11 million in cash. The "jackpot" actually refers to the face value sum of all of the guaranteed annual payments. The big question for lottery winners then is, which is better financially—the lump sum in cash today or the promise of those 30 future payments? The answer to this question is complicated by issues like state and federal income taxes, but we ignore those for now in favor of focusing on the present value of each option.

Example 2.30: In the past, lotteries would allow the winner to choose between the lump sum and 30 equal payments—the first payment happens today and the rest are paid annually over the next 29 years. A 2014 Powerball lottery jackpot was $50 million with a lump-sum cash payout option of $31.2 million. We assume that we have access to an account that will earn the equivalent of 4.5% interest compounded annually. Which is better from a present value point of view: a lump-sum cash payout of $31.2 million or the 30 annual payments?

Note that since the 30 annual payments are all equal, we could work this problem using our explicit formula. However the fact that the first payment actually occurs *today* introduces a slight complication that we will not have to worry about if we use the Excel spreadsheet from Example 2.29. (When we set up that spreadsheet, we included in our payment column the possibility of a payment at time $t = 0$.)

We must first determine the annual payment. This we find by dividing the jackpot by the number of payments: $a(t) = \dfrac{\$50,000,000}{30} = \$1,666,666.67$ for all $0 \le t \le 29$.

Next we enter the 30 payments of $1,666,666.67 in the payment column of our spreadsheet starting at time $t = 0$ and continuing through time $t = 29$. Our spreadsheet with most rows hidden should appear as in Figure 2.39 where the present value is just a stand-in value for now.

	A	B	C
1	Present Value of Irregular Payments		
2			
3	Assumed interest rate, $r =$		4.50%
4	Present value =		$35,000,000.00
5			
6	t	Balance	Future Payments
7	0	$33,333,333.33	$1,666,666.67
8	1	$33,166,666.66	$1,666,666.67
35	28	$24,334,051.19	$1,666,689.67
36	29	$23,762,392.82	$1,666,690.67

FIGURE 2.39   Excel setup for present value of lottery annuity option in Example 2.30.

By using Goal Seek to find the present value that gives a zero balance at year 29, we find that the present value of the annuity option is $28,369,935.58, considerably less than the stated $50 million value of the jackpot. We also note that compared to the lump-sum option, the annuity option is not as attractive. □

In recent years lotteries have changed the way they handle the annuity payout option. The Powerball lottery and others have claimed that over the years winners have expressed concern that due to increases in the cost of living, level annual payments result in a lower standard of living over time. As a result lotteries have started offering an **increasing annuity** where the future annual payments increase by a fixed percentage every year. Powerball uses an increase of 4% every year, while Mega Millions uses an increase of 5% every year. (So if the Powerball annuity payout started with a $1,000,000 payment, the second payment would be $1,040,000.) The annual payments still begin with the first payment today and then continue for the next 29 years, and the face value sum of the annual payments must still equal the reported jackpot. In the exercises the reader is invited to analyze the new system using present value and to perhaps offer a different explanation for the change to increasing annuities.

2.4.4 Section Exercises

For all problems assume access to an account that has an interest rate of $r = 5.5\%$ compounded annually.

1 Find the present value for a future payment of $5000.00 to be made 6 years from today. Explain what this present value represents in a complete sentence.

2 Find the present value of a series of future payments of $400.00 to be made each year for 4 years starting 1 year from today.

3 Suppose you need $50,000 per year to live comfortably in retirement. If you expect your retirement to last for 30 years, determine the nest egg you will need on the day you retire. Explain what your result means in the context of the problem.

4 Use present value to decide which is the better option for you if you are to be the recipient of a stream of future payments:

   a. Option A: $500 every year for 5 years starting a year from today

   b. Option B: $1500 two years from today and $1200 to be paid 6 years from today

   Explain your answer!

5 *Extension*: A **perpetuity** is a series of future payments that continues forever. Use the concept of an equilibrium value to find the present value of a perpetuity that pays $50,000 every year starting 1 year from today.

6 In the past, lotteries would allow the winner to choose between the lump sum and 30 equal payments—the first payment happens today and the rest are paid annually over the next 29 years. The current Powerball lottery jackpot is $40 million with a lump-sum cash payout option of $21.3 million. Assume that you have an account that will earn 3.5% interest compounded annually.

a. What is the annual payment under the annuity option?

b. With a lump-sum cash payout of $21.3 million, which option is better from a present value point of view—the lump sum or the annuity?

c. What would be the required interest rate in order for the lump sum and annuity options to be equivalent?

7 *Extension*: In recent years lotteries have changed the way they handle the annuity payout option. Lotteries now use increasing annuities where the annual payment increases by a fixed percentage every year. Powerball uses an increase of 4% every year, while Mega Millions uses an increase of 5% every year. (So *if* the Powerball annuity payout started with a $1,000,000 payment, the second annual payment would be $1,040,000.) The annual payments still begin with the first payment today and then continue for the next 29 years. The sum of the annual payments must still equal the jackpot.

a. Determine the annual payouts for Powerball if the sum of all 30 annual payouts must equal the jackpot of $50 million.

b. Determine the present value of the annuity option for Powerball.

c. Which annuity is better from a present value point of view: level payments or increasing payments? Note that the answer to this question suggests a different reason for why lotteries may have changed their payout method.

8 Repeat Exercises 7a–c for the Mega Millions game.

9 Given equal jackpots, which game would you prefer to win and why—Powerball or Mega Millions?

2.5 CAR LEASES

There are two main ways that people acquire a new car—by purchasing the car or by leasing the car. When considering a new car, it is almost always advantageous from a purely financial point of view to purchase the car rather than lease. For used cars, though, the situation can be different.

Drivers who do not expect to keep a vehicle very long sometimes opt to **lease** the vehicle rather than purchase it outright. This means that the customer rents the vehicle for a specified period of time (the **term**) for a certain amount of money per month (the **lease payment**). Common terms for new car leases are 2–4 years. At the end of the lease agreement, the customer must return the vehicle to the dealer, and at that time she or he has the option to purchase the vehicle outright for a price that was agreed upon at the signing of the lease. This price is called the **optional lease-end buyout**. The following are some of the pros and cons of leasing that a prospective lessee should consider. Some benefits of new car leases are as follows:

- By always leasing the customer can always drive a new car. New cars come with all of the latest bells and whistles, including the latest safety equipment, and they are generally more trouble-free than older vehicles.
- The monthly lease payment is usually lower than the monthly payment required to purchase the same car. This is because the lessee does not own the car and is only paying for the depreciation in value of the vehicle over the life of the lease.
- New leased vehicles are usually covered under the manufacturer's full factory warranty for the life of the lease. This means the leaseholder will only need to pay for routine maintenance such as oil changes over the term of the lease.
- Customers who lease a new vehicle rather than purchase it generally pay lower sales taxes. This is because they are only taxed on the value of the vehicle's depreciation over the term of the lease.

There are downsides to leasing too:

- Over the long term, leasing new vehicles is very expensive because the lease-holder is always paying for the depreciation at the beginning of a car's life, the time when depreciation is greatest.
- When continually leasing the monthly payments never stop. With purchasing a car, the payments may be higher than an equivalent lease, but they only last for the term of the loan, and then the car is owned outright.
- Car leases limit how much the holder can drive the car. This mileage limit is specified in the lease contract and typically will allow the leaseholder to drive 12,000–15,000 miles per year. Any mileage on the car above that limit will incur a fee when the car is turned in to the dealer, typically 18¢–25¢ per mile over the limit.
- Leases also specify that the car must be returned with only "normal wear and tear" on it. Any damage to the vehicle in excess of normal wear and tear is the responsibility of the leaseholder to fix or pay for.
- Unless the lessee has taken out **gap insurance**, they are exposed to some risk, especially at the beginning of the lease. If a leased car is stolen or totaled, the insurance on the vehicle will pay replacement value to the dealer. Since new cars depreciate so quickly, the replacement value can be less than the amount the leaseholder still owes. The result is that the lessee can owe money for a car they no longer have. Gap insurance provides coverage for the difference between what is owed on the car and what the car is worth, and it is sometimes included in the lease as part of the monthly payment.
- Some leases, particularly those that tout low monthly payments, require a large down payment at signing. Making a large down payment on a lease may be difficult for some customers to accomplish, and if the vehicle is stolen or totaled, the lessee will not recoup any of that money.

In the remainder of this section, we discuss a third alternative for obtaining a vehicle, one that until relatively recently was unknown to most people and difficult to arrange.

2.5.1   Lease Takeovers

An alternative to purchasing or leasing is the **lease takeover**. In a lease takeover situation, the lease seller is someone who has leased a vehicle they no longer want—they may want a new vehicle or they may no longer be able to afford the current lease payments. However, it can be extremely expensive or impossible to just cancel a lease contract. This is where a lease takeover makes sense. A lease buyer is an individual who is willing to take over the lease, that is, assume use of the vehicle and the responsibility for making the lease payments until the end of the lease. This can be a great deal for both parties. The seller gets rid of a vehicle they no longer want or can afford, and the buyer can get a fairly new vehicle on a short-term lease at a good price. Frequently the car is still under factory warranty, the buyer gets to try the vehicle for an extended time (i.e., the remainder of the lease), and at the end of the lease, the buyer can simply return the car to the dealer if he or she does not want to purchase it for the lease-end buyout. Furthermore, the lease seller is sometimes motivated enough to offer a cash incentive to the buyer, which further lowers the cost to the buyer. One caveat is that the finance company for the original lease may charge a fee for the transfer of the lease, so this should be considered when evaluating a lease takeover.

Though it is perhaps not obvious that present value can help us decide when a lease takeover is a good deal, we show how to use present value to our advantage in this situation. By the end of this section, the reader will be able to use Excel and present value to evaluate potential lease takeovers.

The website www.swapalease.com is a site that helps people transfer car leases (Swap-a-Lease, 2015). If someone wants to get out of a lease, she or he can post the terms of the lease on the website in the hopes of transferring it to someone who wants it. That way the lease owner does not have to pay any penalties for breaking the lease, and the person who assumes the lease may end up with a better deal than he or she could get on a new car.

Car leasing is generally not a good move financially, but taking over someone else's lease can be. One reason is that the lease owner has already made any required down payment. Another is that the lease owner may have used far fewer miles than the lease allowed. This can mean that the value of the car at the end of the lease will be higher than predicted because the owner of the lease is paying for more wear on the vehicle than is actually present. Finally, a lease takeover will be even more attractive if the lease owner offers a cash incentive.

Remember that taking over a lease commits the buyer to making the remaining future lease payments and gives the buyer an option to purchase the car at the end of the lease. By comparing the present value of all remaining payments and the buyout to the current value of the car, we can spot lease takeovers that are particularly good deals.

Example 2.31: We first find a lease we are interested in taking over by searching Swapalease.com for whatever kind of vehicle we want. We can also specify particular states to minimize the cost of travel to get the car. A very popular car for leases is the BMW 3 Series. After a quick search we select a gray 2-door 328i that has 6 months remaining on the lease. The car currently has 18,200 miles on it, and we can drive up to 1,967 miles each month without an overmileage penalty. The monthly payment is $585.20, the optional lease buyout is $29,277.00, and the seller is not offering a cash incentive. Is this a good lease to take over from a financial point of view?

The first step is to calculate the present value of the lease takeover assuming that we will purchase the car at the end. In other words, how much cash would we have to deposit into an account today in order to cover all of the payments and the cash buyout at the end? We already have the spreadsheet that will do the necessary calculations for us. However, one modification we will have to make is that since the lease payments are monthly, we have to change our time units to months and use monthly compounding for our account. We assume a rate of 3% for this example; since our payments are monthly, the rate we use in our calculations will be $\frac{3\%}{12} = 0.25\%$. The formula version of the modified spreadsheet is shown in Figure 2.40.

	A	B	C
1	Present Value of a Lea		
2			
3	Assumed interest rate,		0.03
4	Present value =		40000
5			
6	*t*	Balance	Future Payments
7	0	=C4-C7	
8	=A7+1	=B7+$C$3/12*B7-C8	
9	=A8+1	=B8+$C$3/12*B8-C9	

FIGURE 2.40 Excel model for present value of irregular payments with monthly compounding.

Next we enter all of the remaining monthly payments as well as the buyout, which we consider to be the final payment. We then use Goal Seek to find the present value that gives us a zero balance on the month of the optional buyout. The results of a successful Goal Seek are shown in Figure 2.41.

We see from Figure 2.41 that the present value of the lease takeover is $32,331.04. However, knowing the present value of the lease does not tell us whether or not the takeover is a good deal. To decide that we must determine what it would cost to purchase this car today, say, from a dealer. If the present value of the takeover is equal to or perhaps lower than the cost of buying the car outright today, then the takeover is a good deal.

	A	B	C	D	E	F
1	Present Value of a Lease Takeover					
2						
3	Assumed interest rate, r =		3.00%			
4	Present value =		$32,331.04			
5						
6	t	Balance	Future Payments			
7	0	$31,745.84	$585.20			
8	1	$31,240.01	$585.20			
9	2	$30,732.91	$585.20			
10	3	$30,224.54	$585.20			
11	4	$29,714.90	$585.20			
12	5	$29,203.99	$585.20			
13	6	$0.00	$29,277.00			

Goal Seek Status

Goal Seeking with Cell B13 found a solution.

Step
Pause

Target value: 0
Current value: $0.00

OK   Cancel

FIGURE 2.41   Excel result for present value of the lease takeover in Example 2.31.

Pricing Details for a 2012 BMW 3 Series Coupe
328i 2dr Car

Customized True Market Value® Prices

	Trade-In	Private Party	Dealer Retail
National Base Price	$19,186	$20,958	$22,530
Optional Equipment	$4,313	$4,637	$5,466
Power Glass Sunroof	$359	$386	$455
Premium Leather Seating	$567	$609	$719
Adaptive Cruise Control	$965	$1,037	$1,223
6-Speed Shiftable Automatic Transmission	$0	$0	$0
18 Inch Alloy Wheels	$503	$541	$638
8-Way Power Driver's Seat	$139	$150	$176
8-Way Power Front Passenger Seat	$139	$150	$176
Front Sport Seats	$503	$541	$638
Voice Activated Navigation System	$728	$782	$922
harman/kardon speakers	$192	$206	$243
Bluetooth	$162	$175	$206
Driver's Seat Memory	$56	$60	$70
Color Adjustment - Gray	$55	$60	$64
Regional Adjustment - for Zip Code 77024	$25	$27	$29
Mileage Adjustment - 18,200 miles	$1,908	$1,908	$1,908
Condition Adjustment - Outstanding	$999	$1,038	$1,143
Total	$26,486	$28,628	$31,140

Buying a Certified Used Vehicle   Dealer Retail

Certified Used Price $33,344

FIGURE 2.42   Edmunds.com used car evaluation for Example 2.31. Source: Edmunds.com (2015). Reproduced with permission of Edmunds.com.

We can get a pretty good estimate of the current value of the vehicle by using one of many sites on the web for valuing used cars. We need to be careful to include as much information as we can from the lease posting to make sure we get an accurate estimate. Using the used car appraiser through Edmunds.com, we select the year, make, model, mileage, options, and condition listed on the Swap-a-Lease ad. (Note that we are assuming the ad is truthful.) Our results from Edmunds.com are shown in Figure 2.42.

According to Edmunds.com if we were to try to buy this exact car from a private party right now, we would expect to pay around $28,628.00; from a dealer the car would cost $31,140.00 cash today. Comparing these figures to the lease takeover present value of $32,331.04, the takeover appears to not be a particularly good deal. Keep in mind, though, that if we did take over the lease and then decided we did not want to keep the car, we would just turn it in to the dealer in 8 months and walk away. That flexibility is worth something to the buyer. ☐

2.5.2 Section Exercises

1 Choose a vehicle's make and model, and use www.swapalease.com to find a lease takeover for that vehicle. Note that you should refine the search to only include takeovers that include the cost of the optional lease buyout.

   a. Find the present value of the lease takeover by finding the present value of all future payments including the optional lease buyout. You will need to supply your own assumed interest rate and note how you determined it.

   b. Use a website such as www.edmunds.com or www.kbb.com to determine the amount you would have to pay for the same car today if you were to buy it outright. Be sure to include all options when determining the car's value.

   c. Explain in a complete sentence or two whether the lease takeover is a good deal.

2 Find a lease takeover for which the present value of the takeover is lower than the cash price of purchasing the car today.

3

COMBAT MODELS

Up until now our study of discrete dynamical system models has concentrated on **single-compartment models**. These are models for which we have only one quantity of interest such as a population or a loan balance. In this chapter we extend our consideration to **two-compartment models** where we have two interacting populations, this time in the context of combat between two opposing sides.

Throughout this chapter, when we speak of a victory for one side over another, we mean that one side will be completely out of action while the other will have forces remaining. By out of action we do not necessarily mean troops killed or, in the case of a naval battle, ships sunk. Instead we mean that the force has been rendered helpless so that it can no longer inflict damage on the other side. Our notion of victory is a simplistic one. In particular we do not account for the possibility of retreat, surrender, or the fact that an operational goal such as "delaying the enemy" can be achieved by the "losing" side. We also do not distinguish cases of Pyrrhic victory where the cost of battle is so great that the winning side for all practical purposes has also really lost.

Our models progress in complexity and in currency as we move through the chapter. We start with Lanchester's groundbreaking model from 1916, move to the 1953 work of Engel, consider the 1995 work of Hughes, and finally examine the 2013 work of Armstrong.

Models for Life: An Introduction to Discrete Mathematical Modeling with Microsoft® Office Excel®,
First Edition. Jeffrey T. Barton.
© 2016 John Wiley & Sons, Inc. Published 2016 by John Wiley & Sons, Inc.

3.1   LANCHESTER COMBAT MODEL

One of the first mathematical models for analyzing combat was proposed by F. W. Lanchester in 1916 in his book *Aircraft in Warfare: The Dawn of the Fourth Arm* (Engel, 1954). Forty years later Lanchester's work was included in the four-volume collection *The World of Mathematics* (Lanchester, 1956). His model has continued to inspire much discussion and many different modifications. As a testament to Lanchester's influence, Lucas and Turkes write that the Lanchester model is "The most common tool for modeling aggregate attrition (Lucas & Turkes, 2004)." The two-volume work *Lanchester Models of Warfare* by James G. Taylor presents a comprehensive mathematical analysis of Lanchester's model and its variations (Taylor, 1983).

The great strength of the Lanchester combat model and what makes it so compelling is its simplicity. The assumptions inherent in the model, though too severe to be expected to be satisfied in a real battle, nevertheless give rise to important conclusions regarding tactics and strategy. As Niall MacKay notes, "Those who teach military tactics, however, still value Lanchester's model and its generalizations, because, above all, they provoke careful thought about the consequences of the conditions of engagement (MacKay, 2005)." And as Taylor states, "Lanchester-type models are an ideal vehicle for studying combat dynamics because of the relative ease of extracting information from them and the fact that usually no other type of model is better justified (Taylor, 1983)." In this section we present a discrete version of Lanchester's model and see what information we can extract from it.

We begin with two adversaries, Blue and Red, and we represent the numbers of Blue and Red units remaining in the battle at time t as $B(t)$ and $R(t)$, respectively. The meaning of "units" will change with the context of the battle: units could be ships, tanks, soldiers, etc. The basic assumption of the Lanchester model is that a side incurs losses at a rate that is proportional to the size of the enemy's force: the larger the Red force, the more damage it will do to the Blue force and vice versa. We also assume uniformity of units, that is, that all units for each side are equally capable.

In order to complete the model, we introduce a parameter for **fighting effectiveness**, which we define to be the average number of enemy units put out of action by a single opposing unit during each time step. We can think of fighting effectiveness as a kind of overall measure that is affected by things such as quality of training, weapons technology, and experience with the terrain. If we let b represent the fighting effectiveness of a Blue unit and r the fighting effectiveness of a Red unit, then we can represent our combat model with the flow diagram in Figure 3.1. Note that we have two quantities of interest, Blue and Red forces remaining, so we have two ovals, or compartments, in our diagram.

Each time step results in a decrease to both forces proportional to the size of the enemy. The resulting DDS is

$$B(t) = B(t-1) - rR(t-1)$$
$$R(t) = R(t-1) - bB(t-1).$$

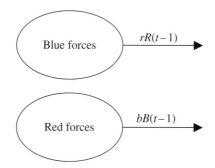

FIGURE 3.1   Flow diagram for Lanchester model.

This is a very different kind of model than the ones we have seen to this point. Not only do we have two dependent variables to track, but the variables depend on each other. We illustrate how the model works in our next example.

Example 3.1: Suppose Blue begins the battle with 50 units, so $B(0) = 50$, and Red begins the battle with 100 units, so $R(0) = 100$. Each Blue unit has a fighting effectiveness of $b = 0.10$, which means that each Blue unit will inflict 0.10 **casualties** (units put out of action) on the Red side per time step. Each Red unit has a fighting effectiveness of $r = 0.20$, which means that each Red unit will inflict 0.20 casualties on the Blue side per time step. After one time step, how many units of each side remain?

We employ the DDS to find both $B(1)$ and $R(1)$. For Blue we have

$$B(1) = B(0) - 0.20R(0)$$
$$= 50 - 0.20 \cdot 100$$
$$= 30.$$

For Red we have

$$R(1) = R(0) - 0.10B(0)$$
$$= 100 - 0.10 \cdot 50$$
$$= 95.$$

We see that after one time step, the Blue side has incurred much heavier casualties than Red. This should agree with our intuition: Red forces outnumbered Blue by 2-to-1 at the outset *and* the Red troops were twice as effective. □

Projecting the battle further into the future quickly becomes tedious by hand, so we enlist the aid of Excel. Fortunately, implementing such a model in Excel is no more difficult than before: we just need a column for each variable. Figure 3.2 shows the

	A	B	C
1	The Lanchester Combat Model		
2			
3	Fighting Effectiveness:		
4	For Blue, b =		0.10
5	For Red, r =		0.20
6			
7	t	$B(t)$	$R(t)$
8	0	50	100
9	1		
10	2		

FIGURE 3.2   Lanchester Excel model setup.

	A	B	C
1	Lanchester Combat		
2			
3	Fighting Effectivene		
4	For Blue, b =		0.1
5	For Red, r =		0.2
6			
7	t	$B(t)$	$R(t)$
8	0	50	100
9	=A8+1	=B8-$C$5*C8	=C8-$C$4*B8
10	=A9+1		

FIGURE 3.3   Lanchester Excel model with formulas displayed.

Excel model setup. In Figure 3.3 we show the Lanchester Excel model with formulas displayed.

Next we use our Excel model to project the battle from Example 3.1 further into the future.

Example 3.2: Use Excel to continue the battle from Example 3.1 to its conclusion. How many remaining units are there for the victor?

In Excel we drag the model equations down until one side is completely eliminated. In this case we see that it takes three time steps for the Red side to completely put Blue out of action, and at the end of the battle, Red will be left with 90.9 units. The results are shown in Figure 3.4.

In some contexts we may wish to simply round to the nearest whole unit, while for others (tanks, ships, etc.), it makes sense to view decimals as partial damage. □

	A	B	C
1	Lanchester Combat Model		
2			
3	Fighting Effectiveness:		
4	For Blue, b =		0.10
5	For Red, r =		0.20
6			
7	t	$B(t)$	$R(t)$
8	0	50	100
9	1	30.0	95.0
10	2	11.0	92.0
11	3	-7.4	90.9

FIGURE 3.4   Lanchester Excel results for Example 3.2.

Example 3.2 brings up an issue in the Lanchester model that we resolve with Excel. Namely, if we drag the model down far enough, in most cases we will end up with negative force levels for one side. Not only does this not make physical sense, but it also has the effect of serving to increase the force level of the enemy, which also does not make physical sense. What we need is a way to "turn off" the model once a force level reaches zero, and we accomplish this in Excel with the "IF" statement.

E.11   IF Statements

In this Excel section we introduce a very useful command in Excel: the IF statement.

The IF function in Excel is a way for the user to tell Excel to do one of two things based on some condition set by the user. The basic structure of the command is "IF(condition X holds, do this, otherwise do that)." The next example illustrates how the IF command works.

Example 3.3: Recall that the absolute value function is always nonnegative: the absolute value of 4 is $|4| = 4$, but the absolute value of -2 is $|-2| = 2$. Set up an "IF" statement in Excel that will take a number stored in column A starting in cell A3 and calculate its absolute value in column B starting in cell B3.

The basic structure of the "IF" statement located in cell B3 should be "IF(cell A3 contains a negative number, report the negative of the number, otherwise just report the number)." Using correct syntax we would enter "=IF(A3<0, -A3,A3)" into cell B3. Once the formula is entered into cell B3, we drag it down as usual. Figure 3.5 shows the formula setup, and Figure 3.6 shows the results.

	A	B
1	IF Statement Ex.	
2		
3	3	=IF(A3<0,-A3,A3)
4	2	=IF(A4<0,-A4,A4)
5	1	=IF(A5<0,-A5,A5)
6	-1	=IF(A6<0,-A6,A6)
7	-2	=IF(A7<0,-A7,A7)
8	-3	=IF(A8<0,-A8,A8)

FIGURE 3.5   Excel IF statement setup for Example 3.3.

	A	B
1	IF Statement Ex.	
2		
3	3	3
4	2	2
5	1	1
6	-1	1
7	-2	2
8	-3	3

FIGURE 3.6   Excel IF statement results for Example 3.3.

Notice that all values in column B are nonnegative as required. As a final note we point out that Excel has a built-in absolute value function; "=ABS(X)" will return the absolute value of the number X. □

In our combat models we want to ensure that we do not allow negative values for force levels. Thus in this context the "IF" command structure should be "IF(a force level would be negative, report '0' for the force level instead, otherwise give the force level as usual)." Since we never know when a force level may turn negative, we must include this "IF" statement as part of our original Excel formulas. Thus instead of entering a formula such as "=B8-$C$5*B8" that could be negative, we enter a formula such as "=IF(B8-$C$5*B8<0,0,B8-$C$5*B8)" that will return a value of 0 if the force level is negative. We then copy the formula down as usual. Figure 3.7 shows the new formula version of our Lanchester model.

	A	B	C
2			
3	Fighting Effectiveness:		
4	For Blue, b =		0.1
5	For Red, r =		0.2
6			
7	t	B(t)	R(t)
8	0	50	100
9	=A8+1	=IF(B8-$C$5*C8<0,0,B8-$C$5*C8)	=IF(C8-$C$4*B8<0,0,C8-$C$4*B8)
10	=A9+1	=IF(B9-$C$5*C9<0,0,B9-$C$5*C9)	=IF(C9-$C$4*B9<0,0,C9-$C$4*B9)
11	=A10+1	=IF(B10-$C$5*C10<0,0,B10-$C$5*C10)	=IF(C10-$C$4*B10<0,0,C10-$C$4*B10)
12	=A11+1	=IF(B11-$C$5*C11<0,0,B11-$C$5*C11)	=IF(C11-$C$4*B11<0,0,C11-$C$4*B11)
13	=A12+1	=IF(B12-$C$5*C12<0,0,B12-$C$5*C12)	=IF(C12-$C$4*B12<0,0,C12-$C$4*B12)

FIGURE 3.7 Modified Lanchester Excel model with formulas displayed.

	A	B	C
1	Lanchester Combat Model		
2			
3	Fighting Effectiveness:		
4	For Blue, b =		0.10
5	For Red, r =		0.20
6			
7	t	B(t)	R(t)
8	0	50	100
9	1	30.0	95.0
10	2	11.0	92.0
11	3	0.0	90.9
12	4	0.0	90.9

FIGURE 3.8 Modified Lanchester Excel model showing nonnegative output.

Next we show the new resulting output in Figure 3.8 using the parameters from Example 3.2.

Note the difference: once the Blue force level would otherwise become negative, the model keeps the level fixed at 0. □

3.1.1 The Fractional Exchange Ratio and Fighting Strength

With the basic Lanchester combat model, it is possible to determine how the battle will end from the very first time step. In fact, as we will see, it is possible to determine the victor before the fighting even starts. The reason is that as soon as one side has an advantage over the other, that advantage grows until the weaker side is defeated.

The key to determining which side has an initial advantage is the **fractional exchange ratio (FER)**, a ratio that compares the relative losses of the two sides. If we let ΔB represent the change in Blue forces and ΔR the change in Red forces, then the relative losses (or proportional change) for the Blue side will be $\frac{\Delta B}{B}$ and the relative losses (or proportional change) for the Red side will be $\frac{\Delta R}{R}$. Then we define the FER to be the ratio of relative losses for Blue to relative losses for Red:

$$\text{FER} = \frac{\Delta B/B}{\Delta R/R}.$$

We examine the FER with a computational example.

Example 3.4: Suppose that initially there are 100 Blue units and 200 Red units and that after one time step, Blue loses 20 units and Red loses 80 units. Calculate and interpret the FER.

We note that Blue has lost $\frac{\Delta B}{B} = \frac{20}{100} = 0.20$, or 20% of its units, while Red has lost $\frac{\Delta R}{R} = \frac{80}{200} = 0.40$, or 40% of its units. The FER is the relative change in Blue divided by the relative change in Red, so $\text{FER} = \frac{20\%}{40\%} = 0.50$. Because the FER is less than one, we see that Blue is losing relatively fewer units than Red. This tells us that Blue is "winning" despite having many fewer units than Red. $\square$

As we will see, the initial advantage for Blue indicated by the FER being less than one will only increase as the battle goes on, resulting in an eventual victory for Blue. If the original FER had been greater than one, that would have indicated that Blue was losing relatively more units than Red and hence would point to a Red victory. The case where the FER equals one means that both sides are sustaining equal relative losses, and in such a case the battle will result in mutual destruction.

According to our DDS, during the first round of fighting, Blue loses $rR(0)$ units so that the initial relative loss for Blue is given by $\frac{\Delta B}{B} = \frac{rR(0)}{B(0)}$. Similarly, the initial relative loss for Red is given by $\frac{\Delta R}{R} = \frac{bB(0)}{R(0)}$. Thus the FER is the ratio of these two quantities:

$$\text{FER} = \frac{rR(0)/B(0)}{bB(0)/R(0)}.$$

Simplifying slightly we can also write

$$FER = \frac{rR(0)}{B(0)} \times \frac{R(0)}{bB(0)},$$

so that our final expression for FER is given by

$$FER = \frac{rR(0)^2}{bB(0)^2}.$$

The next theorem tells us that with the FER calculated as in the last equation, we can determine the eventual victor based purely on the FER.

Theorem 3.1 With the FER defined as previously, we have the following three cases:

1. If $FER < 1$, then Blue will win.
2. If $FER > 1$, then Red will win.
3. If $FER = 1$, then both sides will be put out of action.

We refer the reader to Appendix B for a justification of this result.

Because $FER = \frac{rR(0)^2}{bB(0)^2}$, we can restate Theorem 3.1 slightly to see a very useful way of predicting the course of a battle. With the FER defined as previously, we have the following:

1. If $rR(0)^2 < bB(0)^2$, then Blue will win.
2. If $rR(0)^2 > bB(0)^2$, then Red will win.
3. If $rR(0)^2 = bB(0)^2$, then both sides will be put out of action.

The quantities $rR(0)^2$ and $bB(0)^2$ provide an overall measure of the strength of a force that takes into account both the fighting effectiveness and the number of units. These are known as the **fighting strengths** of Red and Blue, respectively, and our previous work indicates that fighting strength is the key determinant of which side will prevail in a battle: whichever side has the greater initial fighting strength will win. We refer to the situation in Case 3 where both sides have equal fighting strengths as **parity**, and in the case of parity, we get mutual destruction.

Next we provide a quick computational example.

Example 3.5: Suppose that initially there are 100 Red forces with fighting effectiveness $r = 0.10$ and 50 Blue forces with fighting effectiveness $b = 0.30$. Use fighting strength to determine who will win the battle.

Note that this is not an example where we can rely on intuition. Red has twice as many forces as Blue, but Blue's forces are three times as effective as Red's. It is not clear which of these two advantages will win the day. To settle it we calculate the fighting strengths of both sides. The fighting strength for Red is given by $rR(0)^2 = 0.10 \cdot 100^2 = 1000$. The fighting strength for Blue is given by $bB(0)^2 = 0.30 \cdot 50^2 = 750$. Since the fighting strength for Red is larger than the fighting strength for Blue, Red will win the battle. We confirm this prediction with the Excel output given in Figure 3.9. □

	A	B	C
1	Lanchester Combat Model		
2			
3	Fighting Effectiveness:		
4	For Blue, b =		0.30
5	For Red, r =		0.10
6			
7	t	$B(t)$	$R(t)$
8	0	50	100
9	1	40.0	85.0
10	2	31.5	73.0
11	3	24.2	63.6
12	4	17.8	56.3
13	5	12.2	50.9
14	6	7.1	47.3
15	7	2.4	45.1
16	8	0.0	44.4

FIGURE 3.9   Excel confirmation of Example 3.5 results.

The previous example raises an important question: what is the relative benefit of increasing the effectiveness of forces versus increasing their numbers? We examine this question in the following examples.

Example 3.6: Project the course of a battle between two equal forces that are equally effective. In particular assume that $B(0) = R(0) = 100$ and $b = r = 0.20$.

In terms of fighting strengths, note that we have parity between the two forces—each has a fighting strength equal to $0.20 \cdot 100^2 = 2000$. If we enter these values into our Excel spreadsheet model, we get the output shown in Figure 3.10.

We have hidden most of the rows, but the output confirms the mutual destruction for both forces that we expect from parity. □

	A	B	C
1	Lanchester Combat Model		
2			
3	Fighting Effectiveness:		
4	For Blue, b =		0.20
5	For Red, r =		0.20
6			
7	t	$B(t)$	$R(t)$
8	0	100	100
9	1	80.0	80.0
35	27	0.2	0.2
36	28	0.2	0.2

FIGURE 3.10 The Lanchester Excel model confirms mutual destruction for forces at parity.

In our next example we examine the effect of increasing the number of forces for one side.

Example 3.7: Project the course of a battle where forces are equally effective but one side has twice the number of forces. Specifically we use the parameters from Example 3.6 except now we double the initial number of units for Red to $R(0) = 200$.

Of course we expect Red to win now, and our Excel output in Figure 3.11 bears this out. In fact Red will defeat Blue in only three time steps and will have 163.2 units remaining. □

	A	B	C
1	Lanchester Combat Model		
2			
3	Fighting Effectiveness:		
4	For Blue, b =		0.20
5	For Red, r =		0.20
6			
7	t	$B(t)$	$R(t)$
8	0	100	200
9	1	60.0	180.0
10	2	24.0	168.0
11	3	0.0	163.2

FIGURE 3.11 Excel confirmation of Red victory from Example 3.7.

Suppose now that Blue starts a new training program to increase the fighting effectiveness of its units.

Example 3.8: How much would the effectiveness for Blue have to improve in order for Blue to offset Red's numbers advantage from the previous example?

Using Excel for this question is a matter of experimenting with different values for b until Blue achieves parity with Red. Any improvement beyond that point will tip the battle in favor of Blue. As the output in Figure 3.12 verifies, the value we seek is $b = 0.80$.

	A	B	C
1	Lanchester Combat Model		
2			
3	Fighting Effectiveness:		
4	For Blue, b =		0.80
5	For Red, r =		0.20
6			
7	t	$B(t)$	$R(t)$
8	0	100	200
9	1	60.0	120.0
10	2	36.0	72.0
25	17	0.0	0.0
26	18	0.0	0.0

FIGURE 3.12   Result of determining Blue's required fighting effectiveness to achieve parity with Red in Example 3.8.

We can also approach this example using fighting strengths. The fighting strength for the Red side is $rR(0)^2 = 0.20 \cdot 200^2 = 8000$. To achieve parity, the Blue side must increase its fighting effectiveness so that $bB(0)^2 = b \cdot 100^2 = 8000$. Thus we must have $b \cdot 10,000 = 8,000$ or $b = 0.80$. ☐

One way to express the result in the previous example is that when facing a force that is twice as large, the fighting effectiveness for the smaller force must be four times as large as that of the larger force in order to offset the larger force's numbers advantage. A similar calculation shows that if Red has a force three times as large as Blue, Blue's fighting effectiveness must be nine times larger than Red's to offset Red's numbers advantage. To turn this into a general observation, we consider the formula for fighting strength.

Example 3.9: Suppose Red has x times as many forces as Blue. Show that Blue's forces must be x^2 times as effective as Red's in order for Blue to achieve parity with Red.

Parity means that both sides have equal fighting strengths; thus we need to find b such that $bB(0)^2 = rR(0)^2$. Since we assume that Red has x times the number of units as Blue, we have $R(0) = xB(0)$ and the previous equality becomes

$$bB(0)^2 = rR(0)^2 = r(xB(0))^2 = rx^2B(0)^2.$$

Dividing through by $B(0)^2$ gives the result that $b = rx^2$. The presence of the squared terms in the formula for parity implies that Blue must be x^2 times as effective as Red to offset Red having x times the number of forces as Blue. $\square$

In the next section we present a historically important result that also implies Theorem 3.1.

3.1.2   Lanchester's Square Law

Our next result is the discrete-time version of Lanchester's celebrated square law, a discussion of which may be found in (MacKay, 2005). This result allows us to make good approximations to how many forces the winning side will have after the battle.

Theorem 3.2   Lanchester's Square Law: For the basic Lanchester combat model, the following identity holds for all times t:

$$rR(t)^2 - bB(t)^2 = (1-rb)^t \left[rR(0)^2 - bB(0)^2 \right].$$

See Appendix B for a justification of this result.

Because r and b are typically small, we have $(1-rb)^t \approx 1$ as long as the value for t is not too large. Thus $rR(t)^2 - bB(t)^2 \approx rR(0)^2 - bB(0)^2$. This last approximation is very useful for estimating how much of the winning force will remain after a battle is over. We illustrate how to use Lanchester's square law in the following example.

Example 3.10: Suppose that 100 Blue units engage 70 Red units where the Red units are more effective fighters with $b = 0.05$ and $r = 0.09$. Use fighting strength to predict the winner, and use Lanchester's square law to estimate the number of units remaining for the winner.

Note that the fighting strengths are $bB(0)^2 = 0.05 \cdot 100^2 = 500$ for Blue and $rR(0)^2 = 0.09 \cdot 70^2 = 441$ for Red. Because Blue's fighting strength is greater than Red's, Blue will win the battle. With Lanchester's square law, now we can say more. At the end of the battle, the number of Red units will be 0, so we will have

$$rR(t)^2 - bB(t)^2 \approx rR(0)^2 - bB(0)^2$$
$$0 - 0.05B(t)^2 \approx 441 - 500$$
$$-0.05B(t)^2 \approx -59$$
$$B(t)^2 \approx 1180$$
$$B(t) \approx \sqrt{1180} \approx 34.35.$$

Not only do we know that Blue will win, but we now know that Blue will have about 34.35 of its original 100 units remaining. We compare this prediction with the Excel output for the situation given in Figure 3.13.

	A	B	C
1	Lanchester Combat Model		
2			
3	Fighting Effectiveness:		
4	For Blue, b =		0.05
5	For Red, r =		0.09
6			
7	t	$B(t)$	$R(t)$
8	0	100	70
9	1	93.7	65.0
10	2	87.9	60.3
33	25	32.5	1.2
34	26	32.4	0.0

FIGURE 3.13   Excel output for comparison with Lanchester's square law prediction from Example 3.10.

Note that the Excel model verifies that Blue is in fact the winner and that Blue will have 32.4 units remaining at the end of the battle, which is pretty close to our estimate of 34.35. □

In the next section we apply Lanchester's model and square law to a historically significant naval battle: the Battle of Trafalgar.

3.1.3   Strategic Implications and the Battle of Trafalgar

We illustrate the power of the conclusions one can draw from Lanchester's model with a famous naval battle from 1805. For a discussion of this battle using Lanchester's original differential equations model, see *A First Course in Mathematical Modeling* by Giordano, Fox, Horton, and Weir (Giordano, Fox, Horton, & Weir, 2008).

The Battle of Trafalgar in 1805 was a battle between a British Royal Navy force of 27 ships led by Admiral Nelson and a combined French and Spanish naval force of

33 ships led by French Admiral Pierre-Charles Villeneuve. It is famous in part because of the unorthodox strategy that Admiral Nelson successfully employed in securing a very decisive victory for the British. As we will see, Nelson's strategy provides an excellent example of the consequences of Lanchester's square law.

The prevalent naval combat practice at the time was for both forces to form parallel lines of fighting ships during the engagement. Nelson knew that despite the British ship-to-ship advantage in fighting efficiency, his force would not be able to overcome the numbers advantage of the French–Spanish force through fighting effectiveness alone. If he engaged the enemy in a traditional, full force battle, the British would lose. Instead Nelson decided on a "divide and conquer" strategy whereby the British force formed two attacking lines of ships perpendicular to the French–Spanish line (see Fig. 3.14), and he divided Villeneuve's force into three smaller forces, engaging each in succession.

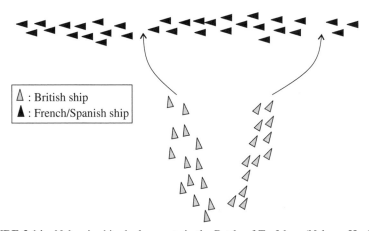

FIGURE 3.14   Nelson's ship deployments in the Battle of Trafalgar (Nelson H., 1805).

In the next example we examine the battle using the Lanchester combat model, and in so doing we lend support to Nelson's belief that a traditional battle would have been disastrous for the British.

Example 3.11: Use the Lanchester model to predict the course of a full force battle between the British and French–Spanish fleets.

With the British forces as Red and the French–Spanish forces as Blue, we assume $R(0) = 27$ and $B(0) = 33$. Like Nelson we also assume that the British were more effective fighters by letting $r = 0.10$ and $b = 0.08$. After setting our parameters to the correct values in Excel, the output shows that the British fleet would have been completely destroyed after 18 time steps, while the French–Spanish fleet would have been left with 12.4 ships at the conclusion of the battle. See Figure 3.15 for the Excel results. □

	A	B	C
1	Lanchester Combat Model		
2			
3	Fighting Effectiveness:		
4	For Blue, b =		0.08
5	For Red, r =		0.10
6			
7	t	$B(t)$	$R(t)$
8	0	33	27
9	1	30.3	24.4
25	17	12.5	0.3
26	18	12.4	0.0

FIGURE 3.15   Model projections of full fleet battle between British and French–Spanish fleets.

We know from a memorandum written by Nelson prior to the battle what he planned to do (Lanchester, 1956) to offset the French–Spanish numbers advantage, though his plan was based on numbers that turned out to be slightly off. Prior to the battle, Nelson believed that his force would comprise 40 ships while Villeneuve's would comprise 46. (So the actual numbers advantage possessed by the French–Spanish fleet turned out to be greater than what Nelson believed.) Nelson planned to split his fleet into three detachments of 16, 16, and 8 ships. The two detachments of 16 were to attack Villeneuve's line: one about a quarter of the way from the rear (about 12 ships in) and the other in the center (see Fig. 3.16).

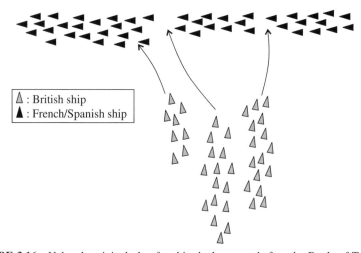

FIGURE 3.16   Nelson's original plan for ship deployments before the Battle of Trafalgar.

Thus a force of 32 British ships would take on the rear 23 ships of the French–Spanish fleet. The mission of the remaining 8 British ships was to attack just ahead of the center of Villeneuve's line to prevent the other 23 French–Spanish ships from attacking the rear. After defeating the rear half of the French–Spanish fleet, the remainder of the entire British force would turn to fight the leading half of Ville-neuve's fleet. Below is an excerpt from Nelson's original memorandum outlining his strategy:

> I should therefore probably make the Second in Command's signal to lead through, about their twelfth Ship from their Rear, (or wherever he could fetch, if not able to get so far advanced); my Line would lead through about their Centre, and the Advanced Squadron to cut two or three or four Ships ahead of their Centre, so as to ensure getting at their Commander-in-Chief, on whom every effort must be made to capture. (Nelson H., 1805)

Next we examine the wisdom of Nelson's strategy using Lanchester's square law.

Example 3.12: Investigate the effect of Nelson's strategy on the fighting strengths for the two sides.

Here the strategy is important, so we use the numbers Nelson assumed rather than the numbers from the actual battle. Thus initially we have $R(0) = 40$ British ships and $B(0) = 46$ French–Spanish ships. With the same fighting effectiveness assumptions as before, we calculate each side's fighting strength. For the British we get $rR(0)^2 = 0.10(40)^2 = 160$ and for the French–Spanish $bB(0)^2 = 0.08(46)^2 = 169.28$. The superior fighting strength of the French–Spanish fleet again suggests they would have won a full force battle.

By dividing the French–Spanish fleet in half, Nelson would reduce its total fighting strength. To confirm this we note that each half of the divided French–Spanish fleet would have a fighting strength of $0.08(23)^2 = 42.32$. To get the divided fleet's total fighting strength, we add the fighting strength of each half together to get $42.32 + 42.32 = 84.64$. Thus by dividing the French–Spanish fleet into two smaller fleets, Nelson could reduce its total fighting strength from 169.28 to 84.64—half of its original.

Nelson's fleet on the other hand was to be divided into an attacking force of 32 (both detachments of 16 attacking one half of the French–Spanish fleet together) and a force of 8. The total fighting strength of Nelson's divided fleet would thus be

$$0.10(32)^2 + 0.10(8)^2 = 102.4 + 6.4 = 108.8.$$

Since the British fighting strength of 108.8 is now greater than the French–Spanish fighting strength of 84.64, Nelson's fleet would now have the advantage. □

Nelson's strategy provides some evidence that he was at least intuitively aware of something akin to Lanchester's square law. For one thing, Nelson's decision to divide the French–Spanish fleet in two equal halves is exactly the point that the square law implies would be the ideal point. (The reader is asked to verify this claim in the exercises.)

In addition, the square law predicts that Nelson's decision to use 32 ships as his main attacking force would have left him with just enough fighting strength to defeat the two 23-ship fleets of the French–Spanish in succession even if Nelson's remaining 8 ships were lost. To see this we perform the same calculation as in Example 3.10, only this time with the numbers from Nelson's plan. The result of those calculations is that after the first battle of 32 versus 23, the British would be left with a fleet of approximately

$$rR(t)^2 - bB(t)^2 \approx rR(0)^2 - bB(0)^2$$
$$0.10R(t)^2 - 0 \approx 0.10 \cdot 32^2 - 0.08 \cdot 23^2$$
$$0.10R(t)^2 \approx 60.08$$
$$R(t)^2 \approx 600.8$$
$$R(t) \approx \sqrt{600.8} \approx 24.5.$$

Even if no ships remained from Nelson's detachment of eight, he would still expect to prevail.

Next we use our Lanchester Excel model to analyze the Battle of Trafalgar as it actually occurred. During the battle Nelson's fleet successfully divided the French–Spanish fleet of 33 into three smaller fleets of 3, 17, and 13 ships (see Fig. 3.14). Nelson dispatched 13 of his 27 ships to destroy the force of 3 first. This initial battle ended quickly. It was followed by all remaining British ships joining forces to fight the French–Spanish force of 17 and then the third and final battle against the force of 13. We now see how the Lanchester model predictions agree with the known results.

Example 3.13: Use the Lanchester Excel model to project the course of the Battle of Trafalgar.

For the first battle we set $R(0) = 13$ and $B(0) = 3$. The Excel results in Figure 3.17 show that the model predicts the battle would end after two time steps with the British retaining 12.6 of their original 13 ships. Here we interpret a decimal value as indicating a partial casualty. This could take the form of either substantial damage to a single ship or very minor damage to several ships.

Next Nelson sends the 14 ships that had been held in reserve together with the 12.6 survivors from the first battle to take on the force of 17 French–Spanish ships. Thus we reset our model so that $B(0) = 17$ and $R(0) = 26.6$. As shown in Figure 3.18, the model again predicts a swift and decisive victory for the British, this time with 5.4 casualties leaving 21.2 ships in the British force.

	A	B	C
1	Lanchester Combat Model		
2			
3	Fighting Effectiveness:		
4	For Blue, b =		0.08
5	For Red, r =		0.10
6			
7	t	$B(t)$	$R(t)$
8	0	3	13
9	1	1.7	12.8
10	2	0.4	12.6
11	3	0.0	12.6

FIGURE 3.17 Lanchester model for the Battle of Trafalgar: the first engagement.

	A	B	C
1	Lanchester Combat Model		
2			
3	Fighting Effectiveness:		
4	For Blue, b =		0.08
5	For Red, r =		0.10
6			
7	t	$B(t)$	$R(t)$
8	0	17	26.6
9	1	14.3	25.2
15	7	0.5	21.2
16	8	0.0	21.2

FIGURE 3.18 Lanchester model for the Battle of Trafalgar: the second engagement.

Finally Nelson's strategy requires the remaining 21.2 British ships to engage the final 13 French–Spanish ships. Again we reset the model to $B(0) = 13$ and $R(0) = 21.2$ with the result that the British win convincingly again with 17.2 of the original 27 ships still remaining. □

The Lanchester model predictions validate Nelson's plan: an almost sure defeat was turned into a decisive victory through the employment of better strategy and tactics. In fact, the results of the actual battle were even more one sided with the British not losing a single ship (Battle of Trafalgar, 2015), though Admiral Nelson was killed. We note that this is not necessarily at odds with our model predictions since 27 ships with substantial damage could result in 17.2 "active units."

In the next section we analyze the Lanchester model's equilibrium points.

3.1.4   Equilibrium Points

In Chapter 1 we learned about equilibrium values: values at which a discrete dynamical system does not change. We extend that idea now to discrete dynamical systems with two dependent variables. In this context we speak of **equilibrium points**, points (B^*, R^*) such that if we plug both values B^* and R^* into our DDS at the same time, the DDS does not change. For the basic Lanchester model, this means an equilibrium point will be any point (B^*, R^*) such that

$$B^* = B^* - rR^*$$
$$R^* = R^* - bB^*.$$

To actually find any equilibrium points, we need to solve the system of equations for B^* and R^*.

It turns out that the basic Lanchester model has only one equilibrium point and that is where both Blue and Red are destroyed. To verify this claim we solve for the values (B^*, R^*) such that

$$B^* = B^* - rR^*$$
$$R^* = R^* - bB^*.$$

The system reduces to

$$0 = -rR^*$$
$$0 = -bB^*.$$

Since r and b must be positive, we see the only solution is

$$0 = R^*$$
$$0 = B^*.$$

Recall, though, that we modified the basic Lanchester model in Excel to make it more realistic. Specifically, we added IF statements that would "turn off" the model once either side's forces became negative. Those IF statements introduce infinitely many equilibrium points into the model, namely, any point where one of the forces becomes zero. At such a point the other force can be any nonnegative number, and the system will still remain constant.

Analytically determining the stability properties of equilibrium points when we have two or more dependent variables involves techniques from linear algebra that are beyond the scope of this text. Instead we will proceed as we did in Chapter 1 and determine stability graphically using Excel. The difference here is that the presence of two dependent variables means we will need a new kind of graph to do so. We introduce this new type of graph in the next section.

3.1.5 Section Exercises

1 Suppose Blue begins a battle with 200 units, so $B(0) = 200$, and Red begins the battle with 100 units, so $R(0) = 100$. Each Blue unit has a fighting effectiveness of $b = 0.10$, and each Red unit has a fighting effectiveness of $r = 0.15$.

   a. Determine by hand the numbers of Blue and Red forces remaining after two time steps.

   b. Determine the eventual victor and how many forces remain at the end of the battle.

2 Suppose Blue initially has 100 units and a fighting effectiveness of $b = 0.25$. Red has a fighting effectiveness of $r = 0.30$. Determine how many units Red needs to achieve parity with Blue.

3 For the parameters given in Exercise 1, determine the following:

   a. The relative losses sustained by each side during the first time step

   b. The fractional exchange ratio (FER)

   c. The fighting strengths of each side

   d. The eventual victor based on the FER

   e. The eventual victor based on the fighting strengths

   f. How do your answers in (d) and (e) compare to the result found in Exercise 1b?

4 Suppose Blue initially has 500 units and a fighting effectiveness of $b = 0.25$. Red initially has 450 units and a fighting effectiveness of $r = 0.30$. Determine the following:

   a. The relative losses sustained by each side during the first time step

   b. The fractional exchange ratio (FER)

   c. The fighting strengths of each side

   d. The eventual victor based on the FER

   e. The eventual victor based on the fighting strengths

5 Suppose Red has 200 units and a fighting effectiveness of $r = 0.20$. If Blue has a fighting effectiveness of $b = 0.25$, how many forces would Blue need initially to completely put Red out of action in the first time step?

6 Suppose that 150 Blue units engage 80 Red units where the Red units are more effective fighters with $b = 0.06$ and $r = 0.10$.

   a. Use fighting strength to predict the winner.

   b. Use Lanchester's square law to estimate the number of units remaining for the winner.

   c. Compare your results in (a) and (b) to the Lanchester Excel model projections.

7 Suppose that 500 Blue units engage 420 Red units where the Red units are more effective fighters with $b = 0.03$ and $r = 0.07$.

    a. Use fighting strength to predict the winner.

    b. Use Lanchester's square law to estimate the number of units remaining for the winner.

    c. Compare your results in (a) and (b) to the Lanchester Excel model projections.

8 Suppose that Blue and Red engage in a battle and that the victor must then fight Green. Blue initially has 500 units and a fighting effectiveness of $b = 0.10$. Red initially has 550 units and a fighting effectiveness of $r = 0.07$. Green initially has 200 units and a fighting effectiveness of $g = 0.08$. Use Lanchester's square law to predict the eventual victor and how many forces the victor has remaining.

9 *Extension*: Suppose there is a three-way battle among Red, Blue, and Green forces. Develop a modification of the Lanchester model for this situation. You may want to introduce additional parameters.

10 *Extension*: Modify the basic Lanchester model to include "foxholes" for one or both of the combatants. Consider a foxhole to be a safe place where a unit can still attack the opposing side but cannot be harmed.

11 *Extension*: Suppose that like Admiral Nelson, Blue aims to employ a divide and conquer strategy against Red in an upcoming battle. Show that dividing the Red forces into two equal halves minimizes the total fighting strength of the two divided forces.

12 *Extension*: Show that the optimal way to divide an opposing force into three smaller forces is to divide the force into three equal forces.

13 *Extension*: Determine the limit for how much a divide and conquer strategy can diminish a force's total fighting strength. To do so consider an initial force of N units divided into N forces of 1 unit each. In other words, each unit must fight the enemy alone.

14 *Extension*: Suppose in the Battle of Trafalgar that Nelson was able to employ the ultimate divide and conquer strategy from Exercise 13 and could fight each French–Spanish ship individually. Determine how large a French–Spanish fleet Nelson could defeat in this scenario.

3.2 PHASE PLANE GRAPHS

A **phase plane graph** provides a convenient way of visualizing the behavior of any two-compartment DDS over time. Recall that at each time t, we get a value for each of our two dependent variables, and we can represent these two values as a single point $(B(t), R(t))$. The idea is that instead of the x-axis representing time as in our usual kind of graph, we put the dependent variable, B, on the x-axis and the dependent variable, R, on the y-axis. Time does not appear explicitly on the graph. However, for every time t, we get a new point $(B(t), R(t))$, and we see the effect of time as the points trace out a **trajectory** in the B,R-plane. We note that if we start a trajectory at an equilibrium point, the trajectory will end up as a single point in the phase plane since such points do not change over time.

We illustrate how to create such a graph in Excel with our next example.

Example 3.14: Recall the parameters from the second engagement in the Battle of Trafalgar: $R(0) = 26.6$, $B(0) = 17$, $r = 0.10$, and $b = 0.08$. Graph the trajectory of the engagement in the B,R-plane.

E.12 Phase Plane Graphs

In this Excel section we show how to create phase plane graphs.

After entering the relevant parameters into Excel, we get the output shown in Figure 3.19.

	A	B	C
1	Lanchester Combat Model		
2			
3	Fighting Effectiveness:		
4	For Blue, b =		0.08
5	For Red, r =		0.10
6			
7	t	$B(t)$	$R(t)$
8	0	17	26.6
9	1	14.3	25.2
10	2	11.8	24.1
11	3	9.4	23.1
12	4	7.1	22.4
13	5	4.9	21.8
14	6	2.7	21.4
15	7	0.5	21.2
16	8	0.0	21.2

FIGURE 3.19 Excel Lanchester output for Example 3.14.

To graph the trajectory of this output in the B,R-plane, we first select *only* the columns for B and R using the thick cross pointer. From there we select our usual kind of graph: a scattergraph with straight lines and markers. Excel will automatically use B for the horizontal axis and R for the vertical axis, so the graph produced will be in the B,R-plane. By selecting the Add Chart Element drop-down from the Chart Design tab, we add axis labels and a chart title to make the graph easier to understand. Our finished graph is shown in Figure 3.20.

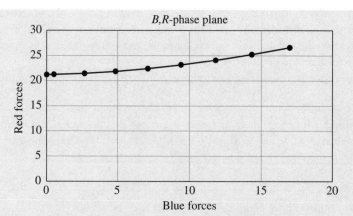

FIGURE 3.20   Excel phase plane graph for Example 3.14.

When viewing a phase plane trajectory in Excel, it is not immediately obviously which way the trajectory is being traced out as time goes on. The direction must be discerned through inspection. If we use our pointer to hover over a point on an Excel graph, a pop-up reveals the coordinates of the point. Figure 3.21 shows what this pop-up looks like.

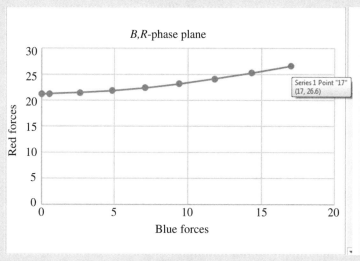

FIGURE 3.21   Determining the starting point of a trajectory on the phase plane graph.

Note that the point in the upper right of the graph is our initial point $(B(0), R(0)) = (17, 26.6)$. Thus the trajectory moves down and to the left over time, eventually ending at a point on the R-axis, indicating a victory for Red with 0 Blue forces remaining. $\square$

To realize the full potential of a phase plane graph, we set it up to graph multiple trajectories on the same graph. In other words we choose many different starting points and let the graph show us the fate of the system for each choice. In this way we get a global view for the behavior of the model.

Example 3.15: Create and analyze a phase plane diagram with multiple trajectories.

E.13 Phase Planes with Multiple Trajectories

In this Excel section we show how to add multiple trajectories to a phase plane graph.

In order to graph multiple trajectories at once, we must first create model output for multiple initial points. Here we select the three separate engagements for the Battle of Trafalgar for our three trajectories. We make columns for each engagement and note that for each of the engagements the Blue and Red forces began with different numbers of ships. The Excel setup for this is shown in Figure 3.22.

	A	B	C	D	E	F	G
1	Lanchester Combat Model						
2							
3	Fighting Effectiveness:						
4	For Blue, $b =$		0.08				
5	For Red, $r =$		0.10				
6							
7	t	B1	R1	B2	R2	B3	R3
8	0	3	13	17	26.6	13	21.2
9	1	1.7	12.8				
10	2	0.4	12.6				

FIGURE 3.22 Excel model setup for graphing multiple trajectories in the phase plane.

Next we copy the model formulas over horizontally making sure to copy the formulas for both Red and Blue simultaneously. Once that is done we copy the four new formulas down, and we are ready to graph the results.

We select all columns (except time) including the headings. We select Insert Scatter from the Charts drop-down menu, but this time we must select "More Scatter Charts" at the bottom of the drop-down. From there we select the style on the right-hand side along with Straight Lines with Markers. Then we click okay, and after some formatting we get the graph in Figure 3.23.

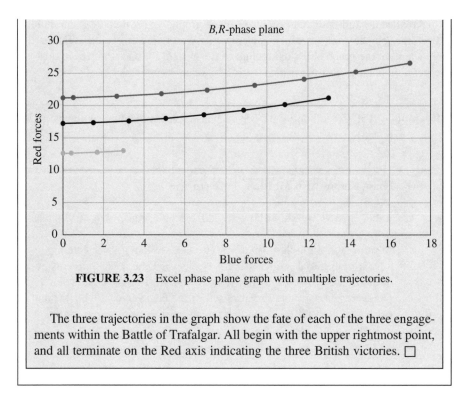

FIGURE 3.23   Excel phase plane graph with multiple trajectories.

The three trajectories in the graph show the fate of each of the three engage-ments within the Battle of Trafalgar. All begin with the upper rightmost point, and all terminate on the Red axis indicating the three British victories. ☐

We conclude this section with a phase plane diagram that shows the trajectories for more than a hundred possible starting points. This diagram uses the fighting effec-tiveness parameters from the Battle of Trafalgar and is given in Figure 3.24.

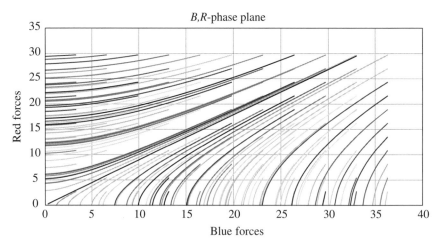

FIGURE 3.24   Excel phase plane trajectories for more than 100 initial points.

Each trajectory represents the model's projection for a battle with a different initial number of forces for Red and Blue. All trajectories that terminate along the Blue axis represent initial force levels that would have led to a French–Spanish victory. All trajectories that terminate along the Red axis represent force levels that would have led to a British victory. Notice that the heavy black diagonal line seems to be a dividing line. Above this line we get British victories and below it we get French–Spanish victories. This line is in fact exactly the set of initial points where the FER equals one. Any point above the line represents an FER greater than one and hence a Red victory. Any point below the line represents an FER less than one and hence a Blue victory. If we start a battle on this line, it will terminate at the origin with mutual destruction.

We can find the equation for the FER = 1 line by using Lanchester's square law. The formula for FER implies that if FER = 1 at the outset of the battle, then $rR(0)^2 = bB(0)^2$. This in turn implies that $rR(0)^2 - bB(0)^2 = 0$. By Lanchester's square law we know that for all t,

$$rR(t)^2 - bB(t)^2 = (1-br)^t \left[rR(0)^2 - bB(0)^2 \right] = 0.$$

Thus for all t we have $rR(t)^2 = bB(t)^2$, so $R(t)^2 = \frac{b}{r}B(t)^2$, which further implies that $R(t) = \pm \sqrt{\frac{b}{r}} B(t)$. However, only the positive root is of interest here.

What we have shown is that as Blue and Red march toward mutual destruction, they do so along the line in the B,R-phase plane that has slope $\sqrt{\frac{b}{r}}$. We note also that this is an unstable situation because any slight advantage gained by one side over the other will tip the entire battle.

3.2.1   Section Exercises

1 Suppose Blue begins a battle with 200 units, so $B(0) = 200$, and Red begins the battle with 100 units, so $R(0) = 100$. Each Blue unit has a fighting effectiveness of $b = 0.10$, and each Red unit has a fighting effectiveness of $r = 0.15$.

   a. Determine the eventual victor by plotting the trajectory of the battle in the B,R-phase plane.

   b. Explain how the result in (a) indicates the eventual victor.

   c. Produce a phase plane diagram showing the fate of several battles for different initial numbers of Blue and Red forces.

   d. Find the FER = 1 line.

   e. Plot a trajectory of mutual destruction in the B,R-phase plane.

2 Suppose Blue begins a battle with 150 units, so $B(0) = 150$, and Red begins the battle with 150 units, so $R(0) = 150$. Each Blue unit has a fighting effectiveness of $b = 0.20$, and each Red unit has a fighting effectiveness of $r = 0.16$.

  a. Determine the eventual victor by plotting the trajectory of the battle in the
  B,R-phase plane.

  b. Explain how the result in (a) indicates the eventual victor.

  c. Produce a phase plane diagram showing the fate of several battles for different
  initial numbers of Blue and Red forces.

  d. Find the FER $= 1$ line.

  e. Plot a trajectory of mutual destruction in the B,R-phase plane.

3.3   THE LANCHESTER MODEL WITH REINFORCEMENTS

In the last section we saw that when two sides have equal fighting strengths, the battle
will result in mutual destruction. However, we also noted that any advantage, however
slight, that one side can achieve over the other in fighting strength will tip the battle
in their favor. Another way to tip the balance of the battle would be for one side to
send reinforcements to the aid of their initial units.

In this section we modify the Lanchester model to allow for reinforcements.
Reinforcements are accounted for by a constant parameter representing the average
number of new units entering the battle at each time step. Let f and w be the respective
numbers of reinforcements for Blue and Red for each time step. Then in our flow
diagram, we represent these reinforcements by adding inward pointing arrows as
shown in Figure 3.25.

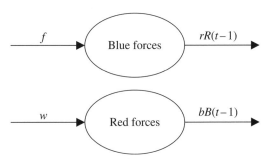

FIGURE 3.25   Flow diagram for Lanchester model with reinforcements.

The corresponding DDS for our new model is given by

$$B(t) = B(t-1) - rR(t-1) + f$$
$$R(t) = R(t-1) - bB(t-1) + w.$$

Incorporating this refinement of the basic Lanchester model in Excel is a matter of
including the two new parameters for reinforcements and adjusting the formulas

for the Blue and Red forces accordingly. The Excel setup and formula versions are shown in Figures 3.26 and 3.27.

	A	B	C	D	E	F
1	Lanchester Model with Reinforcements					
2						
3	Fighting Effectiveness:			Reinforcements:		
4	For Blue, b =		0.30	For Blue, f =		3
5	For Red, r =		0.20	For Red, w =		2
6						
7	t	$B(t)$	$R(t)$			
8	0	75	100			
9	1	58.0	72.0			

FIGURE 3.26 Lanchester with reinforcements Excel setup.

	A	B	C
1	Lanchester Model w		
2			
3	Fighting Effectivene		
4	For Blue, b =		0.3
5	For Red, r =		0.2
6			
7	t	$B(t)$	$R(t)$
8	0	75	100
9	=A8+1	=B8-$C$5*C8+$F$4	=C8-$C$4*B8+$F$5
10	=A9+1	=B9-$C$5*C9+$F$4	=C9-$C$4*B9+$F$5

FIGURE 3.27 Lanchester with reinforcements Excel with formulas displayed.

Next we give a computational example.

Example 3.16: Model the course of a battle if Blue initially has 75 forces and Red has 100. Assume the fighting effectiveness for Blue is $b = 0.30$ and for Red is $r = 0.20$. At each time step 3 units of reinforcements arrive for Blue and 2 for Red.

Here we need to plug all parameter values into Excel and observe the output. We see from the output in Figure 3.28 that Red will win the battle in 13 time steps and will have 28.5 units remaining. □

	A	B	C	D	E	F
1	Lanchester Model with Reinforcements					
2						
3	Fighting Effectiveness:			Reinforcements:		
4	For Blue, b =		0.30	For Blue, f =	3	
5	For Red, r =		0.20	For Red, w =	2	
6						
7	t	B(t)	R(t)			
8	0	75	100			
9	1	58.0	79.5			
20	12	1.6	27.0			
21	13	-0.8	28.5			

FIGURE 3.28 Lanchester with reinforcements Excel output for Example 3.16.

Note that the number of Blue forces turns negative at step 13, which is a physically meaningless result. We can avoid such output by employing an IF statement to turn the model off as we did for the original Lanchester model.

3.3.1 Equilibrium Points and Global Behavior

To find any equilibrium points for the model, we must solve the system

$$B^* = B^* - rR^* + f$$
$$R^* = R^* - bB^* + w$$

for B^* and R^*. After simplifying both equations, we get

$$0 = -rR^* + f$$
$$0 = -bB^* + w.$$

This leads to only one solution:

$$R^* = \frac{f}{r}$$
$$B^* = \frac{w}{b}.$$

The next example shows the basic calculation.

Example 3.17: Find the equilibrium points for the previous example and verify them with Excel.

For Blue we get $B^* = \frac{w}{b} = \frac{2}{0.30} = 6\frac{2}{3}$, and for Red we get $R^* = \frac{f}{r} = \frac{3}{0.20} = 15$. Plugging these values into Excel for our initial numbers of forces, we note that the model does not change from that point. By trying different initial numbers that are slightly higher or slightly lower than the equilibrium values, we see that the equilibrium point appears to be unstable. $\square$

We conclude this section by noting that including an IF statement like the one in the case of the original Lanchester Excel model does not introduce any new equilibrium points here. The continual replenishment on both sides by reinforcements prevents such points from being equilibrium points. For this model a more practical IF statement might be one that completely turns off the model when a side falls below zero. In this case the Excel model would report negative values as zero and would also turn off the addition of reinforcements for both forces. An exercise at the end of this section asks the reader to add such an IF statement to the reinforcements model.

3.3.2   The Battle of Iwo Jima

Real data is difficult to come by when trying to validate a combat model. The nature of warfare makes it very difficult to collect reliable data in the first place, and when it is collected, it is often classified information. One of the few examples of a major historical battle for which we do have data available is the Battle of Iwo Jima, where US forces captured the island of Iwo Jima off the coast of Japan from the Japanese in World War II.

In his paper *A Verification of Lanchester's Law*, J.H. Engel presents a detailed discussion of fitting the Lanchester model to the available data from Iwo Jima with very good results (Engel, 1954). Data provided by Captain Clifford P. Morehouse in *The Iwo Jima Operation* (Morehouse, 1946) indicates that the United States landed 54,000 troops on the first day of the battle and that there were 21,500 Japanese troops on the island. Thus we have $B(0) = 54,000$ and $R(0) = 21,500$. We also know that the United States landed reinforcements during the battle: 6,000 on day 2 and 13,000 on day 5. Note that as we would expect, reinforcements did not arrive at a constant rate as we assumed in the previous section. We need to modify our spreadsheet similar to the way we handled irregular future payments in the present value section of Chapter 2: we give reinforcements their own column and enter any reinforcements by hand on the appropriate day. The Japanese had no reinforcements. Time units are given in days. In Figure 3.29 we show the setup of our Excel spreadsheet with formulas displayed.

We readily see that the United States possessed a substantial numbers advantage over the Japanese both initially and with the reinforcements but that advantage was

	A	B	C	D	E
1	Lanchester Model w				
2					
3	Fighting Effectivene				
4	For Blue, *b* =		0.0088		
5	For Red, *r* =		0.0113		
6				Reinforcements	
7	*t*	*B(t)*	*R(t)*	Blue	Red
8	0	54000	21500		
9	=A8+1	=B8-$C$5*C8+D8	=C8-$C$4*B8+E8		
10	=A9+1	=B9-$C$5*C9+D9	=C9-$C$4*B9+E9		

FIGURE 3.29   Lanchester with variable reinforcements Excel model.

partially offset by the superior fighting effectiveness of the Japanese troops. Engel determined that $b=0.0088$ and $r=0.0113$. Thus each active US troop on average was responsible for killing 0.0088 Japanese troops per day, while each active Japanese troop on average was responsible for killing 0.0113 US troops per day. The Japanese troops were thus $0.0113/0.0088 \approx 1.3$ times as effective as the US troops. However, as we will see in the next example, that effectiveness advantage was not nearly great enough to overcome the superior numbers of the United States.

Example 3.18:   Use the variable reinforcement model with troop data from above to project the course of the Battle of Iwo Jima.

The setup and output of the model simulation (with most rows hidden) are shown in Figure 3.30.

	A	B	C	D	E
1	Lanchester Model with Variable Reinforcements				
2					
3	Fighting Effectiveness:				
4	For Blue, *b* =		0.0088		
5	For Red, *r* =		0.0113		
6				Reinforcements:	
7	*t*	*B(t)*	*R(t)*	Blue	Red
8	0	54,000	21,500		
9	1	53,757	21,025		
10	2	53,519	20,552	6,000	
11	3	59,287	20,081		
12	4	59,060	19,559		
13	5	58,839	19,039	13,000	
14	6	71,624	18,522		
44	36	68,378	164		
45	37	68,376	0		

FIGURE 3.30   Model output for the Battle of Iwo Jima.

The output shows a battle that lasted 37 days with 68,376 US troops remaining and 0 Japanese troops remaining. Since a total of 73,000 US troops participated in the battle, we have a total of 4,624 US fatalities as compared to 21,500 fatalities for the Japanese. These figures agree remarkably well with the actual historical data from the battle: the battle lasted 36 days with 4590 US fatalities, and all Japanese troops killed. Of course we should emphasize that the remarkable fit of the model to the data is due to parameters being estimated after the fact. Due to the nature of warfare, it is very difficult if not impossible to estimate parameters accurately beforehand. □

It is interesting to see how the model-predicted outcome would have been different had the United States not supplied reinforcements.

Example 3.19: Predict the course of the Battle of Iwo Jima had the United States not sent reinforcements.

In this scenario we keep all model parameters the same except that we delete all entries in the reinforcements column for Blue. We find that the eventual outcome of the battle would not have changed, but it would have taken about 49 days for the United States to win and it would have cost nearly 5926 US lives. (See output in Fig. 3.31.)

	A	B	C	D	E
1	Lanchester Model with Variable Reinforcements				
2					
3	Fighting Effectiveness:				
4	For Blue, b =		0.0088		
5	For Red, r =		0.0113		
6				Reinforcements:	
7	t	$B(t)$	$R(t)$	Blue	Red
8	0	54,000	21,500		
9	1	53,757	21,025		
56	48	48,078	318		
57	49	48,074	0		

FIGURE 3.31 Model projection for the Battle of Iwo Jima assuming no reinforcements.

The reinforcements, though not crucial to secure victory, allowed the United States to do so while saving approximately 1300 US soldiers. □

In the next section we turn our attention to a naval combat model due to Hughes.

3.3.3   Section Exercises

1   Model the course of a battle if Blue initially has 250 forces and Red has 230. Assume the fighting effectiveness for Blue is $b = 0.08$ and for Red is $r = 0.10$. At each time step 5 units of reinforcements arrive for Blue and 4 for Red.

    a. Determine by hand the numbers of Blue and Red forces remaining after two time steps.

    b. Determine the eventual victor and how many forces remain at the end of the battle.

2   Model the course of a battle if Blue initially has 100,000 forces and Red has 12,000. Assume the fighting effectiveness for Blue is $b = 0.06$ and for Red is $r = 0.08$. At each time step 200 units of reinforcements arrive for Blue and 50 for Red.

    a. Determine by hand the numbers of Blue and Red forces remaining after two time steps.

    b. Determine the eventual victor and how many forces remain at the end of the battle.

3   Determine the equilibrium point for the situation described in Exercise 1. Confirm the result with Excel.

4   Determine the equilibrium point for the situation described in Exercise 2. Confirm the result with Excel.

5   Use Excel to determine the stability of the equilibrium point in Exercise 3.

6   Use Excel to determine the stability of the equilibrium point in Exercise 4.

7   Blue initially has 100,000 forces and Red has 15,000. Assume the fighting effectiveness for Blue is $b = 0.08$ and for Red is $r = 0.10$. Assuming that Red has no reinforcements, determine how many reinforcements Blue needs per day in order to secure victory.

8   *Extension*: Blue initially has 100,000 forces and Red has 5,000. Assume the fighting effectiveness for Blue is $b = 0.08$ and for Red is $r = 0.10$. Blue wishes to steadily withdraw troops from this battle and deploy them elsewhere.

    a. Assuming that Red has no reinforcements, determine how many troops per day Blue can withdraw while still securing victory.

    b. Determine how many additional Blue casualties result from the withdrawal strategy.

9   *Extension*: Incorporate appropriate IF statements in the Lanchester reinforcements model to turn off the model once one side reaches a force level of 0.

10   *Extension*: Noncombat losses include such things as accident and illness and are often a large part of total overall casualties. Assume that noncombat losses are proportional to the size of the force—the larger the force, the more accidents and illnesses there will be. Modify the basic Lanchester model to include non-combat losses and include an equilibrium analysis.

11 *Extension*: Modify the Lanchester model with reinforcements to include non-combat losses. Implement the model in Excel and carry out an equilibrium analysis.

3.4   HUGHES AIMED FIRE SALVO MODEL

In military terms, a **salvo** is the simultaneous discharge of weaponry. For (nonsubmarine) naval combat, this usually means the firing of surface-to-surface antiship cruise missiles (ASCMs). Thinking of a naval battle as a succession of salvos means we are assuming the battle proceeds in discrete steps. Thus a DDS is the appropriate type of model to use to model naval salvo combat.

In 1995 Wayne P. Hughes Jr. (Hughes, 1995) applies his salvo model, originally presented in his book *Fleet Tactics: Theory and Practice* (Hughes, 1986), to the problem of comparing "the military worth of warship capabilities" (Hughes, 1993). The four attributes included in Hughes' model are (i) number of force, (ii) offensive firepower per unit, (iii) defensive power per unit, and (iv) staying power per unit. We understand offensive firepower to mean the number of well-aimed ASCMs launched per unit per salvo. The defensive power is the number of well-aimed surface-to-air missiles launched per unit per salvo, and so it represents how many enemy ASCMs a unit can destroy before they hit their target. Finally the staying power of a ship is the number of missile hits a ship can sustain before it is put out of action.

The parameters we use to represent each attribute for each side are summarized in Table 3.1.

Considering Blue first, we note that at time t the number of incoming missiles from

TABLE 3.1   Force Attributes for the Hughes Salvo Model

Attribute	Blue	Red
Offensive firepower	b	r
Defensive power	c	s
Staying power	d	u

Red will equal $rR(t-1)$. These are all *potential* hits on Blue, but not all of them will strike. Some of the incoming missiles will be intercepted by Blue, the exact number being $cB(t-1)$. Thus the total number of actual missile hits that Blue will endure with each salvo is given by $rR(t-1) - cB(t-1)$. Since it takes more than one missile hit to put a ship out of action, to find the number of lost units per salvo for Blue, we have to divide the total hits by the number of hits it takes to disable a Blue ship, namely, d. Putting all of this together, we see that the total number of lost units per salvo for Blue will be given by

$$\frac{rR(t-1)-cB(t-1)}{d}.$$

Thus the DDS equation for Blue is

$$B(t)=B(t-1)-\frac{rR(t-1)-cB(t-1)}{d}.$$

Using the same reasoning for the number of Red units, we arrive at the complete DDS for the Hughes salvo model:

$$B(t)=B(t-1)-\frac{rR(t-1)-cB(t-1)}{d}$$

$$R(t)=R(t-1)-\frac{bB(t-1)-sR(t-1)}{u}.$$

To implement the Hughes model in Excel, we need cells for each of our six parameters, and we need to be careful with our cell references when entering the formulas for Blue and Red. In Figure 3.32 we see the Excel setup, followed by a version with the formula for Blue showing in Figure 3.33.

	A	B	C	D	E	F	G
1	The Hughes Salvo Model						
2					**Blue**		**Red**
3	Offensive firepower:			$b=$	10	$r=$	7
4	Defensive power:			$c=$	6	$s=$	6
5	Staying power:			$d=$	5	$u=$	4
6							
7	t	$B(t)$	$R(t)$				
8	0	100	124				
9	1	46.4	60.0				

FIGURE 3.32 Hughes Excel model setup.

Just like for the Lanchester model, we have to be careful to deal only with values that make physical sense: we cannot have a negative number of ships. There is a second restriction for the Hughes model though: the number of ships lost in each salvo must also be nonnegative. Considering the Blue side, for example, we summarize both conditions with the inequality below:

$$0\le\frac{rR(t-1)-cB(t-1)}{d}\le B(t-1).$$

	A	B	C	D	E	F	G
1	The Hughes Salvo Model						
2					**Blue**		**Red**
3	Offensive firepower:			$b =$	10	$r =$	7
4	Defensive power:			$c =$	6	$s =$	6
5	Staying power:			$d =$	5	$u =$	4
6							
7	t	$B(t)$	$R(t)$				
8	0	100	124				
9	1	=B8-($G$3*C8-$E$4*B8)/$E$5					

FIGURE 3.33 Hughes Excel model with formula for Blue units.

The first inequality sign requires that the number of ships lost by Blue is nonnegative, and the second inequality sign ensures that the number of ships lost does not exceed the number of ships Blue has. For the Red side the corresponding inequality would be

$$0 \le \frac{bB(t-1)-sR(t-1)}{u} \le R(t-1).$$

Adding these restrictions to our Excel spreadsheet is slightly more complicated than for the Lanchester model and makes use of nested "IF" statements.

E.14 Nested IF Statements

At each step in our Hughes model, we have to check the two conditions for the number of ships lost by a side: it has to be nonnegative, and it has to be less than the number of extant ships. This means we have to check two conditions and then tell Excel what to do if the conditions are satisfied and what to do if they are not.

In the case of the first condition, if the number of ships lost by Blue is ever negative, then we assume no Blue ships are lost and we enter $B(t-1)$ as the number of Blue ships; otherwise we use the number given by the model. For the second condition, if the number of ships lost by Blue ever exceeds the number of ships Blue has, then we assume all Blue ships are eliminated and enter a "0" for the number of ships for Blue; otherwise we use the number given by the model.

A nested IF statement is an efficient way to get Excel to check multiple conditions by using one IF command inside another. The basic structure is "IF(first condition holds, do this, otherwise check IF(2nd condition holds, do this, otherwise do that))." For our situation we have "IF(# Blue lost < 0, $B(t-1)$, IF(# Blue lost > Blue, 0, use model))." In Figure 3.34 we show the new formula for Blue for our Hughes model.

B(t)
15
=IF(($G$3*C8-$E$4*B8)/$E$5<0,B8,IF(($G$3*C8-$E$4*B8)/$E$5>B8,0,B8-($G$3*C8-$E$4*B8)/$E$5))
=IF(($G$3*C9-$E$4*B9)/$E$5<0,B9,IF(($G$3*C9-$E$4*B9)/$E$5>B9,0,B9-($G$3*C9-$E$4*B9)/$E$5))

FIGURE 3.34 Hughes model formula for Blue with nested IF statements.

We try out the new model with a computational example.

Example 3.20: Consider two fleets whose attributes are given in Table 3.2. Project the course of a battle between the two forces.

TABLE 3.2   **Hughes Salvo Model Parameters for Example 3.20**

Blue		Red	
Parameter	Value	Parameter	Value
$B(0)$	16	$R(0)$	13
b	5	r	6
c	2	s	3
d	5	u	5

We enter all parameters into our Hughes Excel model and drag the model equations down until one side reaches 0. The results are given in Figure 3.35.

	A	B	C	D	E	F	G
1	The Hughes Salvo Model						
2					**Blue**		**Red**
3	Offensive firepower:			$b =$	5	$r =$	6
4	Defensive power:			$c =$	2	$s =$	3
5	Staying power:			$d =$	5	$u =$	5
6							
7	t	$B(t)$	$R(t)$				
8	0	16	13				
9	1	6.8	4.8				
10	2	3.8	0.9				
11	3	3.8	0.0				

FIGURE 3.35 Hughes Excel model output for Example 3.20.

We see that the battle lasts for 3 salvos and Blue is the victor with 3.8 ships remaining. $\square$

3.4.1   The FER and Fighting Strength

Along with the more complicated Hughes model comes a more complicated expression for the FER. Recall that the FER is a ratio of the relative losses sustained by each side. Initially the relative loss that Blue sustains is given by

$$\frac{\Delta B}{B} = \frac{(rR(0)-cB(0))/d}{B(0)} = \frac{rR(0)-cB(0)}{dB(0)}.$$

Similarly the relative loss that Red sustains is given by

$$\frac{\Delta R}{R} = \frac{(bB(0)-sR(0))/u}{R(0)} = \frac{bB(0)-sR(0)}{uR(0)}.$$

The FER is therefore given by

$$\text{FER} = \frac{(rR(0)-cB(0))/dB(0)}{(bB(0)-sR(0))/uR(0)} = \frac{(rR(0)-cB(0))}{dB(0)}\frac{uR(0)}{(bB(0)-sR(0))} = \frac{uR(0)(rR(0)-cB(0))}{dB(0)(bB(0)-sR(0))}.$$

Assuming both $dB(0)(bB(0)-sR(0))$ and $uR(0)(rR(0)-cB(0))$ are positive, we can use the FER as a predictive tool as we did with the Lanchester model: if FER <1, Blue will win; if FER >1, Red will win; and if we have parity at FER $=1$, we get mutual destruction. For the Hughes model the quantities that are akin to fighting strength are $dB(0)(bB(0)-sR(0))$ for Blue and $uR(0)(rR(0)-cB(0))$ for Red. We continue to use the term "fighting strength" in this context despite the fact that a side's fighting strength now depends in part on attributes of the adversary. We still have the result that whichever side has the greater fighting strength at the outset will win the battle, but we do not offer a proof of this result.

3.4.2 Quality versus Quantity: Assessing Warship Attributes

The purpose of the Hughes model is not to predict the course of any actual battle. Rather it is to help weigh the relative importance of (i) force numbers, (ii) offensive firepower, (iii) defensive power, and (iv) staying power. Every force desires more ships with more firepower, more defensive power, and more staying power, but the reality is that no navy can afford to increase all four attributes at the same time. Given finite resources, the question is, how should those resources be allocated? Should a navy build many weaker ships or fewer stronger ships? Should those ships be outfitted with as much offensive firepower as possible, or should more attention be paid to defensive power?

By examining the formula for fighting strength, we see that the force number appears as a squared term (as it does in the Lanchester model). All other parameters appear only as linear terms. This suggests that the single most important determinant of the course of a battle is still force number. Our next example illustrates this point.

Example 3.21: Consider two forces whose attributes are given in Table 3.3. Show that if Red were to double its offensive, defensive, and staying power all at once, Blue could offset all of those gains simply by doubling its number of forces.

Since the forces have identical capabilities and numbers, initially we expect mutual destruction, and our Excel model confirms this. Next we assume that

TABLE 3.3   Hughes Salvo Model Parameters for Example 3.21

Blue		Red	
Parameter	Value	Parameter	Value
$B(0)$	10	$R(0)$	10
b	4	r	4
c	3	s	3
d	2	u	2

◢	A	B	C	D	E	F	G
1	The Hughes Salvo Model						
2					**Blue**		**Red**
3	Offensive firepower:			$b =$	4	$r =$	8
4	Defensive power:			$c =$	3	$s =$	6
5	Staying power:			$d =$	2	$u =$	4
6							
7	t	$B(t)$	$R(t)$				
8	0	20	10				
9	1	10.0	5.0				
15	7	0.2	0.1				
16	8	0.1	0.0				
17	9	0.0	0.0				

FIGURE 3.36   Excel confirmation of result in Example 3.21.

the Red force doubles all of its capabilities. Then $r = 8$, $s = 6$, and $u = 4$. As expected these changes result in a swift and decisive victory for Red. Our claim is that Blue can offset Red's advantage by doubling its numbers. Plugging in $B(0) = 20$ into our spreadsheet confirms this (see Fig. 3.36). □

In our next example we show how to prove the previous result in general using the FER.

Example 3.22: Assume Blue and Red begin at parity so that their fighting strengths are equal. Show that if Red were to double its offensive firepower, double its defensive power, and double its staying power, Blue could offset Red's quality advantage by doubling its number of units.

Consider the fighting strengths for Red, $uR(0)(rR(0)-cB(0))$, and for Blue, $dB(0)(bB(0)-sR(0))$. We begin at parity so that FER = 1 and therefore

$$uR(0)(rR(0)-cB(0)) = dB(0)(bB(0)-sR(0)).$$

If Red doubles all of its quality parameters, the fighting strength for Red becomes $2uR(0)(2rR(0)-cB(0))$, and the fighting strength for Blue becomes $dB(0)(bB(0)-2sR(0))$. Now if Blue doubles only its force number, the fighting strength for Red becomes $2uR(0)(2rR(0)-c2B(0))$, and for Blue it becomes $d2B(0)(b2B(0)-2sR(0))$. Factoring out the 2's from each, we get $4(uR(0)(rR(0)-cB(0)))$, for Red and $4(dB(0)(bB(0)-sR(0)))$. Thus each side has quadrupled its original fighting strength, and we are back to parity. $\square$

The moral of the previous example is that a force that has twice the quality can be matched by a force half as good but twice as big.

In the examples that follow, we begin to get a sense of the relative importance of each of the attributes by examining particular situations. Unfortunately the relative importance of each attribute is highly dependent on the particular force under consideration, so it is difficult to make general claims about them.

Example 3.23: Suppose we know the force attributes for Red and Blue are as given in Table 3.4. Suppose also that Red has the resources to increase any of its quality parameters by two (or two separate attributes by one each). Determine the optimal way Red should make improvements to its fleet.

TABLE 3.4 Hughes Salvo Model Parameters for Example 3.23

Blue		Red	
Parameter	Value	Parameter	Value
$B(0)$	10	$R(0)$	10
b	12	r	8
c	7	s	7
d	3	u	7

	A	B	C	D	E	F	G
1	The Hughes Salvo Model						
2					**Blue**		**Red**
3	Offensive firepower:			$b =$	12	$r =$	9
4	Defensive power:			$c =$	7	$s =$	8
5	Staying power:			$d =$	3	$u =$	7
6							
7	t	$B(t)$	$R(t)$				
8	0	10	10				
9	1	3.3	4.3				
10	2	0.0	3.5				

FIGURE 3.37 Hughes Excel model output for Red's optimal parameter choice in Example 3.23.

As we can verify with Excel, Red is currently at a disadvantage and would quickly lose a battle against Blue. Determining how Red should best allocate its resources is a model simulation question. In other words this is a trial-and-error question that we can answer using our Excel model. Red has six possible ways of allocating the improvements: (1) increase r by 2, (2) increase s by 2, (3) increase u by 2, (4) increase r and s by 1 each, (5) increase r and u by 1 each, and (6) increase s and u by 1 each. Taking each possibility in turn and entering the correct parameter values into Excel, we note the results of each possibility. We show the output in Figure 3.37 for the best outcome, which occurs when r and s are each increased by one. The result of Red's optimal choice is a victory over Blue with 3.5 ships remaining. □

Decisions about how to allocate resources for fleet improvement or construction will be highly force dependent. In other words, changing the initial force attributes of either side can change the result of an example like the preceding one. A navy seeking to optimize its fleet may only be able to do so if it has knowledge of the capabilities of the enemy's fleet.

3.4.3   Equilibrium Points

We begin our discussion of equilibrium points by solving the system

$$B^* = B^* - \frac{(rR^* - cB^*)}{d}$$

$$R^* = R^* - \frac{(bB^* - sR^*)}{u}$$

for (B^*, R^*). The system reduces to the two equations

$$cB^* = rR^*$$

$$sR^* = bB^*.$$

These last two equations imply that $\frac{c}{r} B^* = \frac{b}{s} B^*$, which is certainly true if $B^* = 0$, which in turn would imply that $R^* = 0$. This gives us a trivial equilibrium point at $(B^*, R^*) = (0,0)$.

However if $B^* \neq 0$ we can cancel it from both sides of the equation $\frac{c}{r} B^* = \frac{b}{s} B^*$, and be left with $\frac{c}{r} = \frac{b}{s}$. Then for any $B^* \neq 0$ if $\frac{c}{r} = \frac{b}{s}$, we have an equilibrium point whenever $R^* = \frac{c}{r} B^* = \frac{b}{s} B^*$. In other words when $\frac{c}{r} = \frac{b}{s}$, we end up with an entire line of nonzero equilibrium points.

We should view this last scenario as unlikely. The equality $\frac{c}{r} = \frac{b}{s}$ implies that one of the sides must have more defensive power than offensive firepower, which would be unusual. This scenario also requires that the numbers of ships for Blue and Red are in exactly the right proportion so that neither side can hurt the other.

As in the Lanchester model, the battle only proceeds toward the mutual destruction equilibrium $(B^*, R^*) = (0,0)$ if FER = 1, that is, when the fighting strengths of the two sides are equal. Though the algebra is more complicated, FER = 1 is again represented by a straight line through the origin in the B,R-phase plane. This straight line divides the phase plane into two regions: points above the line result in victory for Red, while points below the line result in victory for Blue. The situation is generally unstable in the sense that any slight advantage given to one side will move the battle off of the FER = 1 line into one of the two regions where victory for that side is assured.

In our Excel implementation of the Hughes model, recall that due to practical considerations, we forced the model to "turn off" if either side's force level became negative. As with the Lanchester model, this introduces equilibrium points along both axes in the B,R-plane: if either side's force level becomes zero, the battle is over.

In our Excel implementation we also forced the model to avoid increases to either side's force level by not allowing the losses for each side to turn negative. We did this by using zero for a side's losses whenever that loss turned negative. In preventing negative losses we introduced another type of equilibrium point: we will get an equilibrium point whenever the losses for both sides are zero.

We examine the equilibrium points introduced by our practical considerations more carefully in the next section.

3.4.4 Overkill and Stalemates

There are two kinds of **overkill** that are important to consider in order to understand the global behavior of the Hughes model. The first is offensive overkill, where one side has more than enough offensive firepower to completely eliminate the enemy in one salvo. The second is defensive overkill, where one side has more than enough defensive power to destroy *all* incoming enemy missiles during a salvo.

If both sides are in an offensive overkill position, then we will get mutual destruction after the next salvo. If both sides are in a defensive overkill position, then the result of the battle is a **stalemate**—an equilibrium point where both sides have forces remaining but neither side is able to harm the other.

The limitations we incorporated with nested "IF" statements in Excel were exactly the requirements that describe overkill positions. Considering the DDS equation for Blue forces, we will have defensive overkill for Blue if

$$\frac{rR(t-1) - cB(t-1)}{d} \leq 0$$

and offensive overkill for Red if

$$\frac{rR(t-1) - cB(t-1)}{d} \geq B(t-1).$$

Thus when

$$0 < \frac{rR(t-1) - cB(t-1)}{d} < B(t-1).$$

some, but not all, of Blue's forces will be eliminated. After some algebra we arrive at the equivalent inequality

$$cB(t-1) < rR(t-1) < (c+d)B(t-1)$$

$$\frac{c}{r}B(t-1) < R(t-1) < \frac{c+d}{r}B(t-1).$$

The fate of the Blue side depends on the *relative* strength of the Red: if the number of Red units ever falls below $\frac{c}{r}B(t-1)$, then Red will no longer be able to harm Blue. On the other hand, if the number of Red units ever rises above $\frac{c+d}{r}B(t-1)$, then Red will completely eliminate Blue during that salvo.

After a similar analysis for the Red forces, we conclude that some, though not all, of the Red forces will be eliminated as long as

$$\frac{s}{b}R(t-1) < B(t-1) < \frac{s+u}{b}R(t-1).$$

The fate of the Red side depends on the *relative* strength of the Blue: If the number of Blue units ever falls below $\frac{s}{b}R(t-1)$, then Blue will no longer be able to harm Red. On the other hand, if the number of Blue units ever rises above $\frac{s+u}{b}R(t-1)$, then Blue will completely eliminate Red during that salvo.

It is useful to rewrite the last inequality so that the Red forces appear in the middle. Assuming positive numbers of forces, we take reciprocals, multiply through by $B(t-1)R(t-1)$, and arrive at

$$\frac{b}{s+u}B(t-1) < R(t-1) < \frac{b}{s}B(t-1).$$

If a battle is in a stalemate situation, what does this mean in terms of the model parameters? In order to have a stalemate, we must have forces remaining on both sides with neither force being able to harm the other. In terms of our inequalities, this means we must have $\frac{b}{s}B(t-1) \leq R(t-1)$ and $R(t-1) \leq \frac{c}{r}B(t-1)$. Taken together these last two inequalities imply that $\frac{b}{s}B(t-1) \leq \frac{c}{r}B(t-1)$, so $\frac{b}{s} = \frac{c}{r}$. This tells us that in order for a stalemate to occur, we must first have $br \leq cs$. In other words the product of the offensive firepower of the two sides must be less than or equal to the product of the defensive powers. Practically speaking this requirement seems very unlikely to occur since offensive firepower is generally greater than defensive power. We note that the case of equality, $\frac{b}{s} = \frac{c}{r}$, produces the equilibrium condition discussed in Section 3.4.3.

In the next example we illustrate an important implication of offensive overkill, especially for forces whose offensive firepower is high compared to their defensive power.

Example 3.24: Consider two forces whose attributes are given in Table 3.5. Analyze a battle between the two forces.

TABLE 3.5   Hughes Salvo Model Parameters for Example 3.24

Blue		Red	
Parameter	Value	Parameter	Value
$B(0)$	2	$R(0)$	4
b	24	r	9
c	16	s	1
d	2	u	1

We plug all model parameters into our Excel model and see that we get mutual destruction after a single salvo. This may come as a surprise given that Blue is the obviously superior force. To quantify how superior, we calculate the fighting strengths of each side. For Blue it is

$$dB(0)(bB(0) - sR(0)) = 2 \cdot 2(24 \cdot 2 - 1 \cdot 4) = 176.$$

For Red the fighting strength is

$$uR(0)(rR(0) - cB(0)) = 1 \cdot 4(9 \cdot 4 - 16 \cdot 2) = 16.$$

Thus Blue has 11 times greater fighting strength than Red, yet Blue is still eliminated in the first salvo. □

This example, based on an example in Hughes (1995), shows that a side with vastly inferior quality can still completely eliminate a vastly superior force. Even though Blue's superior offensive power could destroy the Red fleet several times over, Blue's lack of similar defensive power leaves it vulnerable. Blue's ability to destroy Red many times is for naught since Red still has enough firepower to eliminate Blue once.

This example points out that even a vastly superior force cannot afford to engage an enemy if the superior force has little staying power. In such a case the superior force must also have superior scouting that will enable it to launch an unanswered first salvo. As Hughes writes:

> The unstable circumstance of very strong combat power on both sides relative to their staying power argues under all circumstances in favor of delivering unanswered strikes. First effective attack is achieved by out scouting the enemy. (Hughes, 1995)

3.4.5 Phase Plane Diagram and Combat Implications

A phase plane diagram can help us make sense of all of the inequalities we have been discussing. Recall from the previous section that the fate of $B(t-1)$ is determined by the inequality $\frac{c}{r}B(t-1) < R(t-1) < \frac{c+d}{r}B(t-1)$. We can visualize each end of the inequality as a line through the origin in the B,R-phase plane, the first with slope $\frac{c}{r}$ and the second with slope $\frac{c+d}{r}$. If $R(t-1)$ falls inside the region between these two lines, then Blue will be harmed but not eliminated.

Similarly, we recall that the fate of Red is determined by the inequality $\frac{b}{s+u}B(t-1) < R(t-1) < \frac{b}{s}B(t-1)$. Each of the endpoints represents a line through the origin in the B,R-phase plane, the first with slope $\frac{b}{s+u}$ and the second with slope $\frac{b}{s}$. If $R(t-1)$ falls in the region between these two lines, then Red will be harmed but not eliminated.

The way in which these regions overlap in the phase plane changes depending on the particular choices of parameters. In the examples that follow, we examine some typical scenarios. The first example looks at the case where $\frac{c+d}{r} < \frac{b}{s+u}$.

Example 3.25: Let our parameter choices be given by Table 3.6. Analyze the global behavior of the model using a phase plane diagram.

TABLE 3.6   Hughes Salvo Model Parameters for Example 3.25

Blue		Red	
Parameter	Value	Parameter	Value
b	10	r	9
c	3	s	5
d	4	u	4

Our choices of parameters give us the following four slopes for our lines: $\frac{c}{r} \approx 0.33$, $\frac{c+d}{r} \approx 0.78$, $\frac{b}{s+u} \approx 1.11$, and $\frac{b}{s} = 2$. These four lines through the origin divide the phase plane into five distinct regions (see Fig. 3.38), each corresponding to a different fate for our model.

Working from the bottom up, we consider each region:

- Region I corresponds to points where Red will be completely eliminated in the next salvo with no harm caused to Blue.
- Region II corresponds to points where Red will be completely eliminated in the next salvo but will cause some harm to Blue.

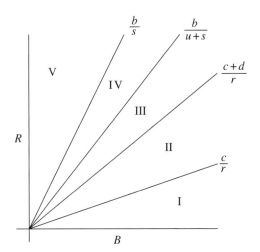

FIGURE 3.38 Phase plane diagram for Hughes model in Example 3.25.

- Region III corresponds to points where both Red and Blue will be completely eliminated in the next salvo.
- Region IV corresponds to points where Blue will be completely eliminated in the next salvo but will cause some harm to Red.
- Region V corresponds to points where Blue will be completely eliminated in the next salvo and Red will not be harmed. □

Our next example examines the case where $\frac{c}{r} < \frac{b}{s+u} < \frac{c+d}{r} < \frac{b}{s}$.

Example 3.26: Let our parameter choices be given by Table 3.7. Analyze the global behavior of the model using a phase plane diagram.

TABLE 3.7   **Hughes Salvo Model Parameters for Example 3.26**

Blue		Red	
Parameter	Value	Parameter	Value
b	7	r	7
c	3	s	5
d	4	u	4

Our choices of parameters give us the following four slopes for our lines: $\frac{c}{r} \approx 0.43$, $\frac{b}{s+u} \approx 0.78$, $\frac{c+d}{r} = 1$, and $\frac{b}{s} = 1.4$. These four lines through the origin divide the phase plane into five distinct regions (see Fig. 3.39), each corresponding to a different fate for our model.

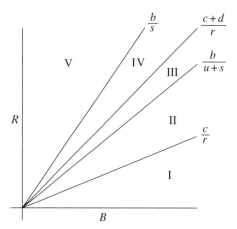

FIGURE 3.39   Phase plane diagram for Hughes model in Example 3.26.

Working from the bottom up, we consider each region:

- Region I corresponds to points where Red will be completely eliminated in the next salvo with no harm caused to Blue.
- Region II corresponds to points where Red will be completely eliminated in the next salvo but will cause some harm to Blue.
- Region III corresponds to points where both Red and Blue will be harmed but not completely eliminated in the next salvo.
- Region IV corresponds to points where Blue will be completely eliminated in the next salvo but will cause some harm to Red.
- Region V corresponds to points where Blue will be completely eliminated in the next salvo and Red will not be harmed.

Region III in this example is interesting because it is the only region where the winner is not determined. In this region we would have to subdivide it further using the FER = 1 line. (The slope of the FER = 1 line in this case is about 0.87, so it is contained in Region III. See Appendix C for a derivation of the line.) Any points in Region III above the FER = 1 line will result in a victory for Red, while any points below the FER = 1 line result in a victory for Blue. □

In the exercises the reader is invited to make a similar analysis for different parameter choices.

3.4.6 Section Exercises

1 Consider two fleets whose attributes are given in Table 3.8. Project the course of a battle between the two forces.

   a. Determine by hand the numbers of Blue and Red forces remaining after one time step.

   b. Determine the eventual victor and how many ships remain at the end of the battle.

TABLE 3.8   Hughes Salvo Model Parameters for Exercise 1

Blue		Red	
Parameter	Value	Parameter	Value
$B(0)$	17	$R(0)$	14
b	4	r	6
c	2	s	3
d	5	u	4

2 Consider two fleets whose attributes are given in Table 3.9. Project the course of a battle between the two forces.

   a. Determine by hand the numbers of Blue and Red forces remaining after one time step.

   b. Determine the eventual victor and how many ships remain at the end of the battle.

3 For the parameters given in Table 3.8, use the FER to determine the eventual winner.

4 For the parameters given in Table 3.9, use the FER to determine the eventual winner.

TABLE 3.9   Hughes Salvo Model Parameters for Exercise 2

Blue		Red	
Parameter	Value	Parameter	Value
$B(0)$	20	$R(0)$	15
b	3	r	5
c	2	s	1
d	4	u	6

5 For the parameters given in Table 3.8, use the fighting strength to determine the eventual winner.

6 For the parameters given in Table 3.9, use the fighting strength to determine the eventual winner.

7 Suppose we know the force attributes for Red and Blue are as given in Table 3.10. Suppose also that Red has the resources to increase any of its quality parameters by 1. Determine the optimal way Red should make improvements to its fleet in order to face Blue.

TABLE 3.10 **Hughes Salvo Model Parameters for Exercise 7**

Blue		Red	
Parameter	Value	Parameter	Value
$B(0)$	20	$R(0)$	15
b	4	r	5
c	2	s	2
d	4	u	6

8 Suppose we know the force attributes for Red and Blue are as given in Table 3.11. Suppose also that Red has the resources to increase any of its quality parameters by a total of 3. Thus Red can increase each parameter by one, one parameter by 3, etc. Determine the optimal way Red should make improvements to its fleet in order to face Blue.

TABLE 3.11 **Hughes Salvo Model Parameters for Exercise 8**

Blue		Red	
Parameter	Value	Parameter	Value
$B(0)$	20	$R(0)$	15
b	4	r	5
c	3	s	2
d	4	u	6

9 Let our parameter choices be given by Table 3.12. Analyze the global behavior of the model using a phase plane diagram.

TABLE 3.12 **Hughes Salvo Model Parameters for Exercise 11**

Blue		Red	
Parameter	Value	Parameter	Value
b	10	r	10
c	3	s	5
d	3	u	5

10 Let our parameter choices be given by Table 3.13. Analyze the global behavior of the model using a phase plane diagram.

TABLE 3.13   **Hughes Salvo Model Parameters for Exercise 12**

Blue		Red	
Parameter	Value	Parameter	Value
b	7	r	6
c	3	s	5
d	5	u	5

3.5   ARMSTRONG SALVO MODEL WITH AREA FIRE

The Hughes salvo model is applicable in naval combat situations where both sides know the location of the other so that all missiles can be assumed to be well aimed. For that reason the Hughes model is referred to as an **aimed fire** model. However, there are situations where enemy locations are only approximately known, and in such cases offensive fire can only target the general area where the enemy is thought to be. Situations in which area fire can arise include sea combat where enemy ships have deployed countermeasures to obscure their location and littoral combat where a sea force engages a land force whose location cannot be precisely determined. In 2013 Armstrong developed an extension of Hughes model to account for this situation (Armstrong, 2013). His model builds on the work of Mahon (2007) and is known as the salvo model with **area fire**.

The idea behind Armstrong's model is to adjust the Hughes model to account for the fact that offensive missiles are no longer well aimed. Instead, there is an element of chance to firing so that each missile has only a certain probability of being on target.

The new parameter that makes this adjustment is the **target area ratio**. The target area ratio is the ratio of the offensive missiles' area of lethality to the total area in which the enemy is known to be located. The higher the target area ratio, the better it is for the attacker. The attacker can increase its target area ratio in two ways: by increasing the destructive capability of its missiles or by increasing the precision with which it knows the enemy's location and thereby decreasing the area where the enemy is known to be. In this way the target area ratio is a measure of both the missile capability and the scouting effectiveness of the attacker.

We let e denote the target area ratio when Blue is attacking and v the target area ratio when Red is attacking. We illustrate the meaning of the target area ratio with an example.

Example 3.27: Suppose Blue knows through scouting that all Red units are located somewhere within an area of 4 mile^2. Each missile that Blue fires has an area of lethality of 0.01 mile^2. Find the target area ratio for Blue.

Here we need to plug in the given values to get the target area ratio for Blue as $e = \dfrac{0.01}{4} = 0.0025$. We can think of e as the proportion of enemy territory that is affected by a single missile. $\square$

Suppose now that there are R Red forces within the area where Red is known to be. Then on average, each missile fired by Blue will hit eR Red units. This idea is illustrated in the example below.

Example 3.28: Suppose Blue's missiles have a lethality area of 0.02 mile^2 and the Red forces are known to be somewhere in an area of 2 mile^2. If there are 100 Red forces, find the average number of Red units affected by each Blue missile.

The target area ratio for Blue is given by $e = \dfrac{0.02}{2} = 0.01$. This means that each missile fired by Blue will affect 1% of the area where the Red forces are located. Then on average each of those missiles will affect 1% of the Red forces. In other words on average each missile will affect $0.01 \times R = 0.01 \times 50 = 0.50$ Red units. $\square$

We make the general observation that with R Red forces and a target area ratio of e for Blue, the number of Red units affected on average by each Blue missile is given by eR. Similarly, if Red is attacking, then the exact same reasoning applies to give us the average number of Blue units affected by each Red missile, namely, vB.

Taking the target area ratios into account, we are now ready to build the area fire salvo model. We use the same parameter variables as for the Hughes model but add to them our target area ratios. These are summarized in Table 3.14.

TABLE 3.14   Force Attributes for the Area Fire Salvo Model

Attribute	Blue	Red
Offensive firepower	b	r
Defensive power	c	s
Staying power	d	u
Target area ratio	e	v

We consider first how the number of Blue units will change with each salvo. The number of Red missiles launched with each salvo is $rR(t-1)$. Because Red only knows the approximate location of the Blue units, each of those missiles would on average only result in $vB(t-1)$ hits on Blue. Thus with each salvo the total number of hits on Blue that would be caused by Red missiles is given by $rR(t-1)vB(t-1) = rvR(t-1)B(t-1)$.

Next we must remember that Blue will be able to shoot down some of these missiles, namely, $cB(t-1)$ of them. That leaves $rvR(t-1)B(t-1)-cB(t-1)$ missiles that will actually strike Blue units. Since it takes d Red missile hits to destroy a Blue unit, we know that with each salvo there will be

$$\frac{rvR(t-1)B(t-1)-cB(t-1)}{d}$$

Blue units destroyed. Finally we arrive at the equation for how the number of Blue units changes with each salvo:

$$B(t) = B(t-1) - \frac{rvR(t-1)B(t-1)-cB(t-1)}{d}.$$

Using the same reasoning for the number of Red units, we arrive at the final DDS for the Armstrong area fire salvo model:

$$B(t) = B(t-1) - \frac{rvR(t-1)B(t-1)-cB(t-1)}{d}$$

$$R(t) = R(t-1) - \frac{beB(t-1)R(t-1)-sR(t-1)}{u}.$$

Implementing the model in Excel is very similar to the Hughes model. The setup for the area fire Excel model is shown in Figure 3.40.

	A	B	C	D	E	F	G
1	Area Fire Salvo Model						
2					**Blue**		**Red**
3	Offensive firepower:			$b =$	8	$r =$	4
4	Defensive power:			$c =$	4	$s =$	2
5	Staying power:			$d =$	6	$u =$	3
6	Target area ratio:			$e =$	0.002	$v =$	0.001
7							
8	t	$B(t)$	$R(t)$				
9	0	200	600				
10	1	200.0	360.0				

FIGURE 3.40   Area fire salvo Excel model setup.

The area fire model equations are similar to the Hughes equations. They are given in Figure 3.41.

As with the Hughes model, we have to modify the model equations in order to ensure that our output is physically meaningful. Just like for the Hughes model, we note that the number of Blue units lost during a salvo cannot exceed the number of existing Blue units (otherwise we get a negative force level), nor can it be negative

8	t	B(t)	R(t)
9	0	150	210
10	=A9+1	=B9-($G$3*$G$6*C9*B9-$E$4*B9)/$E$5	=C9-($E$3*$E$6*B9*C9-$G$4*C9)/$G$5
11	=A10+1	=B10-($G$3*$G$6*C10*B10-$E$4*B10)/$E$5	=C10-($E$3*$E$6*B10*C10-$G$4*C10)/$G$5
12	=A11+1	=B11-($G$3*$G$6*C11*B11-$E$4*B11)/$E$5	=C11-($E$3*$E$6*B11*C11-$G$4*C11)/$G$5
13	=A12+1	=B12-($G$3*$G$6*C12*B12-$E$4*B12)/$E$5	=C12-($E$3*$E$6*B12*C12-$G$4*C12)/$G$5

FIGURE 3.41   Area fire salvo Excel model with formulas displayed.

(otherwise the force level would increase). Thus we have the restriction on the model that

$$0 \leq \frac{rvR(t-1)B(t-1)-cB(t-1)}{d} \leq B(t-1).$$

Similarly from the Red equation, we must have

$$0 \leq \frac{beB(t-1)R(t-1)-sR(t-1)}{u} \leq R(t-1).$$

We take these restrictions into account in Excel using nested "IF" statements that report force levels of 0 whenever the losses in a salvo exceed the number of forces and report the previous force level whenever the losses in a salvo are negative. The structure of the IF statements is identical to that used in the Hughes Excel model.

As a check on our Excel work, we consider the following computational example.

Example 3.29: Model the course of a battle between Blue and Red under the conditions presented in Table 3.15.

TABLE 3.15   Area Fire Model Parameters for Example 3.29

Blue		Red	
Parameter	Value	Parameter	Value
$B(0)$	100	$R(0)$	125
b	10	r	7
c	6	s	6
d	5	u	4
e	0.009	v	0.007

After plugging the required parameters into our Excel model, we get the model projection in Figure 3.42.

Under the given conditions we see that Red is eliminated after 7 salvos, with Blue retaining 97.5 of its original force of 100 or 97.5%. Note that once force levels for Red reach zero, they stay there rather than turning negative. □

	A	B	C	D	E	F	G
1	Area Fire Salvo Model						
2					**Blue**		**Red**
3	Offensive firepower:			$b =$	10	$r =$	7
4	Defensive power:			$c =$	6	$s =$	6
5	Staying power:			$d =$	5	$u =$	4
6	Target area ratio:			$e =$	0.009	$v =$	0.007
7							
8	t	$B(t)$	$R(t)$				
9	0	100	125				
10	1	97.5	31.3				
11	2	97.5	9.6				
12	3	97.5	2.9				
13	4	97.5	0.9				
14	5	97.5	0.3				
15	6	97.5	0.1				
16	7	97.5	0.0				

FIGURE 3.42   Area fire salvo Excel model projections for Example 3.29.

3.5.1 The FER and Fighting Strength

Recall that the FER is a ratio of the relative losses sustained by each side. For the area salvo model, the relative loss sustained initially by Blue is given by

$$\frac{\Delta B}{B} = \frac{(rvR(0)B(0) - cB(0))/d}{B(0)} = \frac{rvR(0)B(0) - cB(0)}{dB(0)} = \frac{rvR(0) - c}{d}.$$

Similarly the relative loss sustained initially by Red is given by

$$\frac{\Delta R}{R} = \frac{(beB(0)R(0) - sR(0))/u}{R(0)} = \frac{beB(0)R(0) - sR(0)}{uR(0)} = \frac{beB(0) - s}{u}.$$

Thus the FER is given by

$$\text{FER} = \frac{(rvR(0) - c)/d}{(beB(0) - s)/u} = \frac{(rvR(0) - c)}{d} \cdot \frac{u}{(beB(0) - s)} = \frac{u(rvR(0) - c)}{d(beB(0) - s)}.$$

Unfortunately the added complication involved in the area salvo model, specifically its nonlinearity, prevents us from making blanket statements about the outcome of the battle based on the FER. Instead we will regard it as an approximate guide: *typically we will have a Red victory if FER > 1 and a Blue victory if FER < 1*. Except under special circumstances (discussed in Section 3.5.5), the case where FER = 1, while it does indicate temporary parity, will not guarantee mutual destruction.

We define fighting strengths as $u(rvR(0)-c)$ for Red and $d(beB(0)-s)$ for Blue and make a similar qualification that whichever side has the greater initial fighting strength will typically win the battle. We note that as a practical matter, fighting strength cannot be negative. If this occurs we interpret the fighting strength as 0.

Note that in the formula for fighting strength, all parameters including force level are raised only to the first power. In other words fighting strength depends linearly on all parameters *including force level*. This is different than the situation in the Lanchester and Hughes models where force levels appear as *squared* terms with the implication that force levels were relatively much more important than the other attributes. As we will see in the next section, this difference has profound strategic implications.

3.5.2 Stealth versus Open Combat

The observation in the previous section that force level is less important in the area fire model than in the Hughes aimed fire model brings us to an interesting question of strategy: should a force engage an enemy in open combat or should it instead try to keep its location hidden even at the expense of not knowing the enemy's exact location? Put another way we are asking whether or not a particular force should prefer to fight with aimed fire as in the Hughes model or with area fire as in the Armstrong model.

We explicate the issue with an example.

Example 3.30: Suppose we have two forces with attributes given in Table 3.16. Use the Hughes salvo model and the area fire salvo model to determine which type of battle Blue should prefer.

TABLE 3.16   Salvo Model Parameters for Example 3.30

Blue		Red	
Parameter	Value	Parameter	Value
$B(0)$	200	$R(0)$	600
b	8	r	4
c	4	s	2
d	6	u	3
e	0.002	v	0.001

Note that the Blue forces have twice the quality of the Red forces in every attribute but that the Red force is three times as large. Plugging all values (except the target area ratio) into the Hughes model gives a swift and decisive victory for Red. We give the results in Figure 3.43.

In fact Red completely eliminates Blue with the first salvo while only sustaining about 133 casualties. In the aimed fire situation, Red's numbers advantage is more than enough to overcome its deficiencies in all other areas.

	A	B	C	D	E	F	G
1	The Hughes Salvo Model						
2					**Blue**		**Red**
3	Offensive firepower:			$b =$	8	$r =$	4
4	Defensive power:			$c =$	4	$s =$	2
5	Staying power:			$d =$	6	$u =$	3
6							
7	t	$B(t)$	$R(t)$				
8	0	200	600				
9	1	0.0	466.7				
10	2	0.0	466.7				

FIGURE 3.43   Hughes model results for Example 3.30.

	A	B	C	D	E	F	G
1	Area Fire Salvo Model						
2					**Blue**		**Red**
3	Offensive firepower:			$b =$	8	$r =$	4
4	Defensive power:			$c =$	4	$s =$	2
5	Staying power:			$d =$	6	$u =$	3
6	Target area ratio:			$e =$	0.002	$v =$	0.001
7							
8	t	$B(t)$	$R(t)$				
9	0	200	600				
10	1	200.0	360.0				
27	18	200.0	0.1				
28	19	200.0	0.0				

FIGURE 3.44   Area fire model results for Example 3.30.

Next we examine the same forces engaging in an area fire situation where neither knows the other's location well enough to aim missiles at particular targets. As
indicated in Figure 3.44, in this case the battle proceeds very differently.

Here Blue is in fact unharmed in the battle while Red is completely eliminated
in a few salvos. In an area fire situation, Blue's advantage in quality easily over-
comes Red's advantage in quantity. □

What this example shows is that high-quality forces—forces with effective weapons, good scouting technology, and highly trained operators—should prefer to engage
the enemy in an area fire situation in order to take full advantage of its superior qualities. Such a force should not risk being detected in order to pinpoint an enemy's location. For forces that have superior numbers but inferior quality, the situation is
reversed. These forces will prefer an aimed fire situation so that they can take full
advantage of their superior numbers. Such forces will do whatever they can to

pinpoint an enemy's location even if it means revealing themselves to the enemy in the process (Armstrong, 2013).

Next we turn our attention to the equilibrium points for the area fire salvo model.

3.5.3   Equilibrium Points

To find equilibrium points for the Armstrong model, we begin by searching for force levels (B^*, R^*) such that

$$B^* = B^* - \frac{rvR^*B^* - cB^*}{d}$$

$$R^* = R^* - \frac{beB^*R^* - sR^*}{u}.$$

Equivalently we need to solve for (B^*, R^*):

$$0 = \frac{rvR^*B^* - cB^*}{d}$$

$$0 = \frac{beB^*R^* - sR^*}{u}.$$

Clearing denominators and factoring out common terms yield the system

$$0 = B^*(rvR^* - c)$$
$$0 = R^*(beB^* - s),$$

which has two solutions: $(B^*, R^*) = (0,0)$ and $(B^*, R^*) = \left(\frac{s}{be}, \frac{c}{rv}\right)$. The first is the trivial point where there are no remaining forces on either side, and the second represents a stalemate where neither side can harm the other. Except under special circumstances (see Section 3.5.5), the course of a battle will not tend toward either equilibrium.

As in the Hughes model, our introduction of IF statements in Excel to prevent physically meaningless results introduces some equilibrium points into the model. In particular, we will have equilibrium points along both axes in the B,R-plane representing complete elimination of one side. Once one side is eliminated, the battle stops. We also introduce equilibrium points in the form of stalemates where neither side can harm the other. In fact as we will see in the following, the area fire salvo model includes an entire region of stalemate equilibrium points in the B,R-plane.

3.5.4   Overkill and Stalemates

Just as with the Hughes model, the area salvo model can exhibit offensive or defensive overkill for one or both sides. A consideration of these will help us more fully understand the behavior of the model.

If both sides are in an offensive overkill position, then we get mutual destruction, and if both sides are in a defensive overkill position, then the result is a stalemate. Considering the DDS equation for Blue forces, we will have defensive overkill for Blue if

$$\frac{rvR(t-1)B(t-1)-cB(t-1)}{d} \le 0$$

and offensive overkill for Red if

$$\frac{rvR(t-1)B(t-1)-cB(t-1)}{d} \ge B(t-1).$$

When

$$0 < \frac{rvR(t-1)B(t-1)-cB(t-1)}{d} < B(t-1),$$

some, but not all, of Blue's forces will be eliminated. After some algebra, and assuming $B(t-1)>0$, we arrive at the equivalent inequality

$$cB(t-1) < rvR(t-1)B(t-1) < (c+d)B(t-1)$$

$$\frac{c}{rv} < R(t-1) < \frac{c+d}{rv}.$$

In contrast to the Hughes model where the fate of the Blue side depends on the *relative* size of the Red force, here it depends on the *absolute* size of the Red force. If the number of Red units ever falls below $\frac{c}{rv}$, then Red will no longer be able to harm Blue, but on the other hand if the number of Red units is ever above $\frac{c+d}{rv}$, then Red will completely eliminate Blue with the next salvo.

After a similar analysis for the Red forces, we conclude that some, though not all, of the Red forces will be eliminated as long as

$$\frac{s}{be} < B(t-1) < \frac{s+u}{be}.$$

The fate of the Red side depends on the *absolute* size of the Blue forces. If the number of Blue units ever falls below $\frac{s}{be}$, then Blue will no longer be able to harm Red, but on the other hand if the number of Blue units is ever above $\frac{s+u}{be}$, then Blue will completely eliminate Red with the next salvo.

In the next example we analyze the implications of a stalemate.

Example 3.31: Suppose a battle is in a stalemate situation. What does this mean in terms of the model parameters?

In order to have a stalemate, we must have forces remaining on both sides with neither force being able to harm the other. In terms of our inequalities, this means we must have $B(t-1) \leq \frac{s}{be}$ and $R(t-1) \leq \frac{c}{rv}$. Taking the largest values for both Red and Blue would give us the point $(B,R) = \left(\frac{s}{be}, \frac{c}{rv}\right)$, our nontrivial equilibrium from Section 3.5.4 and the "maximal stalemate" point. $\square$

Next we revisit Example 3.24 where we showed that under aimed fire (Hughes) conditions, a vastly superior force can be eliminated by an inferior force with slightly higher force levels.

Example 3.32: Consider two forces whose attributes are given in Table 3.17. These are the same attributes as in Table 3.5 except we have the additional parameter for the target area ratio. Analyze an area fire battle between the two forces.

TABLE 3.17   Area Fire Model Parameters for Example 3.32

Blue		Red	
Parameter	Value	Parameter	Value
$B(0)$	2	$R(0)$	4
b	24	r	9
c	16	s	1
d	2	u	1
e	0.1	v	0.05

	A	B	C	D	E	F	G
1	Area Fire Salvo Model						
2					**Blue**		**Red**
3	Offensive firepower:			$b =$	24	$r =$	9
4	Defensive power:			$c =$	16	$s =$	1
5	Staying power:			$d =$	2	$u =$	1
6	Target area ratio:			$e =$	0.1	$v =$	0.05
7							
8	t	$B(t)$	$R(t)$				
9	0	2	4				
10	1	2.0	0.0				

FIGURE 3.45   Area fire model results for Example 3.32.

We get a very different result than we did in the aimed fire case. We see in Figure 3.45 that Blue will eliminate Red in the first salvo with no harm coming to Blue.

With area fire Blue can now take advantage of its superior quality by not engaging Red in a full force, close-range battle. In this case with Red's inferior offensive firepower being dispersed over a large area, Blue has adequate defenses to eliminate all on-target Red missiles. □

3.5.5 Phase Plane Diagram and Combat Implications

We know from our discussion of overkill in Section 3.5.4 that the absolute sizes of the Red and Blue forces (as opposed to the relative sizes) determine the fate of the battle in an area fire situation. Specifically we have already shown that if $B(t-1) \leq \frac{s}{be}$, then no harm will come to Red in the next salvo. Also if $B(t-1) \geq \frac{s+u}{be}$, then Red will be completely eliminated in the next salvo. It is only when $\frac{s}{be} < B(t-1) < \frac{s+u}{be}$ that Red will be harmed but not eliminated.

Similarly we have shown that if $R(t-1) \leq \frac{c}{rv}$, then no harm will come to Blue; if $R(t-1) \geq \frac{c+d}{rv}$, then Blue will be completely eliminated; and if $\frac{c}{rv} < R(t-1) < \frac{c+d}{rv}$, then Blue will be harmed but not eliminated.

We summarize these findings in the phase plane diagram in Figure 3.46.

Note that we have a total of nine regions for the different combinations of the three possibilities each for Red and Blue: unharmed, harmed but not eliminated, and eliminated. Inside the region where Red and Blue are both harmed but not eliminated, it is

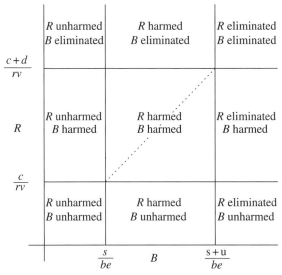

FIGURE 3.46 Phase plane diagram for area fire salvo model. Source: Adapted from Armstrong (2013), figure 1, p. 654. Reproduced with permission from John Wiley & Sons, Inc.

not clear who will win. We get an approximate dividing line for who will win by including the FER = 1 line (pictured as a dashed line in Fig. 3.46), which we find below.

Recall that the line in the B,R-phase plane representing FER = 1 is the line where $\frac{u(rvR(0)-c)}{d(beB(0)-s)} = 1$. We need to solve for $R(0)$ in terms of $B(0)$. Clearing the denominator on the left gives $urvR(0) - uc = dbeB(0) - sd$. Next we collect all constants on the right-hand side to get $urvR(0) = dbeB(0) + uc - sd$. Finally we divide by the coefficient on $R(0)$ to get the equation for our line:

$$R(0) = \frac{dbe}{urv}B(0) + \frac{c}{rv} - \frac{sd}{urv}.$$

As the reader can check by plugging in points, this line is the diagonal line in the phase plane that connects the point $\left(\frac{s}{be}, \frac{c}{rv}\right)$ to the point $\left(\frac{s+u}{be}, \frac{c+d}{rv}\right)$.

It is tempting based on our previous models to claim that this line will mark the boundary between points where Blue will win and points where Red will win. Unfortunately the area fire model is more complicated than that, and we can only regard this line as an approximate boundary.

Though in general the FER = 1 line will not give a clear boundary for the winner in an area fire situation, there is an exception. If the FER = 1 line actually passes through the origin, then we get a couple of unusual results. The first is that the line will be a firm dividing line with victory for Red above it and victory for Blue below. The second is that any point that starts on the line will result in no victory for either side. Depending on the size of the forces, battles starting on the FER = 1 line may end up in mutual destruction, a stalemate (i.e., in Region I), or they may progress ever closer to the nontrivial equilibrium point $\left(\frac{s}{be}, \frac{c}{rv}\right)$.

3.5.6   Section Exercises

1   Suppose Blue knows through scouting that all Red units are located somewhere within an area of 10 mile^2. Each missile that Blue fires has an area of lethality of 0.02 mile^2. Find the target area ratio for Blue.

2   Model the course of a battle between Blue and Red under the conditions presented in Table 3.18.

3   Suppose we have two forces with attributes given in Table 3.19. Use the Hughes salvo model and the area fire salvo model to determine which type of battle Blue should prefer.

4   For the parameters given in Table 3.18, sketch the phase plane diagram that summarizes all possible results. Test each region with an Excel example.

5   For the parameters given in Table 3.19, sketch the phase plane diagram that summarizes all possible results. Test each region with an Excel example.

TABLE 3.18   Area Fire Model Parameters for Exercise 2

Blue		Red	
Parameter	Value	Parameter	Value
$B(0)$	150	$R(0)$	140
b	7	r	6
c	5	s	7
d	4	u	4
e	0.009	v	0.007

TABLE 3.19   Area Fire Model Parameters for Exercise 3

Blue		Red	
Parameter	Value	Parameter	Value
$B(0)$	220	$R(0)$	650
b	7	r	4
c	5	s	2
d	7	u	3
e	0.003	v	0.003

6 *Extension*: Create a salvo model where one side is attacking with aimed fire and the other is attacking with area fire. An example where this situation can arise is with littoral combat where a naval fleet engages an enemy on land. In such a case which side is more likely to engage in area fire and which side in aimed fire? Justify your choice.

7 *Extension*: Modify the basic Lanchester model so that it represents an area fire situation. Carry out a full analysis of the model including numerical experiments, equilibrium analysis, and a phase plane analysis.

4

THE SPREAD OF INFECTIOUS DISEASES

Though modern methods of epidemic control have enjoyed some remarkable successes, particularly in developed parts of the world, infectious diseases are still a significant cause of much suffering around the globe. The mortality and morbidity caused by infectious diseases are certainly more acutely felt in developing countries, but epidemics such as the 1918 flu pandemic that caused an estimated 50 million deaths worldwide, including nearly 700,000 deaths in the United States, serve as reminders that infectious diseases are a truly global concern (Taubenberger & Morens, 2006).

The first known attempt to use a mathematical model to understand the dynamics of an infectious disease was carried out sometime around 1760 by the mathematician and physician Daniel Bernoulli, who was concerned with assessing the risks and advantages associated with variolation against smallpox (Bailey N. T., 1975). In London in 1855 Dr. John Snow, who is considered one of the founding fathers of modern epidemiology, determined through careful study of patterns of cholera cases that the cause of a cholera epidemic was a contaminated water supply found at the Broad Street pump (Bailey N. T., 1975). After gathering statistics on the number and location of cases, Snow convinced the authorities to remove the pump handle in order to prevent new cases (Ramsay, 2006).

The field of mathematical epidemiology was slow in developing until the groundbreaking work of Pasteur (1822–1895) and Koch (1843–1910) who conclusively established a physical basis for the cause of infectious diseases (Bailey N. T., 1975). Until their work—and the work of others such as Snow and William Boyd—infectious diseases such as cholera were believed to be caused by all manner

Models for Life: An Introduction to Discrete Mathematical Modeling with Microsoft® Office Excel®,
First Edition. Jeffrey T. Barton.
© 2016 John Wiley & Sons, Inc. Published 2016 by John Wiley & Sons, Inc.

of strange agents. Once some diseases were conclusively demonstrated to pass from person to person through direct contact, mathematics could be employed to model the process. Progress in developing the first mathematical models for this contact was made by Hamer, Ross, Kermack, McKendrick, Greenwood, Reed, Frost, Soper, and others in the early 1900s. Our initial model is based on the 1927 work of Kermack and McKendrick (1927).

The purpose of mathematical models for the spread of infectious diseases is to help answer questions of practical importance to epidemiologists such as:

1. How fast will the disease spread?
2. How many will be infected?
3. How many will remain uninfected?
4. How long will the epidemic last?

These first questions will serve as a good starting point for developing and testing our models. However, if we stopped there, then the models would be of questionable practical value. Certainly it would be beneficial for planning purposes to know that approximately 150 children are expected to catch measles during an outbreak, but it would be far better if our models could help us decide what to do to minimize the number of infections. The kinds of questions we really want answers to are those like:

1. Should people be vaccinated?
2. If so, how many need to be vaccinated?
3. How should resources be allocated between prevention and treatment?
4. Is there hope of eradication?

Mathematics can be a powerful tool in answering these questions and others. Throughout this chapter, our focus will remain on using mathematical models to answer practical questions—the kinds of questions asked by epidemiologists and health-care practitioners. We concentrate on mathematical successes, but we will also discuss the limitations of the models and mention current problems in which mathematics has some role to play.

4.1 THE S–I–R MODEL

In virtually every text that introduces mathematical models for the spread of infectious diseases, the first model that is presented is the so-called S–I–R model. This model was developed by Kermack and McKendrick (and others) in the 1920s, and it has had a profound and lasting influence on subsequent epidemiological models.

The basis of the model is the partitioning of a population into three distinct classes: those who are *s*usceptible, those who are *i*nfective, and those who have been *r*emoved. Any person who can become infected is considered susceptible. An infective is a person who has the disease and can transmit it to others. Anyone who can no longer contract the disease, either because of immunity after having the disease and recovering or because of death, is considered to be removed.

Our model notation follows naturally from this description. In any population of N individuals, we let S denote the number who are susceptible, we let I denote the number who are infective, and we let R denote the number who have been removed. Since every person in a population must fall into one of these three categories, we have the following fundamental equality:

$$S + I + R = N.$$

Like all of the situations we study, in order to make progress on our model, we must first make simplifying assumptions, and we do our best to make those assumptions explicit so that others may criticize or modify them as they see fit. Initially we assume that our population remains constant and closed during the course of an epidemic. In other words, we assume there are no births, no deaths, and no migration. When epidemics are relatively short in duration—as is the case with colds and influenza—this is not an unreasonable assumption. On the other hand, for diseases like tuberculosis, we will have to do away with this assumption to have any hope of constructing a realistic model. A further assumption of the model is that our population mixes homogeneously. For our purposes this means that people in all three categories are evenly spread throughout the population. Finally, we assume that once a person has the disease and recovers, permanent immunity is conferred; hence, once someone moves from the infective state to the removed state, he or she will remain there for the duration of the epidemic.

Since we will initially be modeling relatively short-lived epidemics, we choose our time units to be in days. Letting time $t = 0$ denote the beginning of an epidemic, we follow our usual notation and write

- $S(t) =$ the number of susceptible people after t days
- $I(t) =$ the number of infective people after t days
- $R(t) =$ the number of removed people after t days

Since we assume that the population remains constant during the course of the epidemic, we know that $S(t) + I(t) + R(t) = N$ for every t.

The structure of the flow diagram for our model is relatively straightforward since an individual can only move from S to I and from I to R. It is shown in Figure 4.1.

FIGURE 4.1 Flow diagram structure for S–I–R model.

Before developing formulas for S, I, and R, we pause to consider how we expect them to behave. Since susceptibles can only leave the category, we should expect the graph for S to only decrease over time. Similarly we should expect the graph for R to only increase because people can only flow into that compartment. Our experience with epidemics is that a few people start off sick and then more and more people get the disease. The number who are infective cannot increase indefinitely though—we have a finite number of people and eventually they move into the R category. With that in mind, we expect the graph for I to increase for a while, hit a peak, and then decrease afterward.

To actually begin building a model based on the flow diagram, we need to label the arrows, that is, we need determine how many people move from one compartment to another each day. This will involve the incorporation of two fundamental parameters based on two properties that have a significant impact on the severity of an epidemic:

1. How easily the disease is transmitted from an infective to a susceptible
2. How long an infective person remains infective

We quantify property 1 by introducing a parameter called the **effective contact rate**, which we denote by a Greek lowercase beta, β. We define β to be the average number of contacts per day that an infective person has that are sufficiently intimate to transmit the disease. What qualifies as a "sufficiently intimate" contact varies from disease to disease. Perhaps a cough in a crowded room is enough to spread the disease. Perhaps a handshake followed by a person rubbing her eyes is enough. In any case we refer to such contacts as **effective contacts**, and β counts how many of them a typical infective person has each day. Thus β is a quantitative measure of how readily the disease is transmitted.

In practice, the parameter β turns out to be extremely difficult to measure directly. It is also a surprisingly complicated number. If we examine the definition, we see that β depends on more than just the etiology of the disease itself. It also depends on societal factors such as hygiene practices, population density, and social customs. For example, we should expect the common cold to have a higher β if an epidemic occurs in a crowded preschool than if an epidemic occurs in a typical office building. Similarly, a disease that can be transmitted through shaking hands would have a much smaller β in a culture where shaking hands is not a customary greeting.

We quantify property 2 by introducing a parameter known as the **duration of infectivity**, and we denote it with a lowercase Greek delta, δ. The duration of infectivity is defined to be the average length of time that an infective person remains infective. Note that δ is not the duration of the *illness* since a person can be infectious before symptoms appear and since symptoms can persist long after a person is no longer infectious. Unlike β the parameter δ can be measured relatively easily for most diseases through direct observation.

We are now ready to label the arrows in our flow diagram, beginning with the arrow from S to I. Since β tells us how many sufficiently intimate contacts each infective person has per day, we know that each day there will be a total of $\beta \times I$ such contacts within the population. Not all of these contacts will result in a new infection, however. Remember that the population is made up of susceptibles, infectives, and those who have been removed all mixed together. Only the sufficiently intimate contacts that occur *with susceptibles* result in new infections. Thus we need to multiply the total number of sufficiently intimate contacts, $\beta \times I$, by the proportion of these that are with susceptible people, $\frac{S}{N}$. Assuming no latent period, the resulting quantity, $\beta I \frac{S}{N}$, will be the number of new infectives that appear each day. In other words, each day there will be $\beta I \frac{S}{N}$ individuals who move from S to I. This gives us the label for the first arrow in our flow diagram.

The number of people who move from infective to removed each day depends on the duration, δ. We illustrate the connection in the next example.

Example 4.1: Suppose that a disease has $\delta = 4$ and that there are 20 infective people. Recall that $\delta = 4$ means that people are infective for an average of 4 days. How many people would we expect to recover the following day?

In the absence of information to the contrary, we assume that the number of infective people can be equally divided into four categories: those who became infective today, those who became infective yesterday, those who became infective 2 days ago, and those who became infective 3 days ago. (If someone became infective 4 days ago or longer, then on average he or she would have recovered by now.) With 20 infective people, we should have roughly 5 people in each category. On average those who became infective 3 days ago will recover and hence be removed by the following day, so we estimate that 5 (or $\frac{1}{4} \times 20$) of our infectives will be removed the following day. $\square$

Following the same reasoning as in the previous example but without particular numbers, each day we expect roughly $\frac{1}{\delta}$ of our infectives to be removed. Thus, each day we will have $\frac{1}{\delta} I$ individuals moving from the infective compartment to the removed compartment. We now complete our flow diagram as shown in Figure 4.2.

FIGURE 4.2   Flow diagram for S–I–R model.

To write down the discrete dynamical system, we proceed as usual, forming an equation for each compartment based on the arrows entering and leaving. For the susceptible compartment, we only have one arrow. Since the arrow is leaving, it represents a subtraction and we have

$$S(t) = S(t-1) - \beta I(t-1) \frac{S(t-1)}{N}.$$

This equation says that the number of susceptibles one day equals the number who were susceptible the previous day minus those who became infective.

For the infective compartment, we must account for two arrows, one representing an addition and one a subtraction:

$$I(t) = I(t-1) + \beta I(t-1)\frac{S(t-1)}{N} - \frac{1}{\delta}I(t-1).$$

Translation: The number of infectives one day is equal to the number of infectives from the day before plus the new infectives minus those who recovered.

Finally, for the removed compartment, we have

$$R(t) = R(t-1) + \frac{1}{\delta}I(t-1).$$

The number in the removed compartment one day is equal to the number who were there the previous day plus the newly removed.

Altogether our S–I–R model is given by the discrete dynamical system

$$S(t) = S(t-1) - \beta I(t-1)\frac{S(t-1)}{N}$$

$$I(t) = I(t-1) + \beta I(t-1)\frac{S(t-1)}{N} - \frac{1}{\delta}I(t-1)$$

$$R(t) = R(t-1) + \frac{1}{\delta}I(t-1).$$

The next step is to create an Excel spreadsheet to allow us to investigate and analyze the model.

Not much is new here as far as Excel goes, but the equations we need to type in are more complicated, so we need to exercise extra caution. The important parameters for a particular epidemic are β, δ, and N, so we store them in their own cells for later experimentation. We will need a column each for the day, the number of susceptibles, the number of infectives, and the number who have been removed. We do not yet have a real situation to consider, so we use stand-in values for β, δ, N, $S(0)$, $I(0)$, and $R(0)$.

At the outset of an epidemic, no one has been removed yet so we set $R(0) = 0$. We choose $N = 1000$ for our total population, and we let $I(0) = 5$ so that initially there are five infectives. In order to keep our total population constant, we need to set $S(0) = N - I(0) = 995$. (No one is removed so everyone who is not infective is susceptible.) Finally we let $\beta = 1.2$ and $\delta = 3$. Note that it is okay for β and δ to not be whole numbers because they represent average values. Before entering any equations, our spreadsheet setup should appear as in Figure 4.3.

When entering the formulas, we need to be careful to refer to the appropriate columns. For example, the formula for $S(t)$ should be entered into cell B9 as "= B8 – $D$3*C8*B8/$D$5." The model with formulas displayed is given in Figure 4.4.

Note that we have automated the finding of $S(0)$ so that we only need to enter the initial number of infectives and Excel calculates the number of susceptibles. After

	A	B	C	D	E
1	S-I-R Model				
2					
3	Effective contact rate, β =			1.2	
4	Duration of infectivity, δ =			3	
5	Total population, N =			1000	
6					
7	t	$S(t)$	$I(t)$	$R(t)$	
8	0	995	5	0	
9	1				
10	2				

FIGURE 4.3   S–I–R Excel model setup.

	A	B	C	D
1	S-I-R Model			
2				
3	Effective contact rat			1.2
4	Duration of infectivit			3
5	Total population, N			1000
6				
7	t	$S(t)$	$I(t)$	$R(t)$
8	0	=D5-C8	5	0
9	=A8+1	=B8-$D$3*C8*B8/$D$5	=C8+$D$3*C8*B8/$D$5-(1/$D$4)*C8	=D8+(1/$D$4)*C8
10	=A9+1	=B9-$D$3*C9*B9/$D$5	=C9+$D$3*C9*B9/$D$5-(1/$D$4)*C9	=D9+(1/$D$4)*C9

FIGURE 4.4   S–I–R Excel model with formulas displayed.

entering the formulas and dragging them down over a period of about 1 month, we arrive at the results shown in Figure 4.5 where we have hidden most rows.

	A	B	C	D
1	S-I-R Model			
2				
3	Effective contact rate, β =			1.2
4	Duration of infectivity, δ =			3
5	Total population, N =			1000
6				
7	t	$S(t)$	$I(t)$	$R(t)$
8	0	995	5	0
9	1	989.0	9.3	1.7
38	30	12.4	0.3	987.4
39	31	12.4	0.2	987.5

FIGURE 4.5   S–I–R Excel results.

The results will be much easier to understand if we graph them. After selecting all desired columns, including time and including column headings, we choose a scatter-plot with straight lines and markers from the "Charts" group of the "Insert" tab. The result is shown in Figure 4.6.

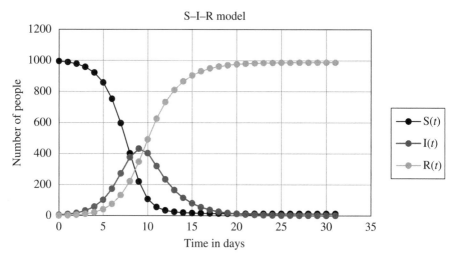

FIGURE 4.6   Graph of S–I–R model over time.

Note that the graphs fit well with our general intuition of how the graphs should look: the graph for S only decreases, the graph for R only increases, and the graph for I increases to a peak and then decreases.

We can use graphs like this to answer some natural questions about the epidemic.

Example 4.2: Given $R(0) = 0$, $N = 1000$, $I(0) = 5$, $\beta = 1.2$, and $\delta = 3$, answer the following questions based on the graph in Figure 4.6:

1. Approximately when was the epidemic at its worst?
2. How many people were sick at the peak of the epidemic?
3. Approximately how long did the epidemic last?
4. Approximately how many people contracted the disease during the course of the epidemic?

We can estimate the values on the graph by inspection, or we can get the exact Excel values by hovering over any point of interest on the graph with our cursor. Just place the cursor on the point of interest and pause. In a brief moment a box will pop up to reveal the exact values of the data point.

The answer to question 1 is the time at which the graph for I hits its peak: day 9. For number 2 we find the value for I at the peak: roughly 432. For number 3 we look for the approximate time corresponding to when the number of infectives reaches 0: roughly day 27. For number 4 we look at the total in the R category at the end of the epidemic because in order for someone to be removed, they must have first gotten sick. This number is roughly 987. □

Even this initial model has met with remarkable success when compared to real data. As we shall see in later sections, the S–I–R model is also important because it helped to make clear some fundamentally important theoretical results that have had far-reaching practical consequences. In the next section, we use the S–I–R model to model a real epidemic.

4.1.1 The Eyam Plague

Once a model has been developed, it is important to test its predictions against known data before trusting its results or considering its use in important decisions. A frequently cited such test for the S–I–R model was carried out by G. F. Raggett using the famous example of the plague that ravaged the village of Eyam, England, from 1665 to 1666 (Raggett, 1982). In this remarkable event the Reverend Mompesson convinced the citizens of Eyam to abide by a self-imposed quarantine in the hopes of sparing nearby villages. Mompesson was also responsible for keeping the detailed records that we use in the discussion that follows.

The story of the Eyam plague has continued to intrigue and inspire people to the present day. The 2001 novel *Year of Wonders* by Geraldine Brooks is a fictionalized account of the Eyam plague. The 1985 children's book *Children of Winter* by Berlie Doherty also recounts the story. Not only has the Eyam plague made its way into popular culture, but it has turned out to provide valuable clues to today's scientists for HIV research. The excerpt below outlines the fascinating link between the Eyam plague and HIV treatment (Frontline: The Age of Aids, 2006):

> In the mid-1990s, after scientists discovered that HIV needs to bind to two receptor proteins, Dr. Stephen O'Brien, chief of the Laboratory of Genomic Diversity at the National Cancer Institute, identified a small number of people who, despite repeated exposures to HIV, had not been infected. He found that these people had a mutated form of one of HIV's receptor proteins, the CCR5 protein on the surface of the CD4 cell. This genetic abnormality doesn't do any harm, and it gives those who have it immunity from HIV.
>
> But how did this particular mutation develop? Testing the DNA of direct descendants of villagers from the British town of Eyam, which went into voluntary quarantine after being infected by the plague in 1665, Dr. O'Brien discovered that those who survived the black death all had the CCR5 mutation. To confirm his theory, he tested people from all different backgrounds for the mutation. Native Americans, Africans, and Asians did not have the mutation at all, but about 14 percent of the descendants of Eyam did, as did people descended from other areas hit hard by the plague.
>
> O'Brien's discovery led to a breakthrough in HIV treatment: the third class of anti-HIV drugs developed, called fusion inhibitors, which bind to the CCR5 protein to protect CD4 cells from HIV.

Many factors contribute to making the Eyam plague a good candidate against which to test our model. We briefly discuss some of these features below and refer the reader to Raggett's article for a fuller treatment (Raggett, 1982):

- The most questionable assumption of the S–I–R model is the assumption of homogeneous mixing of the population. While we cannot hope that this is ever entirely accurate, it is certainly more reasonable for a small village with a population of 350 than, say, a major metropolitan area.
- The S–I–R model assumes that the population is free from population effects like new births or natural mortality. Since the Eyam plague lasted approximately 1 year, this assumption will not be 100% accurate. However, the time span is still short enough that it is not unreasonable to neglect the small number of births or natural deaths that would have occurred during the course of the epidemic.
- The S–I–R model assumes that the population is free from the effects of migration. Here our assumption is valid because of the remarkable self-imposed quarantine undertaken by the village.
- A real obstacle for testing the S–I–R model against historical epidemics is the lack of available data with which to compare it. Thanks to the Reverend Mompesson and some deductive reasoning by Raggett, we have a reasonable collection of data from the Eyam plague. One important, though certainly unfortunate, observation that Raggett uses is that the plague in the seventeenth century was nearly 100% fatal. Thus, the records kept of the deaths due to plague correspond almost exactly to the removed compartment in the S–I–R model; only in this instance there is no recovery (Raggett, 1982).

In the records kept of the Eyam plague, it is apparent that there were actually two outbreaks, the second being roughly twice as severe as the first. It is to this second more severe outbreak that Raggett applies the S–I–R model. The total population at the onset of the second wave was $N = 261$, and Raggett used a duration of infectivity of $\delta = 11$ days (6 days of incubation + 5 days of illness). Table 4.1 includes the Eyam

TABLE 4.1   Numbers of Susceptibles, Infectives, and Removed for the Eyam Plague, 1666, as Recorded by Reverend Mompesson

Numbers of Susceptible, Infective, and Removed during the Eyam Plague			
Date 1666	Susceptibles	Infectives	Removed
June 19	254	7	0
July 3	235	15	11
July 19	201	22	38
August 3	154	29	78
August 19	121	20	120
September 3	108	8	145
September 19	97	8	156
October 4	Incomplete data	Incomplete data	167
October 20	83	0	178

plague data collected by Reverend Mompesson and first published in 1865s *History of Eyam* by W. Wood (1865).

Note that we have $\delta = 11$, $N = 261$, $S(0) = 254$, $I(0) = 7$, and $R(0) = 0$. What remains is to find the value for β that produces the best fit to the Eyam data. Raggett presents a detailed argument for his determination of β that is beyond the scope of this text. However, by using a trial-and-error approach with Excel, we will get very close to what Raggett found. Later in this chapter we confirm our choice with an alternate algebraic method.

E.15 Date Format and Plotting Data Points

We start with our S–I–R model spreadsheet and enter the known parameters. In the cell for $t = 0$, we type in the date "June 19" and recopy the column down from there. Excel should automatically change all time values to date format. We copy all columns down until the date reaches November 20, 1 month past the end of the epidemic. Remember that instead of clicking and dragging, formulas may be copied down by double-clicking on the bottom right-hand corner.

Next we need to create columns for the actual data collected by Raggett. We call these columns *S*-Data, *I*-Data, and *R*-Data to distinguish them from our model-generated numbers. Working from Table 4.1 we type in all of the data at the appropriate date. The spreadsheet setup with all data entered should appear as in Figure 4.7. We have hidden the rows for June 21–July 1 to display more of the data. The value for β is just a stand-in value.

	A	B	C	D	E	F	G
1	S-I-R Model						
2							
3	Effective contact rate, β =			1.2			
4	Duration of infectivity, δ =			11			
5	Total population, N =			261			
6							
7	t	$S(t)$	$I(t)$	$R(t)$	S-Data	I-Data	R-Data
8	19-Jun	254	7	0	254	7	0
9	20-Jun	245.8	14.5	0.6			
21	2-Jul	0.0	117.0	144.0			
22	3-Jul	0.0	106.4	154.6	235	15	11

FIGURE 4.7 Eyam plague Excel setup.

The easiest way to find the best value for β is to graph both the model and the data on the same axes and then experiment with β until the model graph lines up as closely as possible with the data. We select all seven columns through November 20 and insert a scatterplot. Excel will automatically graph the numbers in the data columns as distinct points rather than connected lines. After some formatting changes and with the value for β set at 0.3, the graph should appear as in Figure 4.8.

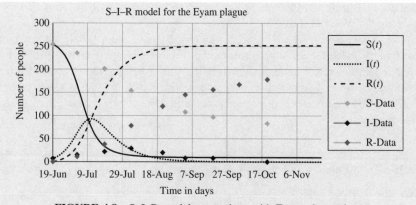

FIGURE 4.8   S–I–R model comparison with Eyam plague data.

Note that we do not yet have very good agreement with the data!

After experimenting with different values for β, we find that β should fall some-where in the range from 0.147 to 0.150. We will use $\beta = 0.149$ as this value seems to give the best visual fit. With this value of β, our graph should appear as in Figure 4.9. The graph shows a remarkably good fit.

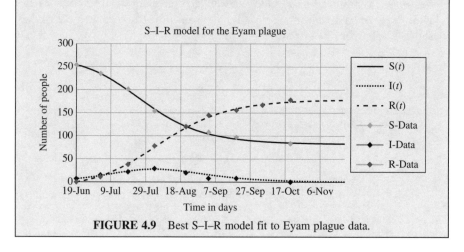

FIGURE 4.9   Best S–I–R model fit to Eyam plague data.

Having determined that $\beta = 0.149$ produces a very good fit for our data, we note that we could now use this value to predict the course of similar plague epidemics for which we may not have data.

We should point out that some have criticized Raggett's results with the Eyam pla-gue as "too good to be true" because bubonic plague is spread from infected rodents via fleas, a fact not accounted for in the model (Brauer & Castillo-Chavez, 2001). However, Murray points out that pneumonic plague is easily transmitted from person to person via coughing, and it is plausible that this was the form of plague in the Eyam outbreak (Murray, 1993). Likewise criticisms of the villagers' decision to quarantine themselves as futile or harmful also seem unfounded given the likelihood of the pneumonic form being prevalent, the possibility of flea transmission via human migration, and the fact that surrounding communities were in fact spared.

In the next section we turn our attention to a single number that provides a measure of how difficult a disease will be to control.

4.1.2   The Basic Reproductive Rate

A fundamental parameter of interest to epidemiologists is the **basic reproductive rate** of a disease. The basic reproductive rate is usually denoted by R_0 (pronounced as "R naught"), and it represents the average number of secondary infections that one infective person would produce in a wholly susceptible population. Finding an expression for R_0 is one of the first things that a modeler will do after creating a model for a disease. For the S–I–R model, we have a relatively simple way of finding R_0, namely, $R_0 = \beta \times \delta$.

The explanation for this formula depends on the definitions of β and δ. Remember that β is the number of effective contacts that an infective has each day and δ is the number of days an individual is available to spread the disease. Thus, $\beta \times \delta$ gives the total number of sufficiently intimate contacts that an infective will have during the period of infectivity. In a population made up entirely of susceptibles, each of these contacts will result in a new infective, and so $R_0 = \beta \times \delta$.

Because it takes into account both the contact rate and the duration of infectivity, R_0 is considered a good measure of how difficult a disease will be to control or eradicate. A disease with a high R_0 value indicates a disease with either a high contact rate so it can be spread easily, a long duration so it can be spread for a long time, or both.

The parameter R_0 is a difficult quantity to measure directly, but it is easier to find than β. We will see later in this chapter how to determine R_0 from real data.

Note that if we know both R_0 and δ, then we can immediately find β by computing $\frac{R_0}{\delta}$. In fact this is the usual practice for finding β: use field data to find R_0 and then deduce β from it.

For the Eyam plague, we were able to determine $\beta = 0.149$ through trial and error. Since we also know $\delta = 11$, we have that $R_0 = 0.149 \cdot 11 = 1.639$. For comparison and for use in later examples, we include in Table 4.2 estimated values for R_0 from a variety of historical epidemics (Anderson & May, 1991).

Note that like β, R_0 depends not only on the disease but also on the society in which the disease is present.

The basic reproductive rate is a fundamental disease parameter that can be estimated from the data collected on the number of infectives during an epidemic. However, its relationship to any particular collection of model parameters depends on the model chosen. We will see that as we refine the basic S–I–R model in the sections that follow, the formula $R_0 = \beta \times \delta$ will need to be updated as well.

In the next sections we use R_0 to develop some important general results.

TABLE 4.2 Estimated Values for R_0 and δ for Several Historical Epidemics

Infection	Location	Time Period	R_0	δ (Days)
Measles	Cirencester, England	1947–1950	13–14	6–7
	England and Wales	1950–1968	16–18	
	Kansas, United States	1918–1921	5–6	
	Ontario, Canada	1912–1913	11–12	
	Willesden, England	1912–1913	11–12	
	Ghana	1960–1968	14–15	
	Eastern Nigeria	1960–1968	16–17	
Pertussis	England and Wales	1944–1968	16–18	7–10
	Maryland, United States	1943	16–17	
	Ontario, Canada	1912–1913	10–11	
Chicken pox	Maryland, United States	1913–1917	7–8	10–11
	New Jersey, United States	1912–1921	7–8	
	Baltimore, United States	1943	10–11	
	England and Wales	1944–1968	10–12	
Diphtheria	New York, United States	1918–1919	4–5	2–5
	Maryland, United States	1908–1917	4–5	
Scarlet fever	Maryland, United States	1908–1917	7–8	14–21
	New York, United States	1918–1919	5–6	
	Pennsylvania, United States	1910–1916	6–7	
Mumps	Baltimore, United States	1943	7–8	4–8
	England and Wales	1960–1980	11–14	
	Netherlands	1970–1980	11–14	
Rubella	England and Wales	1960–1970	6–7	11–12
	West Germany	1970–1977	6–7	
	Czechoslovakia	1970–1977	8–9	
	Poland	1970–1977	11–12	
	Gambia	1976	15–16	
Poliomyelitis	United States	1955	5–6	14–20
	Netherlands	1960	6–7	
HIV (type 1)	England and Wales (male homosexuals)	1981–1985	2–5	
	Nairobi, Kenya (female prostitutes)	1981–1985	11–12	
	Kampala, Uganda (heterosexuals)	1985–1987	10–11	

Source: Anderson and May (1991), table 4.1, p. 70. Reproduced with permission from Oxford University Press.

4.1.3 Threshold Theorems

One of the most important contributions to mathematical epidemiology in the work of Kermack and McKendrick is their so-called threshold theorem, which was one of the first elucidations of the idea of an epidemic threshold (Bailey N. T., 1975). We describe one version of their theorem in the following text, and then we investigate its health practice consequences for vaccination programs.

An epidemic starts to wane once the number of infectives starts to decrease over time. In a closed population this means that the epidemic will die out once the outflows for the I compartment become greater than the inflows. Recall that the only outflow for I is given by the expression $\frac{1}{\delta}I(t-1)$ and the only inflow by $\beta I(t-1)\frac{S(t-1)}{N}$. Our vague notion of what leads to the demise of an epidemic can thus be expressed mathematically as

$$\frac{1}{\delta}I(t-1) > \beta I(t-1)\frac{S(t-1)}{N}.$$

To simplify our notation, we suppress the dependence on t and note that an epidemic begins to die out once $\frac{1}{\delta}I > \beta I\frac{S}{N}$.

We can safely assume that $I > 0$ since otherwise the disease would not be present at all. Thus we can divide both sides of the inequality by I without changing the direction of the inequality signs. We have $\frac{1}{\delta} > \beta\frac{S}{N}$.

We note that β must also be a positive number since otherwise the disease would not be transmissible. Dividing both sides of the inequality by β leaves the direction of the inequality unchanged and gives us $\frac{1}{\beta\delta} > \frac{S}{N}$. Since N is the total population, the right-hand side of the inequality represents the *proportion* of susceptibles in the population. Finally we recall that $R_0 = \beta \times \delta$ so that $\frac{1}{R_0} > \frac{S}{N}$.

We can now summarize all of our work in the preceding discussion with the following version of the Kermack–McKendrick threshold theorem.

Theorem 4.1   Threshold Theorem I: An epidemic begins to die out when the proportion of people who are susceptible drops below the threshold value given by

$$\frac{1}{R_0},$$

where R_0 is the basic reproductive rate of the disease.

If we carry the algebra one step further and multiply both sides of the inequality $\frac{1}{R_0} > \frac{S}{N}$ by N, we have $\frac{N}{R_0} > S$. This provides us with an equivalent version of Threshold Theorem I.

Theorem 4.2   Threshold Theorem II: An epidemic begins to die out when the number of susceptibles drops below the threshold value given by

$$\frac{N}{R_0},$$

where R_0 is the basic reproductive number of the disease and N is the total population size.

To complete our justification of the theorems, we need to tie up one loose end. Namely, how do we know that once the number of susceptibles drops below the threshold, it will stay below the threshold? In other words, we have to be sure that once the threshold is crossed, the epidemic really does die out and cannot rebound.

First we recall the equation for the number of susceptibles:

$$S(t) = S(t-1) - \beta I(t-1)\frac{S(t-1)}{N}.$$

We see that from day to day we find the new number of susceptibles by subtracting a nonnegative number. In other words, the number of susceptibles is always decreasing or constant. Thus once the number of susceptibles falls below a given value, it can never rise above it later.

In the next example we show how to apply the theorem to the Eyam plague.

Example 4.3: Determine the number of susceptibles below which the Eyam plague began to die out.

Here we apply Threshold Theorem II. The theorem tells us that if the number of people who are still susceptible to the plague falls below $\frac{N}{R_0}$, then the epidemic will begin to wane. Using the parameter values from the Eyam plague, we calculate $\frac{N}{R_0} = \frac{261}{1.639} = 159.24$. We see that 159.24 susceptibles are the threshold value, and a quick check of our Excel graph in Figure 4.10 verifies that this indeed was the point at which the plague epidemic began to wane.

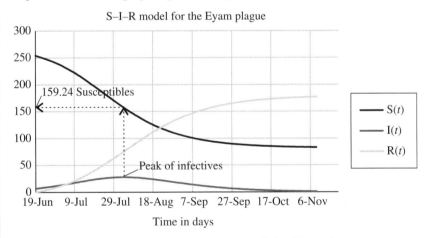

FIGURE 4.10   Threshold theorem applied to Eyam plague.

After locating the value 159.24 on the graph for susceptibles, we note that at the same point in time, the infectives graph is at its peak. Past that point, the epidemic begins to die out. □

The next example emphasizes how Threshold Theorem II can be used to get a sense of how severe a potential epidemic will be.

Example 4.4: Suppose an outbreak of diphtheria occurs in a school that enrolls 200 students. Use Threshold Theorem II to analyze the epidemic.

After consulting Table 4.2, we use $R_0 = 4$ as a reasonable value for a diphtheria outbreak. Since there are 200 students in the school, the theorem tells us that the epidemic will reach its peak when there are $\dfrac{N}{R_0} = \dfrac{200}{4} = 50$ students who remain susceptible. Thus the S–I–R model projects that over—likely well over—150 students will become ill throughout the course of the epidemic. $\square$

In the next section we connect the threshold theorems to vaccination programs and the notion of herd immunity.

4.1.4 Vaccination Programs and Herd Immunity

From the standpoint of our S–I–R model, a vaccination would have the effect of transferring an individual directly from the susceptible compartment to the removed compartment. A triumph of the S–I–R model is that it allows us to estimate how many people need to be vaccinated in order to prevent an epidemic, and it provides mathematical evidence that we need not vaccinate *everyone* in order to do so. In other words the model predicts the existence of an observed phenomenon known as **herd immunity**: the protection of an entire population from a disease achieved by vaccinating some—but not all—members of the population.

The threshold theorems in the previous section tell us that if the number, or proportion, of susceptible individuals falls below a certain value, then there will not be enough transmissions to susceptibles for the disease to progress: the epidemic will begin to die out. Since vaccinations remove people from the susceptible compartment, we can prevent an epidemic from starting by vaccinating enough people to reduce the number of remaining susceptibles to below the threshold. It is perhaps counterintuitive, but we do not need to vaccinate *everyone*. This is very good news, since as Anderson and May put it, "As it will never be possible to immunize every last individual, such indirect effects of 'herd immunity' are crucial in eradication programmes (Anderson & May, 1991)."

The theorem below reports the number of people who need to be vaccinated in order to prevent an epidemic.

Theorem 4.3   Herd Immunity Theorem: In order to prevent an epidemic, the proportion of the population vaccinated must be at least $1 - \dfrac{1}{R_0}$.

The justification for the theorem is not long. We know from Threshold Theorem I that an epidemic will die out if the proportion of susceptibles drops below $\dfrac{1}{R_0}$. Since vaccinations remove people from the susceptible compartment, we must vaccinate enough people that the proportion of remaining susceptibles is less than $\frac{1}{R_0}$. That means that the proportion who are vaccinated must be greater than $1 - \frac{1}{R_0}$. For example, if we need to reduce the proportion of susceptibles to below $\frac{1}{10}$, then we need to vaccinate more than $1 - \frac{1}{10} = \frac{9}{10}$ of the population. $\square$

Example 4.5: Determine the percent of the population that would need to be vaccinated in order to prevent an epidemic of polio similar to the one in the United States in 1955.

To determine the vaccination percentage, we first need to know R_0. For the epidemic described, we estimate from Table 4.2 that $R_0 = 5.5$. Applying the Herd Immunity Theorem gives us $1 - \frac{1}{R_0} = 1 - \frac{1}{5.5} \approx 81.8\%$. If we can manage to vaccinate at least 82% of the population, we will prevent the epidemic. $\square$

In Table 4.3 we repeat Table 4.2, this time including a column for the predicted vaccination percentage that would prevent each epidemic.

Note that the higher the R_0 the harder a disease will be to control through vaccinations. It is also interesting to note that polio has one of the lowest values of R_0 and also happens to be a disease that has successfully been eradicated in the United States.

4.1.5   Equilibrium Points

As discussed earlier in the text, equilibrium values are an important feature of any model. Knowing the equilibrium values and how the model behaves near them gives us some insight into what we should expect from the situation we are modeling. Recall that our definition of an equilibrium point is one where if we plug the values into the model, then we get the same values out of the model. For the S–I–R model, this means that an equilibrium point will be a set of values (S^*, I^*, R^*) at which the model does not change. Thus we need to solve the following system for points (S^*, I^*, R^*):

$$S^* = S^* - \beta I^* \frac{S^*}{N}$$

$$I^* = I^* + \beta I^* \frac{S^*}{N} - \frac{1}{\delta} I^*$$

$$R^* = R^* + \frac{1}{\delta} I^*.$$

After a quick simplification to each equation, we have

$$0 = -\beta I^* \frac{S^*}{N}$$

$$0 = \beta I^* \frac{S^*}{N} - \frac{1}{\delta} I^*$$

$$0 = \frac{1}{\delta} I^*.$$

TABLE 4.3 **The Vaccination Coverage Necessary to Prevent an Epidemic According to the Basic S–I–R Model**

Infection	Location	Time Period	R_0	S–I–R Vaccination %
Measles	Cirencester, England	1947–1950	13–14	92–93
	England and Wales	1950–1968	16–18	94
	Kansas, United States	1918–1921	5–6	80–83
	Ontario, Canada	1912–1913	11–12	91–92
	Willesden, England	1912–1913	11–12	91–92
	Ghana	1960–1968	14–15	93
	Eastern Nigeria	1960–1968	16–17	94
Pertussis	England and Wales	1944–1968	16–18	94
	Maryland, United States	1943	16–17	94
	Ontario, Canada	1912–1913	10–11	90–91
Chicken pox	Maryland, United States	1913–1917	7–8	86–88
	New Jersey, United States	1912–1921	7–8	86–88
	Baltimore, United States	1943	10–11	90–91
	England and Wales	1944–1968	10–12	90–92
Diphtheria	New York, United States	1918–1919	4–5	75–80
	Maryland, United States	1908–1917	4–5	75–80
Scarlet fever	Maryland, United States	1908–1917	7–8	86–88
	New York, United States	1918–1919	5–6	80–83
	Pennsylvania, United States	1910–1916	6–7	83–86
Mumps	Baltimore, United States	1943	7–8	86–88
	England and Wales	1960–1980	11–14	91–93
	Netherlands	1970–1980	11–14	91–93
Rubella	England and Wales	1960–1970	6–7	83–86
	West Germany	1970–1977	6–7	83–86
	Czechoslovakia	1970–1977	8–9	88–89
	Poland	1970–1977	11–12	91–92
	Gambia	1976	15–16	93–94
Poliomyelitis	United States	1955	5–6	80–83
	Netherlands	1960	6–7	83–86
HIV (type 1)	England and Wales (male homosexuals)	1981–1985	2–5	50–80
	Nairobi, Kenya (female prostitutes)	1981–1985	11–12	91–92
	Kampala, Uganda (heterosexuals)	1985–1987	10–11	90–91

The last equation tells us that in order to achieve equilibrium, we must have $I^* = 0$. If $I^* = 0$, then the first two equations each give us $0 = 0$. So if $I^* = 0$, S^* and R^* may be anything as long as $S^* + R^* = N$. Thus we actually have an infinite number of equilibrium points, all sharing the requirement that $I^* = 0$. The only way a disease well modeled by the S–I–R model can achieve a steady state is by disappearing altogether. Epidemics that are well modeled by the S–I–R model will never enter an **endemic** state, where the disease is ever present in the population. As we will see in the next section, the inclusion of births and deaths in the model gives rise to the possibility of endemicity.

4.1.6   Section Exercises

1   Given $R(0) = 0$, $N = 5000$, $I(0) = 5$, $\beta = 0.9$, and $\delta = 5$, produce a graph of the S–I–R model over time, and use it to answer the following questions:

   a. Approximately when was the epidemic at its worst?

   b. How many people were sick at the peak of the epidemic?

   c. Approximately how long did the epidemic last?

   d. Approximately how many people contracted the disease during the course of the epidemic?

2   Given $R(0) = 0$, $N = 3000$, $I(0) = 10$, $\beta = 1.6$, and $\delta = 2$, produce a graph of the S–I–R model over time, and use it to answer the following questions:

   a. Approximately when was the epidemic at its worst?

   b. How many people were sick at the peak of the epidemic?

   c. Approximately how long did the epidemic last?

   d. Approximately how many people contracted the disease during the course of the epidemic?

3   Suppose that during the Eyam plague the duration of infectivity was 10 days rather than 11 so that $\delta = 10$. Graphically determine the corresponding best choice for β to two decimal places of accuracy.

4   Imagine all possible arrows connecting the compartments for S, I, and R on a flow diagram. State the practical meaning of each arrow.

5   Suppose an epidemic has $\beta = 2$ and $\delta = 6$.

   a. Determine the maximum number of new infections a single infective would be expected to cause.

   b. Explain why we do not typically expect a single infective to cause that many new cases.

6   *Extension*: Suppose a disease confers no immunity upon recovery so that those who do recover are immediately susceptible again.

   a. Draw a flow diagram for this situation.

   b. Find the DDS for the situation.

   c. Implement the model in Excel.

   d. Find an example of a disease that is might be well modeled by this new model.

7   *Extension*: Use an IF statement to modify the Excel S–I–R model so that it will not report negative values of susceptibles. (See Chapter 3 for similar examples involving the Lanchester model and the Hughes model.)

8 Use the S–I–R model to model an epidemic of your choice from Table 4.2.

    a. Note the parameter values you used and how you arrived at them.

    b. Produce a graph for the epidemic over a suitable time period.

    c. Apply the threshold theorem to determine how many susceptibles were present when the epidemic began to die out.

    d. Verify the theorem's result by noting the appropriate point on your graph.

    e. Determine how many people were sick at the peak of the epidemic.

    f. Determine how many people in total became ill during the epidemic.

    g. Determine the vaccination percentage required to prevent an epidemic of this disease.

9 Repeat Exercise 10 with a different epidemic.

10 *Extension*: Modify the basic S–I–R Excel spreadsheet to include a one-time round of vaccinations. You should store the vaccination percentage in its own cell for easy experimentation. Use your spreadsheet to verify the result in the Herd Immunity Theorem.

11 *Extension*: An important consideration when proposing a vaccination strategy is the vaccine's **efficacy**, that is, the percentage of vaccines administered that actually achieve the desired immunity. Incorporate a parameter for vaccine efficacy into the modified S–I–R model in Exercise 12.

12 *Extension*: An important consideration when proposing a vaccination strategy is the vaccine's **efficacy**, that is, the percentage of vaccines administered that actually achieve the desired immunity. How does the inclusion of a parameter for vaccine efficacy affect the Herd Immunity Theorem?

13 Is it possible to determine the total number of people who become ill by looking only at the Infectives column in the S–I–R model spreadsheet? Explain.

4.2 S–I–R WITH VITAL DYNAMICS

The S–I–R model we have studied so far is a simple model for describing and predicting the course of an epidemic. We have noted that it is particularly applicable to epidemics that are relatively short lived and occur in small, closed populations. A simplifying assumption of the model is that of a constant population where no births or deaths occur. This assumption is fine if the disease is only present for a short time, but many infectious diseases persist in a population over many years, a time span long enough that any reasonable model must take into account births and deaths. When we include population changes in a model, we say that the model includes **vital dynamics**.

In this section we include vital dynamics, but we do so in a way that will not disturb the model too much: we assume that new births and deaths balance each other exactly, so even though people are dying and being born, the total population, N, does not change.

The inclusion of birth and death rates means we have to be careful about our units. Our model time units are days, whereas most birth and death rates are reported as annual rates. Our solution is to keep days as our units and adjust our birth and death rates to be daily rates. We note that the deaths under consideration here are those due to natural causes and not due to any excess mortality caused by the disease, so it is reasonable to assume that the death rate is the same for all compartments.

Let μ be the daily birth rate for the population, so that every day μN new susceptibles are born. We need to balance new births exactly with deaths, so every day we have μS susceptibles, μI infectives, and μR removed individuals die from reasons other than the disease. We incorporate the changes into the flow diagram given in Figure 4.11.

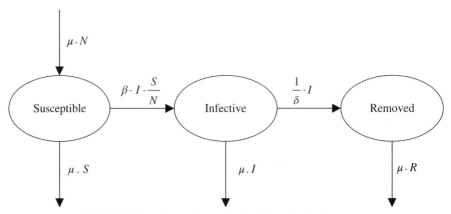

FIGURE 4.11 Flow diagram for S–I–R with vital dynamics.

Accounting for the four new arrows in our DDS gives

$$S(t) = S(t-1) - \beta I(t-1)\frac{S(t-1)}{N} + \mu N - \mu S(t-1)$$

$$I(t) = I(t-1) + \beta I(t-1)\frac{S(t-1)}{N} - \frac{1}{\delta}I(t-1) - \mu I(t-1)$$

$$R(t) = R(t-1) + \frac{1}{\delta}I(t-1) - \mu R(t-1).$$

Creating a spreadsheet for the modified model is now just a matter of making fairly minor alterations to the S–I–R spreadsheet. As with the other parameters, we store μ in its own cell for easy experimentation later. In the next example we highlight some of the differences introduced by the inclusion of vital dynamics.

Example 4.6: Use the S–I–R model with vital dynamics to predict the course of an epidemic of chicken pox by answering the questions below. Assume the following parameters: $\beta = 0.9$, $\delta = 10$, $\mu = 0.0005$, $I(0) = 5$, and $N = 5000$. We note that the value for μ has been set artificially high in an effort to highlight the model behavior on a reasonable time scale.

1. Graph the model predictions over a period of 2 years.
2. Approximately when does the second epidemic wave occur?
3. Approximately how many people were sick at the peak of the second epidemic wave?
4. Give approximate long-term values for S, I, and R.

First we show the initial Excel setup in Figure 4.12. We display the formula for infectives and note that the adjustments to the other formulas are similar.

	A	B	C	D	E	F	G
1	Basic S-I-R Model						
2							
3	Effective contact rate, β =			0.9			
4	Duration of infectivity, δ =			10			
5	Total population, N =			5000			
6	Birth and death rate, μ =			0.0005			
7							
8	t	$S(t)$	$I(t)$	$R(t)$			
9	0	4995	5	0			
10	1	4990.5	=C9+$D$3*C9*B9/$D$5-(1/$D$4)*C9-$D$6*C9				

FIGURE 4.12   Excel S–I–R with vital dynamics setup.

For number 1 we need to drag our formulas down to day 730. (Note that the further down the cursor is pulled, the faster the copying goes.) Once we have done that, we select all of the data we want to include on our graph including the time column and the column headings, select the "Insert" tab, and then select a scatterplot from the "Charts" group. Our completed graph is given in Figure 4.13.

Note that this graph is very different than the basic S–I–R graph. The inclusion of vital dynamics means the susceptible category is continually being replenished. Because of this fresh supply of susceptibles, we get repeated epidemic waves: the disease hits, there is an epidemic, the disease appears to leave, and then once there are enough new susceptibles, we get a new epidemic.

To answer number 2, we see that the second epidemic begins sometime around day 500. For number 3 we scroll down to day 500 or so in our spreadsheet. By examining the values in the infective column, we see that the actual peak of the second epidemic occurred at day 545 with about 161 cases of chicken pox.

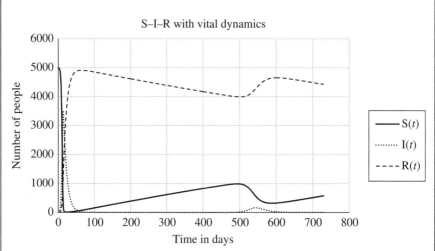

FIGURE 4.13   Graph of S–I–R with vital dynamics over time.

For number 4 we can only get very rough estimates based on our 2-year graph. We can do better by dragging the model down even further, say, to year 10. If we do, we get the graph shown in Figure 4.14.

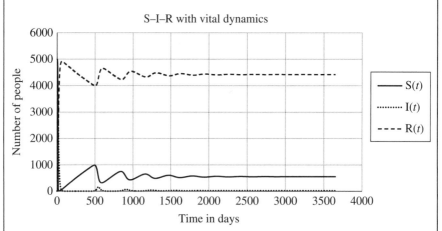

FIGURE 4.14   Long-term model values for Example 4.6.

If we place the cursor over our Excel graph where the graphs have leveled off, we will see the long-term values. These are about 559 for the number of susceptibles, 22 for the number of infectives, and 4419 for the number removed. Again, note the contrast with the basic S–I–R model: with vital dynamics, the disease never completely leaves the population. Instead it settles down into an endemic state. □

As a final remark we point out that even though the graphs for S, I, and R all become constant, there are still people continually moving into and out of the compartments due to new births, new infections, recoveries, and deaths. It is just that all of the processes result in net changes of 0 for each compartment.

4.2.1 The Basic Reproductive Rate

The inclusion of vital dynamics in our model brings about an important change for how we calculate the basic reproductive rate, R_0: we can no longer find it by computing $R_0 = \beta \times \delta$. To see why, recall that this first formula for computing R_0 was based on the observation that

$$R_0 = (\text{Effective contacts per day}) \times (\text{Number of days available to spread the disease}).$$

For the basic S–I–R model, this led naturally to the formula $R_0 = \beta \times \delta$.

Now we restate our observation about R_0 in a subtle but important way:

$$R_0 = (\text{Effective contacts per day}) \times (\text{Number of days spent in compartment } I).$$

Since we now include deaths in our model, the number of days spent in I is no longer the same as the duration, δ. Some infectives will die from natural causes before the duration of infectivity expires; thus on average infectives are not available to spread the disease for the entire duration of infectivity.

What we need is a way to determine on average how long an infective person spends in the infective compartment that takes deaths into account. The solution to this problem comes in the form of the **Waiting Time Principle**, a result that we will use repeatedly throughout the remainder of this chapter.

4.2.2 The Waiting Time Principle

The key to developing the new formula for R_0 is focusing on the *proportion* of people who leave the I compartment each day. The Waiting Time Principle shows that there is a reciprocal relationship between the proportion who leave a compartment each day and how long on average one spends in that compartment.

Theorem 4.4 Waiting Time Principle: Consider a compartment in a discrete dynamical system where a constant proportion of the membership leaves each day. Then the average amount of time someone spends in that compartment (in days) is equal to the reciprocal of this proportion:

$$\text{Average time in compartment} = \frac{1}{\text{Proportion leaving each day}}.$$

Equivalently, we write

$$\text{Proportion leaving each day} = \frac{1}{\text{Average time spent in compartment}}.$$

In practice we use whichever form is more convenient.

We showed that the Waiting Time Principle is at least plausible when we related the duration of infectivity, δ, to the proportion leaving the infective compartment each day, $\frac{1}{\delta}$ (see Example 4.1). Unfortunately a rigorous justification of the result requires both second semester calculus and some probability theory and is beyond the scope of this text. For the interested reader, a proof that assumes the necessary prerequisites is provided in Appendix D.

In the next example we first show how to apply the Waiting Time Principle to the basic S–I–R model to get a familiar result.

Example 4.7: Use the Waiting Time Principle to determine the average time someone spends in the infective compartment for the basic S–I–R model.

According to the flow diagram, the number of infectives leaving the I compartment each day is given by $\frac{1}{\delta}I$. Thus the proportion of infectives leaving the I compartment each day is given by $\frac{1}{\delta}$. According to the Waiting Time Principle, the average amount of time an infective spends in the I compartment is therefore

$$\frac{1}{\text{Proportion leaving } I \text{ each day}} = \frac{1}{1/\delta} = 1 \cdot \frac{\delta}{1} = \delta. \ \square$$

Next we apply the Waiting Time Principle in the less familiar situation of the S–I–R model with vital dynamics.

Example 4.8: Determine on average how long someone spends in the infective compartment for the S–I–R model with vital dynamics.

This example is slightly more complex than the last because infectives leave the compartment via two routes—recovery or death by natural causes. Thus the total number of infectives who leave the compartment each day will be the sum of those who recover and those who die. This total is given by $\frac{1}{\delta}I + \mu I = \left(\frac{1}{\delta} + \mu\right)I$. From this total we can see that the proportion who leaves I each day is therefore given by $\left(\frac{1}{\delta} + \mu\right)$. We are now in position to apply the Waiting Time Principle:

$$\text{Average time spent in } I = \frac{1}{\left(\frac{1}{\delta} + \mu\right)} = \frac{1}{\left(\frac{1 + \mu\delta}{\delta}\right)} = \frac{\delta}{1 + \mu\delta}.$$

Note that because μ and δ are both positive numbers, $\frac{\delta}{1 + \mu\delta} < \delta$. Thus the average time spent in I for the vital dynamics model is less than for the basic S–I–R model. This should agree with our intuition since the inclusion of natural deaths should serve to decrease the average time spent in I. $\square$

Now that we know how long an infective person typically stays in I, we can see how to compute R_0. Using the same idea as before, the number of secondary cases caused by a single infective in a wholly susceptible population is given by

$$R_0 = (\text{Effective contacts per day}) \times (\text{Average time spent in } I)$$

$$= \beta \times \frac{\delta}{1 + \mu\delta}.$$

So for the S–I–R model with vital dynamics, our formula for R_0 is $R_0 = \frac{\beta\delta}{1+\mu\delta}$.

We continue with a computational example.

Example 4.9: Find the basic reproductive rate for the epidemic of chicken pox in Example 4.6.

Recall that the parameters are given by $\beta = 0.9$, $\delta = 10$, and $\mu = 0.0005$. A straightforward application of the formula for R_0 gives us $R_0 = \frac{\beta\delta}{1+\mu\delta} = \frac{0.9 \cdot 10}{1 + 0.0005 \cdot 10} = \frac{9}{1.005} = 8.96$. $\square$

As we have mentioned previously, the parameter β is often the most difficult parameter to find in practice. Typically we find R_0 directly from field data, δ from case studies, and μ from demographic studies. It is frequently up to us to find β as we do in the next example.

Example 4.10: Suppose we know that for an epidemic of mumps, $R_0 = 12$, $\delta = 6$, and $\mu = 0.0003$. Find β.

We take the formula for R_0 and substitute the values we know:

$$R_0 = \frac{\beta\delta}{1+\mu\delta}$$

$$12 = \frac{\beta \cdot 6}{1 + 0.0003 \cdot 6}.$$

Now we solve for β by isolating it on one side of the equation:

$$12 = \frac{\beta \cdot 6}{1.0018}$$

$$12.0216 = 6\beta$$

$$2.0036 = \beta.$$

We see that β is approximately 2 effective contacts per day per infective. $\square$

Finally we show that our new formula for R_0 is really a generalization of the previous one.

Example 4.11: Verify that the new formula for R_0 reduces to the original formula if $\mu = 0$, that is, there are no vital dynamics.

We plug in $\mu = 0$ to get $R_0 = \dfrac{\beta\delta}{1+\mu\delta} = \dfrac{\beta\delta}{1+0\cdot\delta} = \dfrac{\beta\delta}{1} = \beta\delta$. This does indeed agree with our formula for the basic S–I–R model, and it gives us some confidence that the derivation of the new formula was sensible. $\square$

Next we examine the effect of vital dynamics on the threshold theorems.

4.2.3   Threshold Theorems

Just as the inclusion of vital dynamics leads to a more complicated expression for R_0, the development of threshold theorems is also more involved. The process, however, is the same as before: we examine the conditions that would lead to a decrease in I and then follow up with some algebra.

Note as before that the number of infectives will decrease whenever the outflows for I are larger than the inflows. Referring to the flow diagram in Figure 4.11, we see that the inequality representing this condition is given by

$$\frac{1}{\delta}I + \mu I > \beta I \frac{S}{N},$$

or

$$I\left(\frac{1+\mu\delta}{\delta}\right) > \beta I \frac{S}{N}.$$

Dividing through by the positive number I, we get

$$\frac{1+\mu\delta}{\delta} > \beta\frac{S}{N}.$$

Dividing both sides by β yields

$$\frac{1+\mu\delta}{\beta\delta} > \frac{S}{N}.$$

The expression on the left-hand side of the last inequality is the reciprocal of R_0; hence

$$\frac{1}{R_0} > \frac{S}{N}.$$

We now state the threshold theorems for vital dynamics.

Theorem 4.5   **Threshold Theorem I:** An epidemic wanes whenever the proportion of people who are susceptible is below the threshold value given by

$$\frac{1}{R_0},$$

where R_0 is the basic reproductive rate of the disease.

As before we can move the N over to obtain an equivalent version of the theorem.

Theorem 4.6   **Threshold Theorem II:** An epidemic wanes whenever the number of people who are susceptible is below the threshold value given by

$$\frac{N}{R_0},$$

where R_0 is the basic reproductive number of the disease and N is the total population size.

The slightly different language in the statements of the theorems reflects the presence of vital dynamics. Since new births serve to replenish the susceptible compartment, we can no longer conclude that once the susceptibles fall below a certain level, they will stay below that level. In fact we expect the new births to eventually raise the number of susceptibles above the threshold. This observation has implications for potential vaccination programs, as we see in the next section.

4.2.4   Herd Immunity

The threshold theorems shed light on where discussion of a vaccination program should start. We know from these theorems that if the number or proportion of susceptibles can be *maintained* below a certain number, then the epidemic will die out. Thus the goal of any vaccination program will be to vaccinate susceptibles in such a way as to make that happen. Instead of a one-time vaccination campaign as in the basic model, we must now consider a more realistic program where we have ongoing vaccinations of newborns.

Theorem 4.7   **Herd Immunity Theorem:** In order to sustain protection against an epidemic, the proportion of newborns that must be vaccinated is given by

$$1 - \frac{1}{R_0}.$$

Once again the justification is only a couple of lines long. According to the second threshold theorem, we need to maintain the proportion of susceptibles below $\frac{1}{R_0}$. Thus our vaccination program must be such that the proportion vaccinated remains larger than $1 - \frac{1}{R_0}$. $\square$

In its statement and justification, the Herd Immunity Theorem for our modified S–I–R model is very similar to the original. Note, however, that in order to implement a suitable vaccination program, a lot more is involved in the second case. For the basic S–I–R model, the number of susceptibles is always decreasing. This means that a one-time round of vaccinations would be sufficient to halt the epidemic. The inclusion of new births, however, changes this. Since new susceptibles are continually entering the population, our vaccination program must now be *ongoing*. A one-time vaccination spree cannot be sufficient since the number of susceptibles will eventually grow beyond the threshold.

Example 4.12: Assume that pertussis, also known as whooping cough, is a disease that is well modeled by the S–I–R with vital dynamics model. Determine the ongoing vaccination percentage necessary to sustain protection against an epidemic of pertussis. Assume parameter values of $\beta = 1.2$, $\delta = 9$, and $\mu = 0.0003$.

First we find $R_0 = \frac{1.2 \cdot 9}{1 + 0.0003 \cdot 9} = 10.77$. Once we have R_0, we apply the Herd Immunity Theorem to get the required vaccination percentage: $1 - \frac{1}{R_0} = 1 - \frac{1}{10.77} = 90.7\%$. This percentage represents the minimum level of ongoing vaccination coverage necessary to prevent a pertussis epidemic from recurring. □

In the next section we show that the inclusion of vital dynamics has interesting implications for the long-term behavior of our model.

4.2.5 Equilibrium Points

The vital dynamics version of the S–I–R model exhibits some interesting behavior that is not present in the basic model. Namely, we will see that our new model predicts some level of endemicity for most diseases it models. To see this, we examine the equilibrium points. Since endemicity indicates that a disease is ever present in a population, it will appear in our model as a long-term value for I that is nonzero. In other words we will see a stable equilibrium point (S^*, I^*, R^*) where $I^* > 0$.

Recall that finding equilibrium points amounts to finding values at which the DDS remains constant. For our new model, this amounts to solving the system of equations below for (S^*, I^*, R^*):

$$S^* = S^* - \beta I^* \frac{S^*}{N} + \mu N - \mu S^*$$

$$I^* = I^* + \beta I^* \frac{S^*}{N} - \frac{1}{\delta} I^* - \mu I^*$$

$$R^* = R^* + \frac{1}{\delta} I^* - \mu R^*.$$

We are dealing with three equations in three unknowns, so it should not be surprising that the algebra becomes messier. The payoff for our hard work will be the ability to predict the long-term behavior of a potential epidemic simply by knowing the values of our parameters β, δ, μ, and N.

We note that as a first simplification, we have

$$0 = -\beta I^* \frac{S^*}{N} + \mu N - \mu S^*$$

$$0 = \beta I^* \frac{S^*}{N} - \frac{1}{\delta} I^* - \mu I^*$$

$$0 = \frac{1}{\delta} I^* - \mu R^*.$$

On the right-hand side of the second equation, we can factor out I^* to get

$$0 = I^* \left(\beta \frac{S^*}{N} - \frac{1}{\delta} - \mu \right).$$

When a product of two number is 0, at least one of the two numbers must be 0 so we have two cases: $I^* = 0$ or $\beta \dfrac{S^*}{N} - \dfrac{1}{\delta} - \mu = 0$.

If $I^* = 0$ substitution into the S^*-equation allows us to find S^*:

$$0 = -\beta I^* \frac{S^*}{N} + \mu N - \mu S^*$$

$$0 = -\beta \cdot 0 \cdot \frac{S^*}{N} + \mu N - \mu S^*$$

$$0 = \mu N - \mu S^*$$

$$S^* = N.$$

Similarly, substitution into the R^*-equation allows us to find R^*:

$$0 = \frac{1}{\delta} I^* - \mu R^*$$

$$0 = \frac{1}{\delta} \cdot 0 - \mu R^*$$

$$R^* = 0.$$

Thus if $I^* = 0$ we have found that one of our equilibrium points is given by $(S^*, I^*, R^*) = (N, 0, 0)$. In context this equilibrium point corresponds to the disease not being present in the population so everyone remains susceptible.

Next we consider the case where $\beta \dfrac{S^*}{N} - \dfrac{1}{\delta} - \mu = 0$. Solving for S^* gives

$$\beta\frac{S^*}{N} - \frac{1}{\delta} - \mu = 0$$

$$\beta\frac{S^*}{N} = \frac{1}{\delta} + \mu$$

$$S^* = \frac{N}{\beta}\left(\frac{1+\mu\delta}{\delta}\right).$$

Further simplification of S^* yields a concise expression for S^*:

$$S^* = N\left(\frac{1+\delta\mu}{\beta\delta}\right)$$

$$S^* = N\left(\frac{1}{R_0}\right) = \frac{N}{R_0}.$$

To find the corresponding equilibrium value for I^*, we substitute this value for S^* into the S^*-equation, $0 = -\beta I^* \dfrac{S^*}{N} + \mu N - \mu S^*$. We get

$$0 = -\frac{\beta I^*}{N}\left(\frac{N}{R_0}\right) + \mu N - \mu\left(\frac{N}{R_0}\right)$$

$$0 = -\frac{\beta I^*}{R_0} + \mu N - \frac{\mu N}{R_0}$$

$$\frac{\beta I^*}{R_0} = \mu N - \frac{\mu N}{R_0}$$

$$\beta I^* = R_0\mu N - \mu N$$

$$I^* = \frac{\mu N}{\beta}(R_0 - 1).$$

Finally we find the equilibrium value R^* by substituting $I^* = \dfrac{\mu N}{\beta}(R_0 - 1)$ into the R^*-equation, $0 = \dfrac{1}{\delta}I^* - \mu R^*$. This substitution gives us

$$0 = \frac{1}{\delta}\left(\frac{\mu N}{\beta}(R_0 - 1)\right) - \mu R^*$$

$$\mu R^* = \frac{\mu N}{\beta\delta}(R_0 - 1)$$

$$R^* = \frac{N}{\beta\delta}(R_0 - 1).$$

Putting all of our equilibrium values together gives us our first nontrivial equilibrium point for an epidemic:

$$(S^*, I^*, R^*) = \left(\frac{N}{R_0}, \frac{\mu N}{\beta}(R_0 - 1), \frac{N}{\beta\delta}(R_0 - 1) \right).$$

This shows that our model allows for the possibility that over time a disease can enter an endemic state, which is often observed in practice.

Analytically examining the stability characteristics in general of equilibria in a model with three dependent variables is beyond the scope of this text. We only note that for most reasonable parameter choices, this equilibrium point will be stable and represents the long-term values for S, I, and R. We provide evidence for this claim in the following example.

Example 4.13: Recall the chicken pox epidemic from Example 4.6 where $\beta = 0.9$, $\delta = 10$, $\mu = 0.0005$, $I(0) = 5$, and $N = 5000$. Confirm the estimates we made with Excel for the long-term values of S, I, and R using the formulas for our nontrivial equilibrium point.

By examining our Excel graph of the model over a 10-year period, we estimated that the long-term values would be 559 for the number of susceptibles, 22 for the number of infectives, and 4419 for the number removed. We confirm these values by computing the nontrivial equilibrium point $(S^*, I^*, R^*) = \left(\frac{N}{R_0}, \frac{\mu N}{\beta}(R_0 - 1), \frac{N}{\beta\delta}(R_0 - 1) \right)$.

First we compute R_0:

$$R_0 = \frac{\beta\delta}{1 + \mu\delta} = \frac{0.9 \cdot 10}{1 + 0.0005 \cdot 10} = 8.955.$$

This allows us to find $S^* = \frac{N}{R_0} = \frac{5000}{8.955} = 558.34$.

Next we have $I^* = \frac{\mu N}{\beta}(R_0 - 1) = \frac{0.0005 \cdot 5000}{0.9}(8.955 - 1) = 22.10$. Finally we have $R^* = \frac{N}{\beta\delta}(R_0 - 1) = \frac{5000}{0.9 \cdot 10}(8.955 - 1) = 4419.44$. Minor rounding differences aside our Excel work and our algebra have produced the same values for our long-term expectations for the model. ☐

We have finally come to the point where we are ready to connect the theory of our models to real, observable field data. As we show in the next section, R_0 is the key to linking our models to data.

4.2.6   Section Exercises

1   Consider an epidemic of mumps in a city of 1,000,000 similar to the one in Baltimore, Maryland, in 1943. Assume that initially 10 people are infective.

    a. Use the S–I–R model with vital dynamics to model the epidemic; from Table 4.2 use $\delta = 8$, $R_0 = 7$, and $\mu = 0.0006$.

    b. Graph the epidemic over a period of 1 year. Compare your graph to the usual graph from the S–I–R model. How is the new graph different? What accounts for the difference?

    c. Graph the epidemic over a period of 2 years. What new phenomenon do you observe? Explain why it is happening.

    d. Graph the epidemic over a period of 10 years. Describe what you see, paying particular attention to the differences between your new graph and the standard S–I–R graph.

    e. What does Excel predict for the long-term number of infectives? What does this mean practically?

2   For the epidemic in Exercise 1, find the nontrivial equilibrium point. Compare the results of the equilibrium formulas to those predicted by the Excel graph.

3   Repeat Exercise 1 for an epidemic of your choosing. Continue to use the birth and death rate of $\mu = 0.0006$.

4   Repeat Exercise 2 for the epidemic you chose in Exercise 3.

5   *Extension*: Modify the S–I–R with vital dynamics Excel spreadsheet so that it automatically calculates the nontrivial equilibrium values for an epidemic.

4.3   DETERMINING PARAMETERS FROM REAL DATA

This section provides us with a vital link between our models and the real world. In order for any mathematical model to have the potential to be useful, the model parameters must be estimated from real data. For our disease models so far, we have relied on the work of others who have already used real data to provide estimates for R_0 for a variety of diseases.

In this section we present two ways of making such estimates for R_0. Which of the two methods we choose depends on the state of the disease in the population. If the disease is an ongoing epidemic, we show how to use case reports to estimate R_0. On the other hand, if the disease is endemic to the population, we show how to use serosurveys of the population to estimate R_0.

4.3.1 Determining R_0 at the Onset of an Epidemic

Most of the epidemic modeling we have done so far has been retrospective in nature: we have modeled epidemics that have already occurred using parameters already provided to us. This is important to do when constructing a model because comparing a model with past data is fundamental to validating the model. It can give us confidence that what we are doing is reasonable, or it can alert us to the need to modify the model. However, one of the primary purposes of modeling is to predict the course of future events so that we can make decisions about what to do.

When faced with a new epidemic, decisions about what to do must be made quickly. How many people need to be treated? Who do we need to treat? How many hospital beds will we need? How many medical professionals need to be dispatched to the affected area? These are the kinds of questions that epidemic models can help answer, but first the model parameters need to be determined. In particular R_0 must be determined as quickly as possible.

In this section we show how to use **case reports**, that is, reports of numbers of infections, to estimate R_0. Once we have R_0, we then deduce β.

If we examine a typical graph for $I(t)$, we see the familiar shape in Figure 4.15.

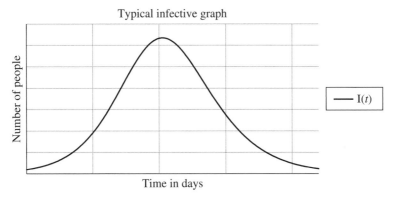

FIGURE 4.15   Typical infectives graph.

If we zoom in on the beginning part of the graph as in Figure 4.16, we get a graph that looks very much like the exponential growth curves from Chapter 1. Thus at the onset of an epidemic, the number of infectives appears to grow roughly exponentially. This is the key to determining R_0 from case reports. If the number of infectives is roughly exponential, then recall from Chapter 1 that there is some positive growth rate, r, for which the DDS is

$$I(t) \approx I(t-1) + rI(t-1).$$

We also have the corresponding explicit formula

$$I(t) \approx (1+r)^t I(0),$$

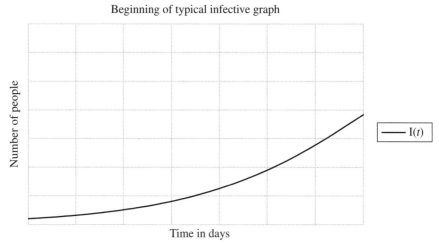

Beginning of typical infective graph

FIGURE 4.16   Beginning of infectives graph is approximately exponential.

where $I(0)$ is the initial number of infectives. It is this explicit formula we use to deduce the growth rate r from case reports—the first step to determining R_0. We outline the method in the next example.

Example 4.14: Returning to the Eyam plague example, we know from the Reverend Mompesson that there were originally seven cases of plague at the onset of the epidemic. We also know that 2 weeks later there were 15 cases. From these two values, we estimate r.

First note that $I(0) = 7$ so that if the number of infectives is approximately exponential, we have $I(t) \approx (1+r)^t 7$. Knowing we have 15 cases in 2 weeks allows us to write $I(14) = 15$ or $15 \approx (1+r)^{14} 7$. To solve for r we write

$$\frac{15}{7} \approx (1+r)^{14}$$

$$\left(\frac{15}{7}\right)^{\frac{1}{14}} \approx 1+r$$

$$\left(\frac{15}{7}\right)^{\frac{1}{14}} - 1 \approx r.$$

Finishing the computation with a calculator gives us $r \approx 0.056$. □

Next we have to figure out how to use our estimate for r to get R_0. To do this we consider how the number of infectives changes on the first day of an epidemic, and we approach the question from two different points of view.

On the one hand, if the number of infectives is roughly exponential at the beginning of an outbreak, then the net increase in infectives the first day is given by the number who are added the first day: $rI(0)$.

On the other hand, we know that during the early stages of an epidemic, virtually everyone in the population is susceptible. Thus the number of new cases caused by the initial infectives during the first δ days should be roughly equal to $R_0 \times I(0)$. At the end of the first δ days, the net change in number of infectives will therefore be $R_0 I(0) - I(0)$ since the original infectives will have been removed. Finally we estimate how the number of infectives increases on day 1 by taking the average increase over the first δ days: $\frac{R_0 I(0) - I(0)}{\delta} = \frac{(R_0 - 1)I(0)}{\delta}$.

We now have two different estimates for the same quantity: the net change in infectives during the first day of an epidemic. We set them equal to each other and solve for R_0:

$$rI(0) = \frac{(R_0 - 1)I(0)}{\delta}$$

$$r = \frac{(R_0 - 1)}{\delta}$$

$$\delta r = R_0 - 1$$

$$1 + \delta r = R_0.$$

At last we have the crucial link! To estimate R_0 for an ongoing epidemic, we use δ as determined by medical researchers, and we use r as determined by case reports. The last step is to compute

$$R_0 = 1 + \delta r.$$

We illustrate the process in the next example.

Example 4.15: Estimate R_0 for the Eyam plague example using the case reports collected by Mompesson.

We recall that for the Eyam plague we have $\delta = 11$. In the previous example we calculated $r \approx 0.056$. The approximate value for the basic reproductive rate is therefore $R_0 \approx 1 + 11(0.056) = 1.616$. $\square$

The next example shows that with R_0 in our possession, we can go a step further and determine the elusive β.

Example 4.16: Use R_0 to estimate β for the Eyam plague.

When we first modeled the Eyam plague with the S–I–R model, we were confronted by the difficulty of having to find β ourselves. Having no other recourse, we resorted to a trial-and-error approach where we experimented with different values of β in our Excel model until we got the S–I–R graph to match the data set. Now that we can determine R_0 from data, we can go one step further and get β as well. We have

$$R_0 = \beta\delta$$

$$1.616 = \beta \cdot 11$$

$$\frac{1.616}{11} = \beta.$$

Finishing with a calculator yields $\beta = 0.147$. This compares favorably with the result of our trial-and-error approach that produced $\beta = 0.149$. $\square$

We now have a reliable, data-driven method for finding model parameters that we can use at the outset of an epidemic when decisions are being made rather than a retrospective trial-and-error method that fits our model to the entire set of data after the epidemic has passed.

4.3.2 Determining R_0 from Serosurveys

If a disease has been in the population for a while and has settled down to an endemic state, then the method of the previous section will not work for determining R_0 because the number of infectives will be roughly constant rather than increasing exponentially. In the endemic case we take advantage of our work in determining equilibrium values for the S–I–R model with vital dynamics.

First we note that if a disease is at or near its (nontrivial) equilibrium—that is, it has settled down into a stable endemic state—then our previous work shows that the proportion of the population who are susceptible is given by

$$\frac{S^*}{N} = \frac{1}{R_0}.$$

This follows immediately from the equilibrium value for the number of susceptibles:

$$S^* = \frac{N}{R_0}.$$

The good news is that the proportion of a population who are susceptible to a disease is something we can physically measure with **serosurveys**, which are blood tests of a population for antibodies to a particular disease. Anyone who has the antibodies is no longer susceptible by virtue of already having had the disease or by vaccination. Conversely, the people who are still susceptible to a disease are precisely those without antibodies in their blood. If we test a large enough number of people, then we will have a good idea of what proportion of the population is still susceptible. Once we have an estimate for the proportion still susceptible, we connect it to the parameter R_0 using the equilibrium value formula for S^*. In general our procedure is as follows:

1. Confirm that the disease is in equilibrium. This would include confirming that there is no ongoing epidemic and that the disease has been present in the population for a long time. If the disease is an ongoing epidemic, we use the method of the previous section instead.

2. Test the blood of a large enough representative sample of the population for antibodies to the disease. What qualifies as "large enough" and "representative" are ideas that a first course in statistics can make precise.

3. Compute the number of people who do not have antibodies to the disease divided by the total number tested. This gives us the proportion of people in the sample who are still susceptible. If we have chosen our sample properly, this proportion is also approximately equal to the proportion of people who are still susceptible in the entire population. In other words,

$$\frac{S}{N} \approx \frac{\text{Number without antibodies}}{\text{Total number tested}}.$$

4. Because we assume that the disease is in equilibrium, we have $\frac{1}{R_0} = \frac{S^*}{N}$,

so

$$\frac{1}{R_0} \approx \frac{\text{Number without antibodies}}{\text{Total number tested}}.$$

5. This last equation gives us the link we need between real, observable data and our mathematical model. To compute the basic reproductive rate for a disease in an endemic state, we run blood tests on a sample of the population and then compute

$$R_0 \approx \frac{\text{Total number tested}}{\text{Number without antibodies}}.$$

In the next example we illustrate this procedure using real data for rubella from The Gambia.

Example 4.17: In 2000 the World Health Organization (WHO) published a report that includes the results of serosurveys for rubella in 45 developing countries, including The Gambia (WHO, 2000). Use the WHO reported data to approximate R_0 for rubella.

From the serosurveys administered in The Gambia in 1966, 1971, and 1976, we know that at the time approximately 94% of the population had antibodies for rubella in their blood. Thus 6% of the population remained susceptible, or 0.06 as a proportion. Our work from before tells us that $\frac{1}{R_0} \approx 0.06$ so $R_0 \approx 16.67$.

We compare our result to Table 4.2, which reports that R_0 for rubella in The Gambia in 1976 was between 15 and 16. The fact that our estimate is slightly outside of the range reported in Table 4.2 could be due to rounding in the WHO serosurvey data. If, for example, the actual percentage of those testing positive for rubella antibodies had been rounded from 93.6%, we would have calculated $\frac{1}{R_0} \approx 0.064$ and $R_0 \approx 15.63$, which is in the reported range. Another possible source of error may be that our serosurvey data covers a 10-year period from 1966 to 1976, while the specified R_0 in Table 4.2 is specific to 1976. $\square$

Recall that in Section 4.2.5 we carried out some fairly involved algebra to produce a formula for the equilibrium value for S. This section shows us the payoff. Our equilibrium analysis in combination with blood tests done in the field provides us with a useful way of determining R_0 from data for endemic diseases.

4.3.3   Ebola Virus Disease

Ebola virus disease (EVD) is a severe infectious disease that is transmitted through contact with an infected person's bodily fluids. Cases have been reported in countries all over the world, though most recent cases have occurred in West Africa. Symptoms of EVD include fever, fatigue, muscle pain, and headache. As the disease progresses, muscle and abdominal pain, diarrhea, vomiting, and unexplained hemorrhaging can also occur (CDC, 2014). Ebola is fatal in approximately 50% of cases on average, though fatality rates between 25 and 90% have been observed in past outbreaks (WHO, 2015).

In March 2014 the largest Ebola epidemic in history began in Guinea in West Africa. Within 2 months of the outbreak, epidemics also began in bordering nations Sierra Leone and Liberia. According to the CDC's Ebola update on January 30, 2015, these three countries have seen a combined total of 22,124 cases and 8,829 fatalities (CDC, 2015b). The January 30 update also reported some hopeful news: it was the first week since June 29, 2014, that there were fewer than 100 reported new cases of EVD in the three countries combined.

Prevention and control measures include reducing wildlife to human transmission through the proper cooking of meat, reduction of human to human transmission via separating healthy from infected, the wearing of gloves while treating infected, hand washing after contact, and quick and safe burial of the deceased. While two potential vaccines are currently being tested for safety in humans, no approved vaccines are available yet (WHO, 2015).

Currently the only treatments are the treatment of specific symptoms and administering of fluids to keep the patient hydrated. Other potential treatments are in development (WHO, 2015). Once an infective recovers, immunity to EVD is known to last at least 10 years, possibly longer or for life (CDC, 2014).

The first case of EVD ever diagnosed in the United States was on September 26, 2014, in Dallas, Texas. Nineteen days later there were two more cases—two nurses who tended to the first victim. In what follows we apply the S–I–R model with vital dynamics to predict what could have happened in Dallas if nothing were done.

We must keep in mind that our model is based on assumptions that are unrealistic for a city like Dallas—the population is not closed, nor is it well mixed. Ebola is also not particularly well modeled by the S–I–R model because it has a long **incubation period**, the time from infection to when symptoms develop. For Ebola humans are not infectious until symptoms develop, and the incubation period can range between 2 and 21 days. The S–I–R model does not take this into account, but the reader is asked to modify the S–I–R model to include an incubation period in the exercises.

Despite all of these objections, we can still derive some interesting results by using the S–I–R model with vital dynamics to explore "what-if" questions about Ebola in the United States.

Those who have contracted EVD are contagious as soon as symptoms appear. Estimates for the duration of infectivity for EVD vary depending on the effected country. This variation is due in part to different burial practices. Patients who die from EVD are still contagious after they die, and in countries where burial rituals involve close contact with the deceased, the duration of infectivity extends beyond death. Common estimates for the duration are between 6 and 10 days, and because we are dealing with a potential outbreak in the United States, we expect the duration to be on the low end of this range. Thus we let the duration be $\delta = 6$ days. In our next example we determine R_0.

Example 4.18: Determine R_0 for the spread of EVD in Dallas, Texas.

We treat the Dallas cases as an ongoing epidemic where $I(0) = 1$ and $I(19) = 2$. As in Section 4.3.1, we assume that the number of infectives grows exponentially at the beginning of an epidemic. This allows us to estimate the growth rate, r. We have

$$I(19) = (1 + r)^{19} I(0)$$
$$2 = (1 + r)^{19} \cdot 1$$
$$2^{1/19} = 1 + r$$
$$0.0372 = r.$$

Once we know r, we can find R_0 from the formula

$$R_0 \approx 1 + \delta r = 1 + 6 \cdot 0.0372 = 1.223.$$

Note that this is a very low value for R_0. It tells us that in the United States where virtually everyone is susceptible to the disease, a single infective would cause about 1.2 new cases of Ebola on average. This number is reasonable compared to estimates for R_0 that range from 1.5 to 2.5 from past epidemics depending on the affected country. Our value also lends support to the notion that with proper treatment and control methods, Ebola is actually a relatively easy disease to control compared to a disease like measles whose R_0 is 12–18. ☐

In the next example we determine the remaining required parameters and model the course of an uncontrolled outbreak of Ebola in Dallas.

Example 4.19: Use the S–I–R model with vital dynamics to predict the course of an uncontrolled outbreak of EVD in Dallas.

We can deduce β from R_0, but first we need to find μ. Based on data from the Institute for Health Metrics and Evaluation (IHME), a reasonable value for the daily birth and death rate in Dallas County, Texas, is $\mu = 0.000035$ (IHME, 2010).

For β we use the model's equation for R_0 to get

$$R_0 = \frac{\beta\delta}{1 + \mu\delta}$$

$$1.223 = \frac{\beta \cdot 6}{1 + 0.000035 \cdot 6}$$

$$1.2233 = \beta \cdot 6.$$

Thus $\beta = 0.204$. Finally we find N, the population of Dallas, which according the most recent US Census is about $N = 1,200,000$ (U.S. Census, 2010).

With all of our parameters in place, plug them into our Excel model and run our projection. We present a graph of the projected epidemic in Figure 4.17.

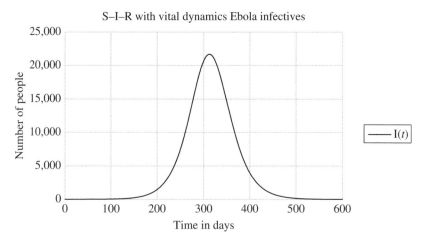

FIGURE 4.17 Dallas, Texas, projected uncontrolled Ebola epidemic.

The peak of the initial epidemic occurs on day 312 with 21,676 people sick, and the initial epidemic ends after about 630 days. ☐

As we mentioned at the beginning of this section, we should be skeptical about the validity of the results from the last example. Though the example is an interesting exercise, the simplifying assumptions we made in developing our S–I–R models in general will not be satisfied in an Ebola epidemic.

4.3.4   Section Exercises

1   Suppose a mysterious flu-like illness is spreading on a college campus of 20,000 students. Originally only 1 student was infective, but by the time school officials became concerned 1 week later there were 50 cases. If the duration of infectivity for the disease is known to be 5 days and $\mu = 0.0005$, determine R_0 for the disease.

2   Continuing Exercise 1, determine the effective contact rate, β, for the new disease.

3   Consider a disease like measles that has a very high R_0. Using Table 4.2, in this case assume that $R_0 = 15$ and $\delta = 7$. Estimate the growth rate of the number of infectives at the beginning of a measles epidemic.

4   Suppose that pertussis is endemic to a population where $\mu = 0.0005$ and that there is no ongoing epidemic. A serosurvey for antibodies to pertussis is administered to a representative sample of 4514 members of the population. Of the 4514 surveyed, 4190 tested positive for antibodies. Estimate R_0 for pertussis in this population.

5   Continuing Exercise 4, determine the effective contact rate, β, for pertussis in this population. Use Table 4.2 to estimate the duration of infectivity.

6   Project the course of an uncontrolled outbreak of Ebola in your own hometown if originally one person was infected.

7   *Extension*: One reason the S–I–R model with vital dynamics is not the most appropriate model for Ebola is that EVD has a considerable incubation period (~10 days) during which time the person who has the disease is asymptomatic and cannot transmit it. A better choice of model would be an S–E–I–R model with vital dynamics, where we add a category for those who are *exposed* but not yet infectious.

   a. Draw a careful flow diagram for an S–E–I–R model with vital dynamics.

   b. Give the DDS for the model.

   c. Implement the model in Excel.

   d. How does the course of the potential epidemic in Dallas change when modeled by the S–E–I–R model with vital dynamics?

4.4   S–I–R WITH VITAL DYNAMICS AND ROUTINE VACCINATIONS

For the basic S–I–R model, vaccinations have the effect of moving a susceptible person straight into the removed compartment. Because there were no vital dynamics in that model, our vaccination was simply a "one-time campaign" that we assumed occurred before the onset of the epidemic. With the inclusion of a constant supply of new births, our vaccination program can be made more realistic since we now think of vaccinations of newborns on a routine or ongoing basis as they enter the susceptible population.

The inclusion of a routine vaccination program changes the dynamics of our model as well as its equilibrium points. To introduce the vaccination of newborns, we let ρ (a Greek lowercase rho) denote the proportion of newborns who are vaccinated. Instead of heading into the susceptible compartment, newborns who are vaccinated move directly into the removed compartment.

Since the number of new births each day is given by μN, each day vaccinations are sending $\rho\mu N$ newborns directly into R. The remainder $\mu N - \rho\mu N = (1-\rho)\mu N$ enter S. We incorporate these observations into the flow diagram in Figure 4.18.

Our discrete dynamical system changes accordingly to become

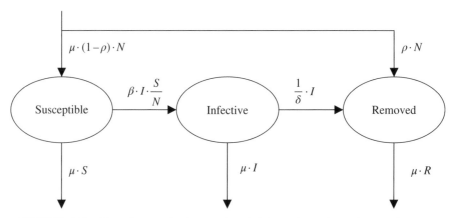

FIGURE 4.18   Flow diagram for S–I–R with vital dynamics and ongoing vaccinations.

$$S(t) = S(t-1) - \beta I(t-1)\frac{S(t-1)}{N} + (1-\rho)\mu N - \mu S(t-1)$$

$$I(t) = I(t-1) + \beta I(t-1)\frac{S(t-1)}{N} - \frac{1}{\delta}I(t-1) - \mu I(t-1)$$

$$R(t) = R(t-1) + \frac{1}{\delta}I(t-1) - \mu R(t-1) + \rho\mu N.$$

Next we provide a computational example that lends support to the Herd Immunity Theorem.

Example 4.20: Recall the chicken pox epidemic from Example 4.6. Determine the newborn vaccination proportion required to eventually eradicate the disease; confirm the result with Excel.

Our parameter assumptions from Example 4.6 were $\beta = 0.9$, $\delta = 10$, $\mu = 0.0005$, $I(0) = 5$, and $N = 5000$. To find the vaccination proportion, we first must find R_0. Because the model includes vital dynamics, we find

$$R_0 = \frac{\beta \cdot \delta}{1 + \mu \cdot \delta} = \frac{0.9 \cdot 10}{1 + 0.0005 \cdot 10} = 8.96.$$

According to the Herd Immunity Theorem, the proportion of newborns we need to vaccinate must be at least $1 - \frac{1}{R_0} \approx 0.89$.

Next we test this result with Excel. Including the vaccination proportion as its own parameter, we need to update the formulas for the susceptible and removed compartments. Figure 4.19 shows the setup with the formula for the removed category displayed.

	A	B	C	D	E	F	G	H
1	S-I-R with Vital Dynamics and Ongoing Vaccinations							
2								
3	Effective contact rate, β =			0.9				
4	Duration of infectivity, δ =			10				
5	Total population, N =			5000				
6	Birth and death rate, μ =			0.0005				
7	Vaccination proportion, ρ =			0.60				
8								
9	t	$S(t)$	$I(t)$	$R(t)$				
10	0	4995	5	0				
11	1	4989.0	9.0	=D10+(1/$D$4)*C10-$D$6*D10+$D$7*$D$6*$D$5				

FIGURE 4.19 Excel S–I–R with vital dynamics and vaccinations setup.

One final modification we make is to account for vaccinations having been ongoing before the start of the epidemic. Thus we assume that before the onset of the chicken pox epidemic, only 11% of the 5000 in the population was susceptible. As a result we start with $R(0) = 0.89 \cdot 5000 = 4450$, and $S(0) = 0.11 \cdot 5000 - I(0) = 550 - 5 = 545$. With the vaccination proportion set at $\rho = 0.89$, we get the graph for the epidemic shown in Figure 4.20.

Note that the infectives decrease from the outset and tend to 0. We leave it to the reader to check that if we set the vaccination proportion to 0.88 instead, it is not quite enough, and the number of infectives will increase at the outset. $\square$

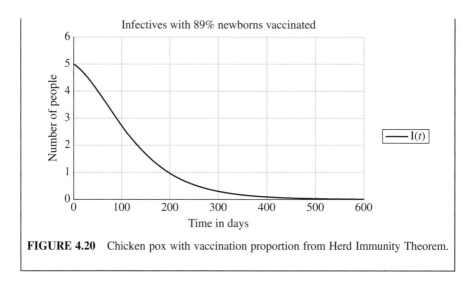

FIGURE 4.20   Chicken pox with vaccination proportion from Herd Immunity Theorem.

In the next section we show that the inclusion of ongoing vaccinations does not change the formula for computing the basic reproductive rate.

4.4.1   The Basic Reproductive Rate

Since we have a new model, we need to determine R_0 from the model parameters. We rely on the same general observation as before:

$$R_0 = (\text{Effective contacts per day}) \times (\text{Average time spent in } I).$$

We note that the inclusion of a vaccination program has left the infective compartment unchanged in our flow diagram and DDS. Hence the average time spent in I is the same as before, and so is the expression for R_0, namely,

$$R_0 = \frac{\beta\delta}{1+\mu\delta}.$$

As we see in the next section, it is a different story for equilibrium points: vaccinations do change them.

4.4.2   Equilibrium Points

As before, finding the equilibrium points amounts to solving the system of equations

$$S^* = S^* - \beta I^* \frac{S^*}{N} + (1-\rho)\mu N - \mu S^*$$
$$I^* = I^* + \beta I^* \frac{S^*}{N} - \frac{1}{\delta}I^* - \mu I^*$$
$$R^* = R^* + \frac{1}{\delta}I^* - \mu R^* + \rho\mu N$$

for S^*, I^*, and R^*.

We note that as a first simplification, we have

$$0 = -\beta I^* \frac{S^*}{N} + (1-\rho)\mu N - \mu S^*$$

$$0 = \beta I^* \frac{S^*}{N} - \frac{1}{\delta} I^* - \mu I^*$$

$$0 = \frac{1}{\delta} I^* - \mu R^* + \rho \mu N.$$

Note again that the I^*-equation is unchanged by the vaccination program; this means that the algebra begins just as it did for the vital dynamics model. Rather than repeat it here, we refer to that earlier work and note that if we start with the I^*-equation, we will find two cases. Either $I^* = 0$ or $S^* = \frac{N}{R_0}$.

Suppose $I^* = 0$. Direct substitution of $I^* = 0$ into the S^*-equation and R^*-equation yields the equilibrium point

$$(S^*, I^*, R^*) = ((1-\rho)N, 0, \rho N).$$

This point indicates that when no disease is present, the population will be split into the two compartments, S and R, with S containing the unvaccinated individuals and R containing the vaccinated. Note that we get the expected proportion in each compartment based on our vaccination coverage.

Next suppose $S^* = \frac{N}{R_0}$. To find I^* we substitute $S^* = \frac{N}{R_0}$ into the S^*-equation, $0 = -\frac{\beta I^*}{N} S^* + (1-\rho)\mu N - \mu S^*$. We get

$$0 = -\frac{\beta I^*}{N} \left(\frac{N}{R_0} \right) + (1-\rho)\mu N - \mu \left(\frac{N}{R_0} \right)$$

$$0 = -\frac{\beta I^*}{R_0} + (1-\rho)\mu N - \frac{\mu N}{R_0}$$

$$\frac{\beta I^*}{R_0} = (1-\rho)\mu N - \frac{\mu N}{R_0}$$

$$\beta I^* = R_0(1-\rho)\mu N - \mu N.$$

Thus we have

$$I^* = \frac{\mu N}{\beta}(R_0(1-\rho) - 1).$$

Finally, we find R^* by substituting $I^* = \dfrac{\mu N}{\beta}(R_0(1-\rho)-1)$ into the R^*-equation, $0 = \dfrac{1}{\delta}I^* - \mu R^* + \rho\mu N$. We find

$$\mu R^* = \frac{1}{\delta}I^* + \rho\mu N$$

$$R^* = \frac{1}{\mu\delta}I^* + \rho N$$

$$R^* = \frac{1}{\mu\delta}\left(\frac{\mu N}{\beta}(R_0(1-\rho)-1)\right) + \rho N.$$

Thus we have

$$R^* = \frac{N}{\beta\delta}(R_0(1-\rho)-1) + \rho N.$$

Note that if $(1-\rho)$ is small enough, that is, if ρ is close enough to 1, then the equilibrium value for I can be negative. This of course makes no physical sense. Under such conditions the number of infectives will be understood to be 0, and the disease will be eradicated.

So far in our work, we have taken for granted that we *ought* to be vaccinating, and our main concern has been how many we should vaccinate. In the next section we see that under some circumstances vaccinations may not be the right course of action.

4.4.3   Life Expectancy and Average Age at Infection: A Complication for Vaccination Programs

For some diseases the health consequences of infection depend on the age at which an individual contracts the disease. This age dependence must be taken into account when deciding on a vaccination program because an unintended consequence of such programs is that they tend to raise the average age at which people contract the disease (Nelson B., 2005). In this section we use our ongoing vaccination model to examine the mathematical consequences of such a program on the age at infection.

When people enter the susceptible category at birth, the age at which they become infected is the amount of time they spend in the susceptible compartment. Thus by **average age at infection**, we mean the average amount of time someone spends in the susceptible compartment. To compute a sensible estimate for this age, we must first assume that the disease is in equilibrium as we did when finding R_0 from serosurveys.

Since we aim to find the average time someone spends in the susceptible compartment, we apply the Waiting Time Principle once again. To do so we must first find the proportion of susceptibles who leave the compartment each day. Once we do that, the waiting time is straightforward to compute. From the flow diagram in Figure 4.18, we know that the total number of susceptibles who leave the compartment each day is the sum of those who become infective and those who die. This total is given by

$$\beta I \frac{S}{N} + \mu S = \left(\beta \frac{I}{N} + \mu \right) S.$$

It is tempting to say that the proportion of susceptibles who leave each day is therefore given by $\left(\beta \frac{I}{N} + \mu \right)$. The trouble with this approach is that the quantity I changes over time, but we need a single value for the proportion. We get around this obstacle by assuming that the disease has been present in the population long enough to be in equilibrium. Thus we use

$$I^* = \frac{\mu N}{\beta} (R_0(1-\rho)-1)$$

as our single fixed value for I. As long as we are at equilibrium, the proportion of susceptibles removed from S each day will therefore be equal to

$$\left(\beta \frac{I^*}{N} + \mu \right) = \left(\beta \frac{(\mu N/\beta)(R_0(1-\rho)-1)}{N} + \mu \right)$$
$$= (\mu(R_0(1-\rho)-1) + \mu)$$
$$= \mu R_0(1-\rho).$$

Finally, we appeal to the Waiting Time Principle and compute the average age at infection, which we denote by A:

$$A = \frac{1}{\mu R_0(1-\rho)}.$$

We observe that the average age at infection depends on several factors: the population's natural death rate, the ease with which the disease can spread, and the proportion of newborns who are routinely vaccinated.

Considering our formula for A, we can go a step further and note that μ can be thought of as the removal rate from the entire population. Thus by the Waiting Time Principle, the reciprocal of μ is equal to the average length of time that someone spends in the population. Put another way, the reciprocal of μ represents the average **life expectancy**, which we denote by L, of someone in the population. We capture this relationship as

$$L = \frac{1}{\mu},$$

and we note that this is the relationship used to estimate μ for the Ebola example in the previous section. We can also use this relationship to rewrite our expression for the average age at infection as

$$A = \frac{L}{R_0(1-\rho)}.$$

As mentioned previously the average age of infection depends on the vaccination proportion, ρ. Since $(1-\rho)$ appears in the denominator, the higher the vaccination proportion, the closer $(1-\rho)$ will be to 0 and thus the higher the age at infection will be. That is, the more people we vaccinate, the older people will typically be when they contract the disease. The trouble with this is that some diseases are much more serious when contracted later in life. In such cases a vaccination program can actually increase the chances for serious complications from a disease.

In the next section we consider a disease for which the age at infection is a crucial factor when considering a vaccination program.

4.4.4 Rubella and Congenital Rubella Syndrome in The Gambia

In the United States it is recommended that all children be given the mumps, measles, and rubella (MMR) vaccine—the first dose at age 12–15 months and the second dose at age 4–6 years (CDC, 2012). Rubella, also known as German measles, is particularly interesting because it is an example of a disease whose seriousness depends on when it is contracted. Rubella is generally a fairly mild disease with one important exception. If a pregnant woman contracts the disease during the first trimester of pregnancy, there is an 80% chance that her baby will be born with **congenital rubella syndrome** (CRS). Complications for infants with CRS include cataracts, mental retardation, deafness, and cardiac defects (Anderson & May, 1991). We now use our vaccination model to analyze the effect that a Rubella immunization program aimed at newborns might have in The Gambia.

First we must find values for all of the model parameters. We have from serosurvey data (see Example 4.17) that $R_0 = 16.67$, and from Table 4.2 we may take the duration of infectivity to be $\delta = 11.5$ days. The life expectancy, L, for The Gambia is difficult to know precisely. Various estimates place it at 45–55 years. We will use $L = 50 \cdot 365 = 18,250$ days.

Currently there is not a routine vaccination program for rubella in The Gambia. Should there be?

With no vaccinations our model predicts that the average age at infection for rubella in The Gambia is about $A = \frac{L}{R_0(1-\rho)} = \frac{18,250}{16.67(1-0)} = 1,095$ days, that is, 3 years old. This estimate agrees with the one given in Anderson and May of 2–3 years (Anderson & May, 1991). Thus in The Gambia where there is no vaccination program and virtually everyone gets rubella, there are very few complications due to CRS because almost all women of childbearing age have already had the disease.

We contrast this with what would happen under a routine newborn vaccination program. Suppose we were able to successfully vaccinate 80–85% of all infants in The Gambia. Then our new average age at infection would be between $A = \frac{L}{R_0(1-\rho)} = \frac{18,250}{16.67(1-0.80)} = 5,475$ and $A = \frac{18,250}{16.67(1-0.85)} = 7,300$ days, or between 15 and 20 years old.

Now we can see the problem with vaccinations here. By vaccinating infants we will succeed in dramatically reducing the number of cases of rubella; however, those who *do* contract the disease will now be between 15 and 20 years old. The new, higher age at infection means that while we decrease the total cases of rubella, we may simulta-neously *increase* the number of the most serious cases—pregnant women who pass CRS to their babies. Thus vaccinating infants against rubella in The Gambia might very well be a bad idea—a result that is certainly not obvious at the outset.

This last example highlights the power of mathematical modeling. Even though the S–I–R model and its variants may not be accurate for predictions involving large populations or long time periods, they can still play an important role by highlighting issues to consider before taking action.

4.4.5   Section Exercises

1   According to the World Health Organization, vaccination coverage for measles in the United States was 91.9% in 2014. Using values for R_0 from Table 4.2 and life expectancy estimates from an Internet source, estimate the average age at infection for measles in the United States.

2   According to the World Health Organization, vaccination coverage for pertussis in Nigeria was 41% in 2012. Using values for R_0 from Table 4.2 and life expectancy estimates from an Internet source, estimate the average age at infection for pertus-sis in Nigeria.

3   Our minimum vaccination coverage required to prevent an epidemic has been based on the Herd Immunity Theorem. A second way to think of disease eradica-tion would be to vaccinate enough people so that the long-term value for the num-ber of infectives equals 0. In other words, we could vaccinate so that $I^* = 0$. Show that the vaccination level implied by this requirement turns out to be the same as that given in the Herd Immunity Theorem.

4   Another way to approach the idea of disease eradication is to vaccinate enough people that the average age at infection is greater than the life expectancy of the population. Show that the vaccination level produced by this approach is the same as that given by the Herd Immunity Theorem.

5   *Extension*: A reason that we do not actually vaccinate newborns in practice is that newborns are temporarily protected from many diseases by antibodies that are passed on from the mother. This phenomenon is known as **maternal antibody protection**, and if a vaccine is administered too soon after birth, maternal antibo-dies can interfere with the vaccine working properly. Suppose that maternal

antibody protection lasts 90 days. Modify the S–I–R model with vital dynamics and routine vaccinations to include a compartment, M, for maternal antibody protection.

a. Give a flow diagram for the M–S–I–R model with vital dynamics and routine vaccinations.

b. Give the DDS for the model.

c. Implement the model in Excel.

5

DENSITY-DEPENDENT POPULATION MODELS

If we project exponential population growth far enough into the future, it eventually becomes unrealistic because the population continues to grow without bound. We do not really expect the Yellowstone grizzly population to reach, say, a billion bears; yet, if we rely uncritically on our exponential model, eventually that is what we will see.

Real populations seldom exhibit exponential growth for long. Certainly there are many examples where populations do grow exponentially for a time, but both experience and common sense tell us that eventually the growth must taper off. As overcrowding develops, resources like food, water, and shelter become more and more scarce, diseases spread more easily, and as a consequence it becomes more difficult for the population to continue growing. Models that take these growth-limiting effects into account are said to be **density dependent**.

5.1 THE DISCRETE LOGISTIC MODEL

We begin by assuming that for any population there is a maximum number that a given environment can support. This maximum number is called the **carrying capacity**, and we follow convention by denoting this number by K. We should note that the carrying capacity depends both on the particular species and on the particular environment in which it is found. A small pond, for example, will have a smaller carrying capacity for goldfish than a large lake—it is not just the goldfish themselves that

Models for Life: An Introduction to Discrete Mathematical Modeling with Microsoft® Office Excel®,
First Edition. Jeffrey T. Barton.
© 2016 John Wiley & Sons, Inc. Published 2016 by John Wiley & Sons, Inc.

determine the carrying capacity. Similarly, a lake will have a larger carrying capacity for minnows than for catfish.

Our task in this section is to build a DDS that models a population when we know its growth is limited by the carrying capacity of its environment. Before we attempt to write down the DDS, we give some thought to the features that such a model should have.

The notion of a carrying capacity implies that:

1. The growth rate of the population should decline as the population nears the carrying capacity.
2. The growth rate should be 0 if the population reaches the carrying capacity.

In the nineteenth century Pierre Verhulst was the first to formulate such a model (Verhulst, 1838). Here we present the discrete version of Verhulst's model: the **discrete logistic growth** model.

Recall that the DDS for exponential growth is given by $P(t) = P(t-1) + rP(t-1)$, where the growth rate r is assumed to be the same regardless of how large or small the population is. The idea in Verhulst's model is to replace the fixed growth rate, r, with an expression that varies in accordance with properties 1 and 2 above.

Consider the basic exponential growth model where r does not depend on the population. Then the graph of the growth rate versus population should appear as in Figure 5.1. Note that the graph for r is a horizontal line because the growth rate remains constant regardless of how large the population is.

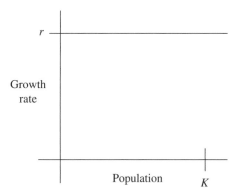

FIGURE 5.1   Growth rate versus population for exponential model.

A more realistic graph would show a decline in the growth rate with increasing population to reflect the effects of overcrowding, and it would indicate a growth rate of zero if the population reaches the carrying capacity, K. The simplest such graph is given in Figure 5.2: a straight line that starts with a maximum growth rate of r and decreases to a growth rate of 0 at the carrying capacity.

The graph is an important step, but to accomplish our ultimate goal of a DDS for logistic growth, we need to find the equation of this straight line: the line that connects

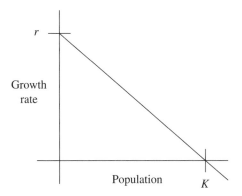

FIGURE 5.2 Growth rate versus population for discrete logistic model.

the points $(0, r)$ and $(K, 0)$. To do so we recall the slope–intercept form of a line: $y = mx + b$, where m is the slope of the line and b is the y-intercept. The formula for finding the slope is $m = \dfrac{\text{rise}}{\text{run}} = \dfrac{y_2 - y_1}{x_2 - x_1}$, where (x_1, y_1) and (x_2, y_2) are any two points on the line. Using the only two points we know, we find $m = \dfrac{0 - r}{K - 0} = -\dfrac{r}{K}$.

We already know where the line crosses the y-axis, namely, at $b = r$. Finally, we can write down the equation of the line that represents our varying growth rate:

$$y = -\frac{r}{K}x + r,$$

which we can also write as $y = r - \dfrac{r}{K}x = r\left(1 - \dfrac{x}{K}\right)$. Remembering that x represents population and y the growth rate, we arrive at the formula

$$\text{Growth rate} = r\left(1 - \frac{P}{K}\right).$$

To distinguish the constant r from our varying growth rate formula, we now refer to r as the **intrinsic growth rate** of the population. This is the maximum possible growth rate the population would experience under ideal conditions. With our new, more realistic notion of a varying growth rate in place, we can write down the DDS for logistic growth:

$$P(t) = P(t-1) + r\left(1 - \frac{P(t-1)}{K}\right)P(t-1).$$

It is worth comparing this new DDS to the exponential growth model. The logistic model still says that "to get from 1 year's population to the next, add a percentage of the previous population." The important difference is that the percentage now depends on how large the population is relative to the environment's carrying capacity.

Next we apply the discrete logistic model to the population of Antarctic baleen whales.

5.1.1   Antarctic Baleen Whales

Baleen whales, also known as great whales, are whales that feed by filtering food through baleen plates in their upper jaw. Examples of baleen whales are the blue whale, fin whale, and sei whale. Due to overfishing, baleen whale populations in the Antarctic declined to dangerously low levels in the mid-1900s. On December 2, 1946, the International Whaling Commission (IWC) was formed to

> provide for the proper conservation of whale stocks and thus make possible the orderly development of the whaling industry, provide for the complete protection of certain species; designate specified areas as whale sanctuaries; set limits on the numbers and size of whales which may be taken; prescribe open and closed seasons and areas for whaling; and prohibit the capture of suckling calves and female whales accompanied by calves. (International Whaling Commission, n.d.)

Prior to 1963 the IWC used the **blue whale unit** (BWU) as its unit in setting whale quotas (Clark, 1985). In these units we have 1 blue whale = 1 BWU, 1 fin whale = 1/2 BWU, and 1 sei whale = 1/6 BWU. In retrospect the IWC's choice of units was a regrettable one as it was not specific enough to protect any particular species and led to the further depletion of baleen whales by overfishing.

Based on IWC estimates we adopt an initial population in 1985 of 75,000 BWU, an intrinsic growth rate of $r = 5\%$, and a carrying capacity of $K = 400,000$ BWU for the Antarctic baleen whale fishery (Clark, 1985). It is worth taking a few moments to understand the units here. Saying that the carrying capacity is 400,000 BWU means that the environment could support as many as 400,000 blue whales, or 800,000 fin whales, or 2,400,000 sei whales, or any combination of the three species that does not exceed the 400,000 BWU threshold.

We begin with a computational example.

Example 5.1: Assume that in 1985 population we have $P(0) = 75,000$, $r = 5\%$, and $K = 400,000$. What would the discrete logistic growth model predict for the 2015 abundance of baleen whales in the Antarctic fishery?

This problem is a straightforward application of the discrete logistic model. The main difficulty will be typing in the formula correctly into Excel. Recall that the DDS for this model is

$$P(t) = P(t-1) + r\left(1 - \frac{P(t-1)}{K}\right)P(t-1).$$

We store all parameters in their own cells and then refer to them using absolute addressing. We show the Excel setup with the population formula displayed in Figure 5.3.

	A	B	C
1	Discrete Logistic Model		
2			
3	Intrinsic growth rate, r =		5.0%
4	Carrying capacity, K =		400,000
5			
6	t	Population	
7	0	75,000	
8	1	=B7+$C$3*(1-B7/$C$4)*B7	

FIGURE 5.3   Discrete logistic Excel model setup.

Once we have entered the formula, we copy it down to year 30 (2015) and observe the results shown in Figure 5.4 with most rows hidden.

	A	B	C
1	Discrete Logistic Model		
2			
3	Intrinsic growth rate, r =		5.0%
4	Carrying capacity, K =		400,000
5			
6	t	Population	
7	0	75,000	
8	1	78,047	
36	29	197,121	
37	30	202,120	

FIGURE 5.4   Discrete logistic Excel model output for Example 5.1.

We see that the model predicts a population of 202,120 BWU for 2015. Notice that even after 30 years (assuming no harvesting), the baleen whale population is still well below its carrying capacity. This example illustrates that great whale populations are in general very slow to recover from overfishing. □

In the next example we consider how long it would take for the baleen whale population to grow to near its carrying capacity.

Example 5.2: How long under the conditions in Example 5.1 would it take the baleen whale population to reach 390,000 BWU?

This question can be answered by copying our formulas down far enough. Based on our model projections, it would take approximately 102 years—over a century—for the whales to recover to near their carrying capacity, even with full protection from the IWC. A graph of the population over this time period is presented in Figure 5.5.

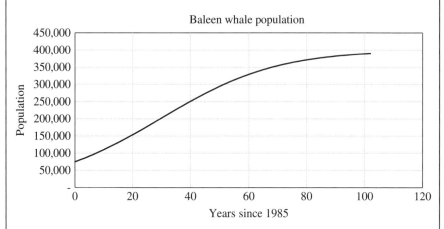

FIGURE 5.5   Model predictions for baleen whale population from Example 5.2.

Notice the S shape of the graph; this is the characteristic shape of logistic growth. Observing this shape in a graph of data is often a clue that a logistic model is an appropriate one to try. ☐

5.1.2   Equilibrium Values

Recall that an equilibrium value for a population is a value P^* where the model does not change. In the next example we find the equilibrium values for the discrete logistic model.

Example 5.3: Using the same parameters as in Example 5.1, find the equilibrium values for the baleen whale model.

We need to find values P^* such that

$$P^* = P^* + 0.05\left(1 - \frac{P^*}{400,000}\right)P^*.$$

Subtracting P^* from both sides gives us

$$0 = 0.05\left(1 - \frac{P^*}{400,000}\right)P^*.$$

When a product of real numbers is equal to 0, at least one of the factors must equal 0. Thus we must have either $P^* = 0$ or $\left(1 - \dfrac{P^*}{400,000}\right) = 0$. Solving the second equation gives us $P^* = 400,000$. What we have shown is that the population will be at equilibrium if it becomes extinct or if it reaches its carrying capacity. $\square$

The results from Example 5.3 are not a special case. For the discrete logistic model, we always have two equilibria: extinction and carrying capacity, a fact the reader is invited to verify in the exercises.

As we will see later in this chapter, it is not possible to make a general claim about the stability of the equilibrium values for the discrete logistic model because the stability of the equilibrium values actually depends on the parameter values. We can, however, determine the stability of the equilibrium values for particular cases using Excel as we did in Chapter 1.

Example 5.4: Determine the stability of the equilibrium values from Example 5.3.

Recall that a stable equilibrium value is one where if the population starts off of the equilibrium, the population will tend back toward it. On the other hand, an unstable equilibrium is one where if the population starts off of it, the population will tend to get further away from it. By using Excel to graph population projections for several different initial populations on the same graph, we can determine the stability of an equilibrium value by observing the behavior of the populations. (See Section 1.5.2 for a refresher on stability and how to create an Excel graph to test for it.) We produce such a graph in Figure 5.6.

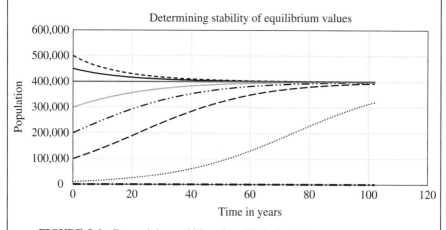

FIGURE 5.6 Determining stability of equilibria for baleen whale population.

First note that the constant populations at $P(0) = 0$ and $P(0) = 400,000$ confirm that we do indeed have equilibrium values at those points. Also note that populations just above 0 tend to grow away from 0. Thus $P^* = 0$ is an unstable equilibrium. On the other hand, all other populations tend toward the carrying capacity, $P^* = 400,000$. Populations that start above 400,000 decline toward it, and populations that start below 400,000 grow toward it. Thus the equilibrium value $P^* = 400,000$ is a stable equilibrium. $\square$

In the next section we provide an alternative method for classifying equilibria graphically.

5.1.3   Cobweb Diagrams

We saw in Chapter 1 and in the previous section that we can determine the stability of equilibrium values graphically by selecting several different initial populations and observing whether the graphs all seem to approach or move away from a given equilibrium. In this section we introduce a new kind of graph called a **cobweb diagram** that is commonly used to determine stability. An advantage of using a cobweb diagram is that once we understand how to construct and interpret one, stability can be determined simply by inspecting the graph without reference to any particular initial population.

We begin by considering a logistic growth model where the intrinsic growth rate is $r = 0.80$ and the carrying capacity is $K = 100$. We know that the DDS for the population is given by

$$P(t) = P(t-1) + 0.80\left(1 - \frac{P(t-1)}{100}\right)P(t-1).$$

We define a function, f, using the DDS: $f(x) = x + 0.80\left(1 - \frac{x}{100}\right)x$. Computing the population from one time to the next is equivalent to evaluating our function f at x where x is the population at a given time. The function f is called the **reproduction function**— given a population, x, it computes the next, $f(x)$.

The graph of the reproduction function is called the **reproduction curve**, and it provides a useful way of doing DDS calculations graphically. If we graph the function $f(x) = x + 0.80\left(1 - \frac{x}{100}\right)x$, we can evaluate it at any population x by starting at x on the horizontal axis and traveling vertically up to the graph of f. Once we are on the graph, we estimate the function value by glancing at the vertical axis. If we want to know what the population will be the day after that, we take our current estimate, locate it on the x-axis, and again travel vertically up to the graph so we can estimate the new population value. In this way we can at least estimate the population as far into the future as we like by repeating the steps.

The cobweb diagram introduces a clever graphical device that makes this process of graphical iteration much easier. In fact, it will allow us to project our population into the future without having to lift our pencil from the page. The idea is to graph the reproduction function $f(x) = x + 0.80\left(1 - \dfrac{x}{100}\right)x$ along with the line $y = x$ on the same axes. We point out that an equilibrium value is a value x^* such that $f(x^*) = x^*$. In this context such a point is often referred to as a **fixed point** of f. Graphically, fixed points correspond to points where the line $y = x$ and f intersect. Thus, if we have not already found our equilibrium values, we can estimate them easily by inspecting our new graph.

Example 5.5: Show that the equilibrium values for a logistic growth model with intrinsic growth rate $r = 0.80$ and carrying capacity $K = 100$ correspond to the fixed points of the reproduction function.

Algebraically we compute the value of the reproduction function at each equilibrium point. For the carrying capacity, we get

$$f(100) = 100 + 0.80\left(1 - \frac{100}{100}\right)100 = 100.$$

For the equilibrium at zero, we get

$$f(0) = 0 + 0.80\left(1 - \frac{0}{100}\right)0 = 0.$$

In both cases we have $f(x^*) = x^*$.

To approach the problem graphically, we first must graph the reproduction function and the line $y = x$ on the same axes.

E.16   Graphing Functions

Graphing functions in Excel requires a slightly different approach than graphing a DDS. We will need a column for the x-variable and a column for the function values or y-variable. For the x-variable, we need to decide on the range we want to graph. In this case we will graph x-values from $x = 0$ to $x = 110$. We also need to decide how many points to graph. Somewhere around 100 points should be plenty and is not difficult to do in Excel. We want our points to be evenly spaced, so starting at $x = 0$ we use all whole number values of x up to 110. The Excel setup with most rows hidden and formulas displayed is given in Figure 5.7.

Next we enter the formula for $f(x)$. The result is shown in Figure 5.8.

Note how the formula for $f(x)$ refers to the x-variable column to get population values, rather than from the "previous cell" as was the case for the DDS. The last step before graphing is to include a third column for the line $y = x$. This is done in Figure 5.9.

	A	B
1	Graphing Functions	
2		
3	x	f(x)
4	0	
5	=A4+1	
6	=A5+1	
113	=A112+1	
114	=A113+1	

FIGURE 5.7 Graphing functions in Excel setup.

	A	B
1	Graphing Functions	
2		
3	x	f(x)
4	0	=A4+0.8*(1-A4/100)*A4
5	=A4+1	=A5+0.8*(1-A5/100)*A5
6	=A5+1	=A6+0.8*(1-A6/100)*A6
113	=A112+1	=A113+0.8*(1-A113/100)*A113
114	=A113+1	=A114+0.8*(1-A114/100)*A114

FIGURE 5.8 Graphing functions in Excel with formula for $f(x)$.

	A	B	C
1	Graphing Functions		
2			
3	x	f(x)	y = x
4	0	=A4+0.8*(1-A4/100)*A4	=A4
5	=A4+1	=A5+0.8*(1-A5/100)*A5	=A5
6	=A5+1	=A6+0.8*(1-A6/100)*A6	=A6
113	=A112+1	=A113+0.8*(1-A113/100)*A113	=A113
114	=A113+1	=A114+0.8*(1-A114/100)*A114	=A114

FIGURE 5.9 Graphing multiple functions in Excel.

Finally we are ready to graph. We select all three columns, including the column headings, and select a scattergraph with straight lines but no markers. The result is given in Figure 5.10.

Note that the two intersection points for the graphs of the reproduction function and the line $y = x$ correspond exactly to the fixed points for f. □

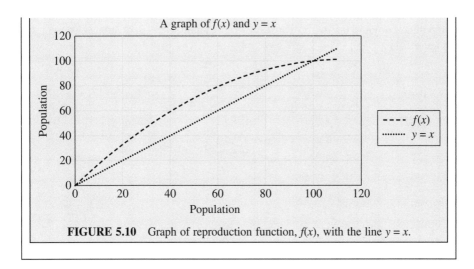

FIGURE 5.10   Graph of reproduction function, $f(x)$, with the line $y = x$.

We are now interested in what happens to our population under repeated iteration of the function, f. We begin with some initial population x_0 and note that the population the following year will be $x_1 = f(x_0)$, which corresponds to the y-value on the graph of f above the point x_0.

Now we come to the clever part of our graphing technique: instead of having to locate the new population x_1 on the x-axis, we simply move *horizontally* from the point $(x_0, f(x_0))$ over to the graph of the line $y = x$. When we hit the line $y = x$, we will be at a point whose y-coordinate is $x_1 = f(x_0)$ (because we have moved horizontally) and hence whose x-coordinate is exactly the value we want, namely, x_1. Now to find the next population $x_2 = f(x_1)$, we simply move vertically to the graph of f again. We can carry out as many iterations (i.e., population values) as we like simply by repeating the process of moving horizontally to $y = x$ to get the next x-coordinate and then vertically to f to get the next population. If this process brings us closer and closer to an equilibrium value, then that value is stable. If it moves us further away, the equilibrium is unstable. Occasionally we will have a value that we approach from one side but move away from on the other. Such an equilibrium will be called **semistable**.

Example 5.6: Use a cobweb diagram to examine the logistic growth model from the previous example where $r = 0.80$ and $K = 100$.

We start with an initial population $x_0 = 60$ and observe how the population changes with each iteration.

The first step is to evaluate the function at the initial population. When we do, we get a population of $f(60) = 60 + 0.8\left(1 - \dfrac{60}{100}\right)60 = 79.2$ for the following year. The 79.2 becomes our new x-value $x_1 = 79.2$ to which we apply the function f again. Graphically this is equivalent to moving right until we hit the line $y = x$

and then moving up to the function graph again. The first step is shown in Figure 5.11.

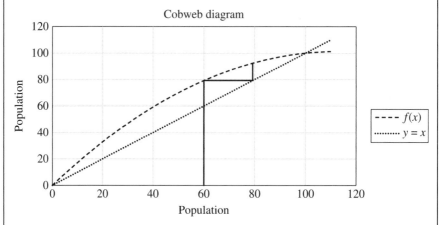

FIGURE 5.11   Iterating the reproduction function using cobwebbing.

The next graph shows several iterations of the cobwebbing method. Notice how the points are tending toward the fixed point $x^* = 100$ (Fig. 5.12).

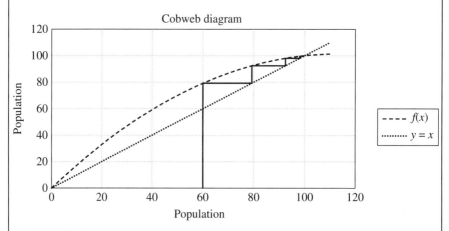

FIGURE 5.12   Several iterations of the reproduction function using cobwebbing.

The next graph in Figure 5.13 shows the result of starting our cobweb at the initial point $x_0 = 20$.

Note that this graph also shows points tending toward $x^* = 100$. This indicates that the fixed point, or equilibrium value, at 100 is stable. At the same time we have shown that the fixed point or equilibrium value at 0 is unstable since our populations are all tending away from it. ☐

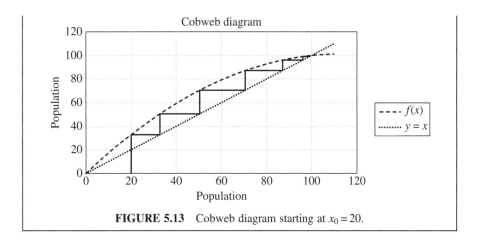

FIGURE 5.13   Cobweb diagram starting at $x_0 = 20$.

The steps for creating cobweb diagrams in Excel are somewhat more involved than for our usual types of graphs. The interested reader can find the details in Appendix E and further information in Gurney (2004).

Next we use cobwebbing to reinforce our work on a previous example.

Example 5.7:   Consider again our affine model for the white-tailed deer population in Chapter 1. We assume that the population is growing by 26% each year and that 780,000 deer are harvested per year. Use a cobweb diagram to verify that the equilibrium value $P^* = 3,000,000$ is unstable.

The DDS for this example is $P(t) = P(t-1) + 0.26P(t-1) - 780,000$; hence our reproduction function f is $f(x) = x + 0.26x - 780,000$. The cobweb diagram in Figure 5.14 shows that no matter the initial population (one below equilibrium, one above), we will always tend *away* from the equilibrium value. This confirms our earlier work showing that $P^* = 3,000,000$ is an unstable equilibrium. □

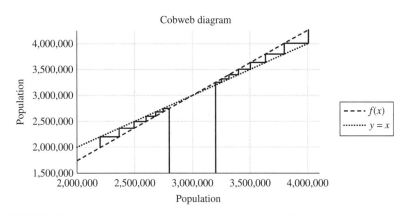

FIGURE 5.14   Cobweb diagram showing an unstable equilibrium for Example 5.7.

5.1.4 Section Exercises

1 Recall that for baleen whales we assume that $r = 5\%$ and $K = 400,000$. If in 1985 the population was $P(0) = 75,000$, determine how long it will take for the population to reach 300,000 blue whale units.

2 Use the formula for the growth rate for the logistic model to find the population at which the absolute (as opposed to percentage) population growth would be largest. Confirm your result by noting this population on a graph of the discrete logistic model.

3 Reveal a shortcoming in the discrete logistic model by choosing a very large initial population.

4 *Extension:* For the general discrete logistic model, find the smallest initial population that will result in a negative population during the next time step.

5 *Extension:* For the general discrete logistic model, find the population that results in the largest absolute population growth during the next time step.

6 For the general discrete logistic model, show algebraically that the two equilibrium values are K and 0.

7 Explain in a complete sentence or two why the values you found in Exercise 6 make sense intuitively.

8 For the baleen whale example, determine the stability of the two equilibrium values by producing an appropriate graph.

9 Find the estimates for the intrinsic growth rate and carrying capacity of the earth's human population. Be sure to cite your source(s). Use the current world population along with the parameters you found to predict the world's population over the next 50 years. Display your predictions on a graph.

10 Suppose in the discrete logistic model that $r = 0.10$ and $K = 50,000$.

  a. Show that if $P(0) > 20,000$, then the population will move away from 20,000.

  b. Explain why 20,000 is not an unstable equilibrium value.

5.2 LOGISTIC GROWTH WITH ALLEE EFFECTS

Our logistic growth model incorporated more realism into our population models by putting a limit on how large a population can eventually get, and we called this limit the carrying capacity. In this section we incorporate a similar observation from many real populations: as a population declines, eventually it can become too small to recover and will crash. This limit on how small a viable population can be is known as the **sustainability threshold**, the level below which the population will go extinct.

Some reasons why we see sustainability thresholds in real populations include members having more difficulty finding mates and being more vulnerable to predators. Such effects are known as **Allee effects** after Warder Clyde Allee, a zoologist who first observed them in the 1930s (Drake & Kramer, 2011).

Let S denote the sustainability threshold for a population. Our new model must reflect the fact that if $P(t-1) < S$, then the population will decline. Stated in terms of the population growth rate, we have that if $P(t-1) < S$, then the population growth rate should be negative. Together with the features of the logistic growth rate, we now seek a model whose growth rate satisfies:

1. The growth rate is negative if the population is less than S.
2. The growth rate is positive if the population is between S and K.
3. The growth rate is negative if the population is larger than K.

The simplest function that accomplishes all three requirements is a downward-pointing parabola, or quadratic. We give the graph for such a growth rate in Figure 5.15.

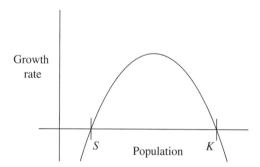

FIGURE 5.15   Growth rate versus population for Allee effects model.

To form the quadratic for the new growth rate, we force a root to appear at $P(t-1) = S$ by multiplying the original logistic growth rate by the factor $\left(\frac{P(t-1)}{S} - 1\right)$. Thus the quadratic growth rate for our new model is given by

$$r\left(1 - \frac{P(t-1)}{K}\right)\left(\frac{P(t-1)}{S} - 1\right).$$

Note that this growth rate satisfies the three requirements above.

Using the new growth rate in place of the logistic version gives us the DDS for the new model:

$$P(t) = P(t-1) + r\left(1 - \frac{P(t-1)}{K}\right)\left(\frac{P(t-1)}{S} - 1\right)P(t-1).$$

We call this new model the **logistic model with Allee effects** or simply the Allee effects model. Here we still refer to r as the intrinsic growth rate even though in the new model it is no longer the maximal growth rate.

In the next example we use the Allee effects model in order to investigate the population of giant pandas, an endangered species that has become a symbol for wildlife conservation efforts. It is a naturally very slow growing population with reproductive females typically producing 1 cub every 2 years (WWF, 2015). This slow growth makes the giant panda vulnerable to negative effects from poaching and from the removal of reproductive pandas for placement in zoos (Carter, Ackleh, Leonard, & Wang, 1999).

The giant panda is unusual in that its diet is made up almost entirely of 1 food source: bamboo. In order to find enough bamboo to eat, giant panda populations continually migrate in search of new sources. Thus a major source of stress on the giant panda is not only habitat depletion but also the closing off of migration routes due to development (Carter, Ackleh, Leonard, & Wang, 1999).

At low levels giant panda populations have been shown to experience Allee effects related to lack of mate selection (Moller & Legendre, 2001). This helps explain the difficulty that captive breeding programs for the giant panda have experienced, and it indicates that our Allee effects model is a reasonable choice as a starting model for the giant panda population. Though we will consider the wild giant panda population as a whole, it is really comprised of several subpopulations, each of which could be modeled individually.

Estimates of giant panda population parameters are difficult to make. For parameter values for r and K, we rely primarily on the work of Carter, Ackleh, Leonard, and Wang (1999), and we have to make our own estimate for K based on their data. We use a combination of the data in Carter and in Wang, Li, and Pan (2001) to estimate the sustainability threshold, S. For population estimates, we use those provided by the World Wildlife Fund (WWF, 2014).

Example 5.8: For the wild giant panda population, we let $r = 0.92\%$, $K = 16,000$, and $S = 480$. Population estimates put the giant panda population at approximately

▲	A	B	C
1	Allee Effects Model		
2			
3	Intrinsic growth rate, r =		0.0092
4	Carrying capacity, K =		16000
5	Sustain. thresh., S =		480
6			
7	t	Population	
8	0	1100	
9	=A8+1	=B8+$C$3*(B8/$C$5-1)*(1-B8/$C$4)*B8	
10	=A9+1	=B9+$C$3*(B9/$C$5-1)*(1-B9/$C$4)*B9	

FIGURE 5.16   Allee effects Excel model setup.

1100 in 1977. Use the Allee effects model to project the giant panda population in the year 2004.

We begin by starting with our Excel model for the discrete logistic model and storing the sustainability threshold in its own cell. We then enter the formula for the Allee effects model. The result is shown in Figure 5.16.

Next we copy our model equations down to year $t = 27$ (2004), hide the rows we do not need to see, and display the result in Figure 5.17. We see that the model predicts about 1627 pandas for the year 2004. $\square$

	A	B	C
1	Allee Effects Model		
2			
3	Intrinsic growth rate, r =		0.92%
4	Carrying capacity, K =		16,000
5	Sustain. thresh., S =		480
6			
7	t	Population	
8	0	1100	
9	1	1112.2	
34	26	1596.5	
35	27	1627.3	

FIGURE 5.17   Allee effects model projections for giant panda population.

A 2004 survey of the giant panda population estimated the population at about 1600 pandas. Thus our model seems to have generated a reasonable prediction over the time period from 1977 to 2004. The most recent census of the panda population indicated that the giant panda population continued to increase to a level of 1864 pandas in 2014. In the exercises the reader is invited to compare a model prediction with this most recent estimate.

In the next section we show that the inclusion of the sustainability threshold in our model introduces a new equilibrium value.

5.2.1   Equilibrium Analysis

To find the equilibrium values, we must find values for P^* that leave the DDS unchanged. In this case we must find values for P^* such that

$$P^* = P^* + r\left(1 - \frac{P^*}{K}\right)\left(\frac{P^*}{S} - 1\right)P^*.$$

This last equation is equivalent to $0 = r\left(1 - \frac{P^*}{K}\right)\left(\frac{P^*}{S} - 1\right)P^*$. We are now in a familiar situation where we have a product of terms that equals zero so that at least one of the terms must be zero. This leads to three possible equilibrium values. If $\left(1-\dfrac{P^*}{K}\right)=0$, then $P^*=K$; if $\left(\dfrac{P^*}{S}-1\right)=0$, then $P^*=S$; and otherwise $P^*=0$. Thus the three equilibria for the Allee effects model are the carrying capacity, the sustainability threshold, and extinction.

In the next example we determine the stability of the equilibrium values for the giant panda population from Example 5.8.

Example 5.9: Assuming the same parameters as in Example 5.8, determine the stability of the equilibrium values for the panda Allee effects model.

We proceed by using Excel to create a graph for the panda model starting at several different initial populations. Because of the wide spread between sustainability and carrying capacity, it is difficult to produce a graph that clearly shows both at the same time. We start by examining the equilibrium at the carrying capacity of 16,000. The result is given in Figure 5.18.

The graph indicates populations converging to the carrying capacity so we have a stable equilibrium there.

Figure 5.19 focuses on the behavior near the sustainability threshold of 480.

This graph shows populations moving away from 480 so we have an unstable equilibrium there. Note that we also see populations below 480 crashing to extinction. This gives us a stable equilibrium at 0, which is different from the behavior exhibited by the discrete logistic model without Allee effects. □

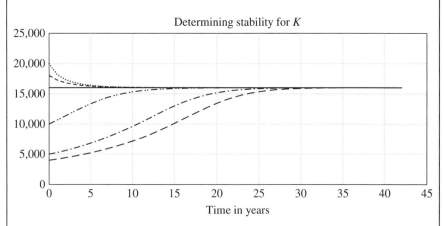

FIGURE 5.18 Determining stability for giant panda equilibrium at carrying capacity.

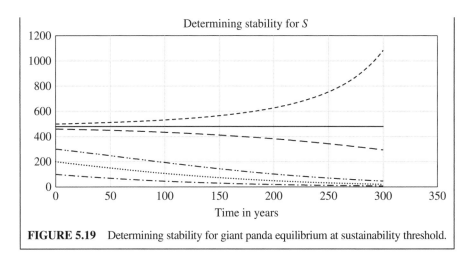

FIGURE 5.19 Determining stability for giant panda equilibrium at sustainability threshold.

As is the case for the discrete logistic model, we cannot make general claims about the stability of the equilibrium values for the Allee effects model. Different parameter choices can lead to different results, so it is always a good idea to test the particular situation of interest with Excel.

5.2.2 Section Exercises

1 The most recent survey of the panda population yielded an estimate of 1864 pandas in 2014. Compare this estimate to what our model from Example 5.8 projects.

2 Suppose that the Antarctic baleen whale population would crash if it falls below 30,000 BWU. Use the Allee effects model to graph the population from 1985 to 2015.

3 Suppose that the giant panda population did not experience Allee effects and was instead well modeled by the discrete logistic model.

   a. Using the same parameters as in Example 5.8, project the giant panda population in 2020 using the discrete logistic model.

   b. Compare the projection in (a) to the projection for 2020 where Allee effects are incorporated.

   c. Determine by examining the Allee effects DDS what, specifically, accounts for the difference.

4 Determine the panda population at which the population experiences its largest absolute growth.

5 Reveal a shortcoming in the Allee effects model for the giant panda population by finding populations that will result in a negative population projection for the next time step.

6 Investigate the effect that setting $K = S$ has on the discrete logistic model with Allee effects.

7   Analyze the graph of the growth rate function for the logistic model with Allee effects. Why is this graph not realistic for some population values?

8   *Extension*: For the general discrete logistic model with Allee effects, determine the population that will result in the largest absolute increase in the population for the next time step.

5.3 LOGISTIC GROWTH WITH HARVESTING

In the previous section we refined the logistic model by including the effect of a sustainability threshold. In this section we modify the logistic model again but in a different direction. We examine logistic growth with harvesting in the context of a fishery model, and we consider two different harvesting strategies. The first is **constant take** harvesting. Here we imagine fishers having a goal (or a government-set limit) for the number of fish they take each day, regardless of how long it takes them to do so. In this situation we have a constant *number* of fish that will be harvested each day. The second type of harvesting is **constant effort** harvesting. Here we have fishers who can only fish for, say, 8 h per day, and so the catch will vary depending on how abundant the fish are. In this situation we will have a constant *percentage* of available fish harvested each day rather than a constant number.

First we examine the constant take situation.

5.3.1   Constant Take Harvesting

For this scenario, we set h as the constant number of fish harvested each time period. Then we modify our logistic model in a familiar way:

$$P(t) = P(t-1) + r\left(1 - \frac{P(t-1)}{K}\right)P(t-1) - h.$$

The introduction of harvesting has an interesting and telling effect on the equilibrium values and overall behavior of the population.

5.3.2   Equilibrium Analysis

The algebra for finding the equilibrium values is more involved than before and requires the **quadratic formula**. Recall that the quadratic formula states that if $ax^2 + bx + c = 0$, then $x = \frac{-b \pm \sqrt{b^2 - 4ac}}{2a}$.

We start with

$$P^* = P^* + r\left(1 - \frac{P^*}{K}\right)P^* - h$$

and immediately simplify to $0 = r\left(1 - \frac{P^*}{K}\right)P^* - h$. After collecting like terms and clearing the coefficient on $(P^*)^2$, we arrive at the quadratic equation $(P^*)^2 - KP^* + \frac{Kh}{r} = 0$.

With $a=1$, $b=-K$, and $c=\dfrac{Kh}{r}$, the quadratic formula gives us the solution

$$P^* = \frac{K \pm \sqrt{K^2 - 4\dfrac{Kh}{r}}}{2}.$$

We get two distinct equilibrium values if the discriminant, $K^2 - \dfrac{4Kh}{r}$, is greater than 0, one unique equilibrium value if the discriminant equals 0, and no equilibrium values if the discriminant is less than 0. Thus the value for h that makes the discriminant equal to 0 represents a harvesting number where the model's behavior changes dramatically. By setting the discriminant equal to 0 and solving for h, we see that this harvesting number is $h = \dfrac{rK}{4}$.

In the next example we explore the practical implications of our analysis for the Antarctic baleen whale population.

Example 5.10: Recall that for our baleen whale population, we have $r = 0.05$ and $K = 400,000$. We assume constant harvesting of the population at a level of $h = 3000$ per year. Perform an equilibrium analysis on the model.

According to our algebra, we should have two equilibrium values:

$$P^* = \frac{400,000 \pm \sqrt{400,000^2 - 4(400,000 \cdot 3,000 / 0.05)}}{2}.$$

This gives us the values $P^* = 326,491.1$ and $P^* = 73,508.9$. In Figure 5.20 we show these two equilibrium values along with population graphs for several different initial populations.

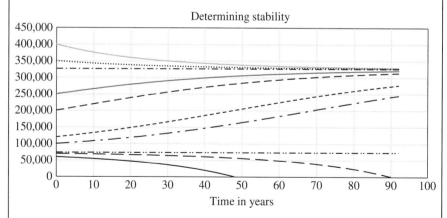

FIGURE 5.20   Model behavior when $h = 3000$.

Note that the larger equilibrium appears to be stable, while the smaller equilibrium appears to be unstable. Fishing at a constant level has, in effect, introduced a sustainability threshold into the population even though there was not one in the original model. If the population of whales dips below 73,508.9, it will go extinct. □

Next we increase the harvesting level to 4000 BWU per year.

Example 5.11: Examine the effect of increasing the harvesting level to 4000 BWU per year.

The overall behavior of our model does not change. We still get two equilibrium values: a larger one that is stable and a smaller one that is unstable. The particular values change, however. With $h = 4,000$ we get

$$P^* = \frac{400,000 \pm \sqrt{400,000^2 - 4(400,000 \cdot 4,000/0.05)}}{2}.$$

This gives us the two values $P^* = 289,442.7$ and $P^* = 110,557.3$. The graph, given in Figure 5.21, looks similar, but if we look closely we see an important difference: the equilibrium values have gotten closer together since the unstable value has gotten larger and the stable value has gotten smaller. This means that there is less room for error with a larger harvesting number. If our population were to fall below about 110,000, it would not recover. □

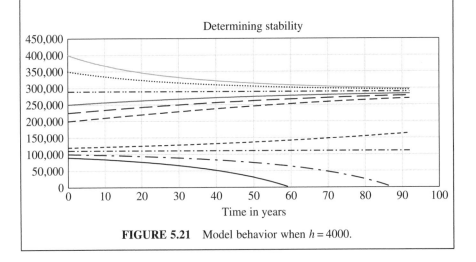

FIGURE 5.21 Model behavior when $h = 4000$.

A natural question at this point is whether there is a harvesting level above which the population is doomed. We show what our model says about this question in the next example.

Example 5.12: Examine the effect of increasing the harvesting level to 5000 BWU per year.

With a harvesting level of 5000 per year, we get qualitatively different behavior. Here we have only one equilibrium value, $P^* = 200,000$. When we graph model projections for several different initial populations (see Fig. 5.22), we see that populations above 200,000 tend to decrease down to 200,000, while populations below 200,000 move away from 200,000 and crash.

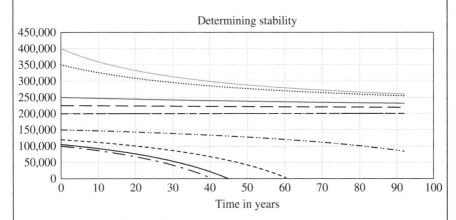

FIGURE 5.22   A bifurcation occurs when $h = 5000$.

This is our first example of a **semistable equilibrium**, and it is not a good situation for the population of whales. On the one hand, if the population starts out above 200,000, it seems safe since it will never fall below it. On the other hand, if we overfish by even a little bit or if there is some other minor disturbance that causes the population to drop below 200,000, it will eventually crash. □

An example of such a dramatic fishery collapse occurred in 1972 when a warming El Niño and overfishing combined to cause the anchoveta anchovy fishery to collapse off the coast of Peru. This had formerly been one of the largest fisheries in the world yielding 13 million tons of fish at its peak (Schwartzlose et al., 1999).

Finally we look at the case where the harvest level is $h = 6000$.

Example 5.13: Examine the consequences of a harvest level of $h = 6000$.

In this case the model has no equilibrium, and we see the consequences of this in the graph in Figure 5.23. With such a high harvest level, the population will crash regardless of the initial population. □

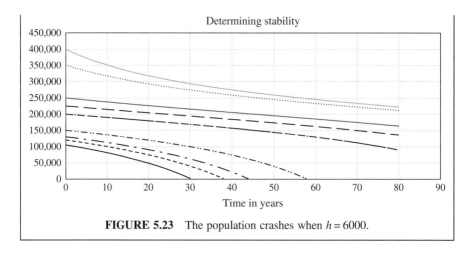

FIGURE 5.23   The population crashes when $h = 6000$.

The harvest level $h = 5000$ is a special kind of parameter value known as a **bifurcation point**. A bifurcation point is a parameter value at which the qualitative behavior of a model changes, for example, with a change in the number or types of equilibria present. In the baleen whale model, the bifurcation at $h = 5000$ marked a place where the model went from having two equilibria (one stable, one unstable) to one semistable equilibrium. Any harvest level above $h = 5000$ yields no equilibria and a crashing population.

5.3.3   Maximum Sustainable Yield for Constant Take

We have seen in the previous section that harvesting too much from a population can have devastating consequences. A natural next question is to ask "How much should we harvest?" An answer to this question that has been used by many fisheries is based on the concept of **maximum sustainable yield** (**MSY**). The MSY is the largest number that can be harvested from a population while allowing the population to sustain itself in the long term.

In order for a harvest to not decrease the overall population, the most that can be harvested in any time step is the amount of growth the population undergoes. The MSY will occur where this growth is the largest. As the next example shows for the discrete logistic model, the MSY occurs at a population size equal to half the carrying capacity.

Example 5.14: Find the MSY for the baleen whale example.
    We have $r = 0.05$ and $K = 400,000$. The DDS for the model is

$$P(t) = P(t-1) + 0.05\left(1 - \frac{P(t-1)}{400,000}\right)P(t-1).$$

Note that the part of the DDS that gives the absolute change in population from 1 year to the next is $0.05\left(1-\frac{P(t-1)}{400,000}\right)P(t-1)$. We need to find the maximum value for this change.

Thinking of the change as the function $g(x)=0.05\left(1-\frac{x}{400,000}\right)x$, we graph g to find its maximum. Figure 5.24 shows the graph for g.

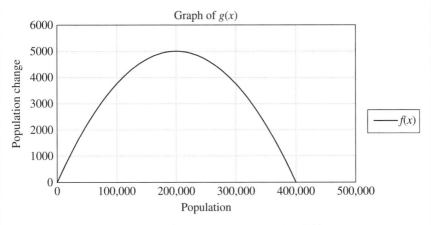

FIGURE 5.24   The MSY occurs at $h = 5000$.

Note that the highest point on the graph occurs at a population of $x=200,000$, with a value of $g(200,000)=5,000$. Thus the largest possible sustainable harvest level for the baleen whale population is MSY $=5000$. This is the largest possible harvest level for the baleen whale population that would prevent a decrease in the population long term. $\square$

Note that the value found in Example 5.14 is the bifurcation point where the model changes from having two equilibria to having one semistable equilibrium. If the whale population began over 200,000, then a harvest level at MSY would keep the population level above 200,000 indefinitely. However, as we pointed out previously, this puts the whale population in a precarious position since any perturbation below 200,000 would mean eventual extinction for the whales.

Finally we note that the population value for MSY occurred exactly halfway between the two equilibrium values of extinction and the carrying capacity. This is not a coincidence. The function representing the change in the population, g, is a downward-pointing parabola with roots at $x=0$ and $x=K$. The vertex of a parabola occurs midway between the two roots, so our maximum will always occur at a population of $x=\frac{K}{2}$. Then the MSY will always be equal to

$$\text{MSY}=g\left(\frac{K}{2}\right)=r\left(1-\frac{K/2}{K}\right)\frac{K}{2}=r\left(\frac{1}{2}\right)\frac{K}{2}=\frac{rK}{4}.$$

While the MSY may initially appear to be an attractive method for determining a harvest level, in practice it is generally not a good idea. As we note previously, the MSY harvesting level is a bifurcation point that introduces a semistable equilibrium value into the model. If the population falls below this value, then the MSY harvest number will cause the population to collapse. Given the difficulties involved with obtaining accurate population estimates and the possibility of outside factors that could negatively affect population levels, MSY should be used with caution and only in conjunction with strict population monitoring.

5.3.4 Constant Effort Harvesting

In this section we examine an alternate method of harvesting. Instead of setting a quota, we set a limit on the fishing *effort* expended. As an example of this kind of control, rather than allowing as many boats as necessary to catch a particular number of fish, we could restrict the number or length of time that boats can fish. If we only allow, say, 10 boats to fish for 2 weeks no matter the population, then the catch will not be constant. It will instead be based on how easy it is for those boats to find fish and hence how abundant the fish are. Consequently, we associate constant effort fishing with a harvest level that corresponds to a proportion of the fish available.

We assume now that we have restricted fishing effort so that a certain percentage of the fish population is harvested in a given time step. We denote this percentage by e and we modify our logistic model to reflect this change:

$$P(t) = P(t-1) + r\left(1 - \frac{P(t-1)}{K}\right)P(t-1) - eP(t-1).$$

As we did for the constant take case, we examine the equilibria of the model next.

5.3.5 Equilibrium Analysis

To find the equilibrium values, we solve $P^* = P^* + r\left(1 - \frac{P^*}{K}\right)P^* - eP^*$ for P^*. This reduces to solving $0 = r\left(1 - \frac{P^*}{K}\right)P^* - eP^*$. Collecting like terms and factoring give us $0 = \left(r - \frac{rP^*}{K} - e\right)P^*$.

One equilibrium value is therefore $P^* = 0$ (extinction). The other we find by solving $0 = r - \frac{rP^*}{K} - e$ for P^*. This yields $P^* = K\left(1 - \frac{e}{r}\right)$.

As before we examine the model graphically for a variety of choices for the parameter e and different initial populations.

Example 5.15: Investigate the effects of constant effort harvesting on the baleen whale population if $e = 0.01$.

Our DDS becomes $P(t) = P(t-1) + 0.05\left(1 - \frac{P(t-1)}{400,000}\right)P(t-1) - 0.01P(t-1)$. We know from our work above that we have two equilibrium values, one at extinction

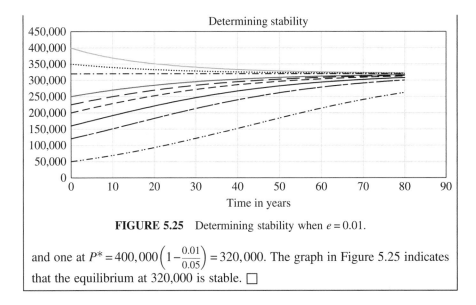

FIGURE 5.25   Determining stability when $e = 0.01$.

and one at $P^* = 400,000\left(1 - \dfrac{0.01}{0.05}\right) = 320,000$. The graph in Figure 5.25 indicates that the equilibrium at 320,000 is stable. $\square$

Unlike the constant take case, the constant effort model does not introduce a sustainability threshold. Furthermore, increasing the harvesting level to 2.5% lowers the equilibrium to 200,000, but unlike the constant take case, the equilibrium is still stable. In other words, with constant effort, the baleen whale population can withstand occasional setbacks due to slight overfishing or natural disasters. It is only when the effort rate equals the population's intrinsic growth rate that we have guaranteed extinction.

5.3.6   MSY for Constant Effort

We noted in Section 5.3.3 that the MSY for a population under the constant take model occurs at a population of $\dfrac{K}{2}$ and equals $\dfrac{rK}{4}$. We use these two facts to set an effort level, e, that produces the MSY when the population is equal to $\dfrac{K}{2}$. What we need is to find e such that

$$eP = e\frac{K}{2} = \frac{rK}{4}.$$

Solving for e yields $e = \dfrac{r}{2}$. Thus the MSY for the constant effort model occurs when fishing effort is one half of the population's intrinsic growth rate.

Example 5.16: Find the fishing effort that produces the MSY for the Antarctic baleen whale fishery.

   This is a straightforward calculation because we already have $r = 0.05$. Thus $e = \dfrac{0.05}{2} = 0.025$. $\square$

The result of using a fishing effort of $e = \frac{r}{2}$ is an equilibrium value of

$$P^* = K\left(1 - \frac{e}{r}\right) = K\left(1 - \frac{r/2}{r}\right) = \frac{K}{2}.$$

This is the same equilibrium value that we found in the constant take model when using MSY to set our harvest level. However, the situation here is much different. In the constant take case, the resulting equilibrium value was semistable: any fluctuation of the population below the equilibrium would cause the population to crash. In this case, however, the equilibrium remains stable. Fluctuations of the population of whales below the equilibrium will not cause the population to crash. Instead the lower population would result in a leaner harvest because the constant effort model harvests a proportion of the available catch. This allows the population to recover to the desired MSY population level.

In the next section we return to our study of the discrete logistic model. We have mentioned that we cannot make general claims about the stability of its equilibrium values because stability depends on the choice of parameters. In the following we will examine just how different the model behavior can be for different parameter values.

5.3.7 Section Exercises

1 Suppose a fishery has an intrinsic growth rate of $r = 0.10$ and a carrying capacity of $K = 1,000,000$.

   a. Find the maximum sustainable yield for the fishery.

   b. Assume that a constant take strategy is used to harvest fish at the MSY. What happens to the population over time if it starts at carrying capacity?

   c. Assume that a constant effort strategy is used to harvest fish at the MSY. What happens to the population over time if it starts at carrying capacity?

2 For the fishery described in Exercise 1, answer the following.

   a. Assume that a constant take strategy is used to harvest fish at the MSY. What happens to the population over time if it starts at half of its carrying capacity?

   b. Assume that a constant effort strategy is used to harvest fish at the MSY. What happens to the population over time if it starts at half of its carrying capacity?

3 For the fishery described in Exercise 1, answer the following.

   a. Assume that a constant take strategy is used to harvest fish at the MSY. What happens to the population over time if it starts at 400,000?

   b. Assume that a constant effort strategy is used to harvest fish at the MSY. What happens to the population over time if it starts at 400,000?

4 *Extension:* For the fishery described in Exercise 1, the MSY was set using the assumed intrinsic growth rate of $r = 0.10$. Suppose a mistake was made when estimating the intrinsic growth rate and that it actually turns out to be $r = 0.08$. Investigate the consequences of this mistake for both the constant take and constant

effort models if the MSY harvest levels were set using the mistaken value of $r = 0.10$.

5 *Extension*: For the fishery described in Exercise 1, the MSY was set using the assumed carrying capacity of $K = 1,000,000$. Suppose that a mistake was made when estimating the carrying capacity and that it actually turns out to be $K = 800,000$. Investigate the consequences of this mistake for both the constant take and constant effort models if the MSY harvest levels were set using the mistaken value of $K = 1,000,000$.

6 In light of Exercises 3, 4, and 5, discuss which strategy seems to be riskier for managing a fishery: constant take harvesting or constant effort harvesting.

7 For the constant take harvest model, show that if there is a semistable equilibrium value, then it occurs at a population equal to half the carrying capacity.

8 Show that for the constant effort model extinction occurs whenever $e > r$. Explain why this makes intuitive sense.

5.4 THE DISCRETE LOGISTIC MODEL AND CHAOS

Over the last few decades, the discrete logistic model has been the subject of dozens of scholarly articles by mathematicians and other scientists. One of the reasons for all of the interest is the fact that even this relatively simple deterministic model displays some remarkable properties—properties that have potentially interesting or even alarming implications for biological systems (May & Oster, 1976). The single most interesting, and initially disturbing, property of the model is that it is one of the first examples of a simple model from which **chaos** emerges. In this section we present a few illustrative numerical examples of what is meant by chaos. A full mathematical treatment of the material that follows may be found in many sources, including May R. M. (1976).

As the intrinsic growth rate, r, in the discrete logistic model is allowed to increase, the model begins to show more and more complex behavior. In the following discussion, we examine this behavior by (i) fixing the carrying capacity at $K = 1000$ and (ii) graphing the logistic model over time for several different initial populations.

Under what we would probably call "normal" conditions, the graph of the discrete logistic model forms an S shape. In Figure 5.26 we show a typical discrete logistic graph for $r = 0.10$ with three initial populations that are all close together: 100, 120, and 140.

Important features to notice are that (i) the carrying capacity, 1000, is a stable equilibrium and (ii) the graphs corresponding to the different initial populations start out close together and remain close together over time.

Next, in Figure 5.27 we examine the same graph but with $r = 0.50$ This time we see that the graphs are steeper, but observations (i) and (ii) from before still hold.

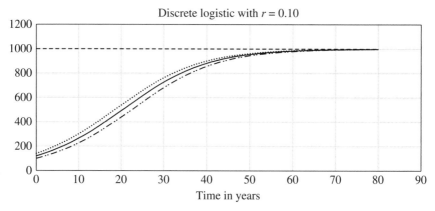

FIGURE 5.26   Typical discrete logistic S-shaped graph when $r = 0.10$.

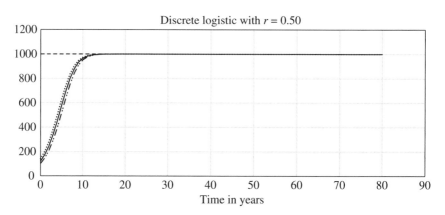

FIGURE 5.27   Discrete logistic graph when $r = 0.50$.

Next we try a much larger growth rate $r = 2.1$ and present the resulting graph in Figure 5.28. This is our first indication that something unusual is going on. Instead of approaching the carrying capacity and staying there, our population settles down into what is called a **stable cycle of period two**. This means that the population eventually follows a pattern where it is bouncing back and forth between two constant values, in this case the values 1129 and 824. With such a large growth rate—the population more than doubles every time step—the population experiences **overshoot**. This happens when a population temporarily exceeds its carrying capacity. It then overcorrects to below carrying capacity, and the cycle repeats.

If we increase r to $r = 2.5$, we find a stable cycle of period four as shown in Figure 5.29. Here the population settles down into a pattern where it cycles through the same four values: 1225, 536, 1158, and 701. Again we see that increasing r has led to an increase in the complexity of the model behavior. The model is bifurcating as we increase r.

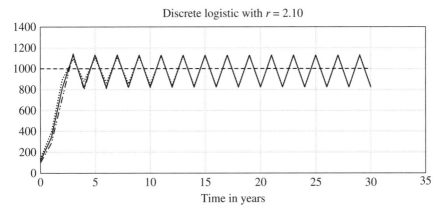

FIGURE 5.28 Discrete logistic graph when $r = 2.1$.

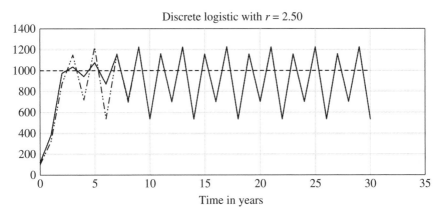

FIGURE 5.29 Discrete logistic graph when $r = 2.55$.

If we increase r to 2.55, we get a stable cycle of period eight, though the values become difficult to distinguish on the graph.

A remarkable fact about the behavior of the discrete logistic model that is not easy to prove is that we can arrange to get a stable cycle whose period is *any* power or two by continuing to increase r in small increments (May R. M., 1978).

Even more remarkable is that once we increase r to be greater than 2.570, we get chaos (May R. M., 1978). As an example of what we mean by chaotic behavior, we show the result of taking $r = 2.8$ on our graph in Figure 5.30. In this graph there is no discernible pattern: it is chaotic in a literal sense. We also see that our initial populations, which start out close together, end up far apart. The model is exhibiting what is known as **sensitive dependence on initial conditions**, also known as the **butterfly effect**, whereby small changes in the model input result in large changes in the output. This is one of the defining characteristics of chaotic behavior.

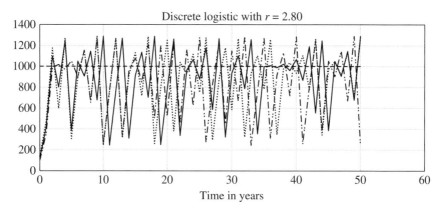

FIGURE 5.30   Discrete logistic graph when $r = 2.8$.

TABLE 5.1   The Behavior of the Discrete Logistic Model for Increasing Values of r

Value of Intrinsic Growth Rate	Population Behavior
$0 < r < 2.000$	Tends to K
$2.000 < r < 2.449$	Cycles through 2 values
$2.449 < r < 2.544$	Cycles through 4 values
$2.544 < r < 2.564$	Cycles through 8 values
$2.564 < r < 2.570$	Cycles through 16, then 32, then 64, etc.
$2.570 < r < 3.0$	Chaotic behavior: cycles of arbitrary period or aperiodic behavior
$r > 3.0$	Guaranteed extinction

Source: Adapted from May R. M. (1976), table 2.3, p. 14. Reproduced with permission from John Wiley & Sons, Inc.

We note that if r is greater than 2.570, we do not necessarily get chaotic behavior. In fact just about anything can happen. We can get a stable cycle of any period, or we can get behavior that never repeats. The exact cutoffs for when one behavior stops and another starts are difficult to determine analytically, but May has done so. His results are presented in Table 5.1.

5.5   THE RICKER MODEL

We based the discrete logistic model on the observation that the growth rate for a species will not remain constant in the long term; instead the growth rate should decline as the population grows. The simplest way to represent a declining growth rate is with a negatively sloped straight line as shown in Figure 5.2. However, the scale of that graph hides an important flaw in this representation, namely, that if we *continue* the straight line, the growth rate eventually falls below −1. This means that our model would predict more than a 100% decrease in the population, which does not make physical sense. Even under the worst circumstances, the limit on our negative growth rate should be −1.

In this section we modify the discrete logistic model to prevent the possibility of a growth rate less than −1. The result is called the Ricker model, named after W. C. Ricker who developed it. The Ricker model is an historically important model that has been used in the management of the Pacific salmon (Clark, 1985).

We represent the idea behind the Ricker model with a graph for the population growth rate similar to the one in Figure 5.2. We still begin the graph at the intrinsic growth rate, and we still require that at the carrying capacity, K, the growth rate should be 0. The important difference in the new graph is that the growth rate can never fall below −1 regardless of how large the population grows. Instead we have a horizontal asymptote at the value −1 (see Fig. 5.31).

Note that the graph satisfies the three requirements: (1) the growth rate is equal to the intrinsic growth rate when $P = 0$; (2) the growth rate is equal to 0 when the population is at carrying capacity, $P = K$; and (3) no matter how large the population gets, the growth rate never falls below −1.

The challenging part of developing the Ricker model is finding a function for the growth rate whose graph matches the curve presented in Figure 5.31. There are in fact many different curves with the correct shape, but the one that Ricker employed is an exponential function. Specifically, we seek a function of the form $ae^{-bP(t-1)} - 1$, where a and b are positive constants that we select to fit to our graph. The negative exponent forces the exponential term to level off at 0 as the population grows larger and larger, and the subtraction of 1 translates the graph down so that it levels off at the required −1 instead.

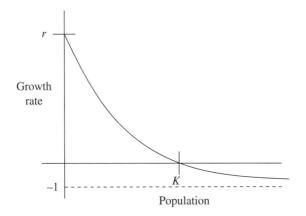

FIGURE 5.31   Growth rate versus population for Ricker model.

What remains is to determine the constants a and b. We accomplish this by using the two known points on our graph to set up two equations for our two unknowns. The two known points are the same ones we used when developing the logistic equation: $(0,r)$ and $(K,0)$. Our two equations for the new model are then $r = ae^{-b \cdot 0} - 1$ and

$0 = ae^{-b \cdot K} - 1$. Since $e^0 = 1$, the first equation tells us that $a = 1 + r$. We plug this information into the second equation and solve for b:

$$0 = (1+r)e^{-bK} - 1$$
$$1 = (1+r)e^{-bK}$$
$$e^{bK} = (1+r)$$
$$bK = \ln(1+r).$$

The result is $b = \dfrac{\ln(1+r)}{K}$.

We now plug in our values for a and b to get the Ricker model for growth rate as a function of population:

$$(1+r)e^{-(\ln(1+r)/K)P(t-1)} - 1.$$

Hence the discrete dynamical system for the Ricker model is

$$P(t) = P(t-1) + \left[(1+r)e^{-(\ln(1+r)/K)P(t-1)} - 1 \right] P(t-1).$$

In this DDS the leading $P(t-1)$ cancels to give us the slightly simpler version:

$$P(t) = (1+r)P(t-1)e^{-(\ln(1+r)/K)P(t-1)}.$$

E.17 Exponential and Logarithmic Functions

Implementing the Ricker model in Excel requires the use of two of Excel's built-in mathematical functions: the exponential function and the natural logarithm. The syntax for raising the number e to a power is "=EXP(number)," while the natural logarithm of a number is written as "=LN(number)." The formula version of our Excel Ricker model is given in Figure 5.32 with stand-in values for the intrinsic growth rate and carrying capacity.

	A	B	C	D
1	Ricker Model			
2				
3	Intrinsic growth rate, r =		0.20	
4	Carrying capacity, K =		10,000	
5				
6	t	Population		
7	0	1000		
8	1	=(1+$C$3)*B7*EXP(-LN(1+$C$3)*B7/$C$4)		
9	2	1383.9		

FIGURE 5.32 Ricker Excel model setup.

Salmon populations are particularly well suited for discrete models because salmon have nonoverlapping generations. The size of any particular generation then depends on the size of the previous generation. In fact it was trying to predict salmon populations that originally led Ricker to develop his model (Ricker, 1954). In the next example we apply Ricker's model to a salmon population using parameter values estimated in *Ricker Salmon Model* (Phaser, 2015). These parameter estimates are based on approximately 30 years of salmon population data collected by the US Army Corps of Engineers at the John Day Dam (U.S. Army Corps of Engineers, 2004).

Example 5.17: Use the Ricker model to project a salmon population over the next 15 years if the population starts at 150,000. Assume $r = 0.61$ and $K = 477,700$.

Once we input the relevant parameters, we need to drag our model down to year 15 and observe the results. The model predicts that the salmon population will grow to 477,598 fish, which is very near carrying capacity. A graph for the population is provided in Figure 5.33. □

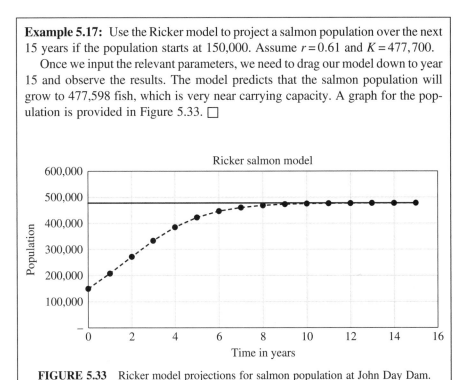

FIGURE 5.33   Ricker model projections for salmon population at John Day Dam.

We note that for these particular parameter values, the carrying capacity once again appears to be a stable equilibrium. We verify that the carrying capacity is in fact always an equilibrium for the Ricker model in the next section.

5.5.1 Equilibrium Analysis

To find any equilibrium values, we must solve the following for P^*:

$$P^* = (1 + r)P^* e^{-(\ln(1+r)/K)P^*}.$$

We must either have $P^* = 0$ (extinction) or $P^* \neq 0$ in which case we divide both sides by P^* to see that we must solve

$$1 = (1+r)e^{-(\ln(1+r)/K)P^*}.$$

Moving the exponential over, we get

$$e^{(\ln(1+r)/K)P^*} = (1+r).$$

Taking the natural logarithm of both sides yields

$$\frac{\ln(1+r)}{K}P^* = \ln(1+r)$$

$$\frac{P^*}{K} = 1.$$

The end result is $P^* = K$.

We have shown that just like the discrete logistic model, the only equilibrium values for the Ricker model are when the population is extinct or at its carrying capacity.

5.5.2 Section Exercises

1 Use the Ricker model to project the salmon population from Example 5.17 over 20 years if the population starts at 50,000.

2 Determining the maximum sustainable yield based on the Ricker model is more complicated than it is for the discrete logistic model. However, while the MSY for the Ricker model will not occur at a population equal to exactly $\frac{K}{2}$, the value $\frac{K}{2}$ will produce a good approximation. For the salmon fishery in Example 5.17, approximate the MSY based on the Ricker model.

3 *Extension:* Modify the Ricker model to include constant take harvesting using the MSY from Exercise 2. Investigate the effects of this harvesting level by projecting the fate of several different initial salmon populations.

4 *Extension:* Modify the Ricker model to include constant effort harvesting using the MSY from Exercise 2. Investigate the effects of this harvesting level by projecting the fate of several different initial salmon populations.

5 *Extension:* Show that under a constant effort harvesting level of e, the Ricker model's positive equilibrium value becomes

$$P^* = K\left(1 - \frac{\ln(1+e)}{\ln(1+r)}\right).$$

Note the similarity with the constant effort harvesting equilibrium for the discrete logistic model.

6 *Extension*: The Ricker model fixed a shortcoming of the discrete logistic model by ensuring that the growth rate for the model never falls below -1. Ricker used an exponential function to accomplish this, but others have suggested different functions. One famous example is known as the **Beverton–Holt model**. In the Beverton–Holt model, a rational function of the form $\frac{a}{b+P(t-1)}-1$ where a and b are constants is used in place of an exponential to prevent a growth rate less than -1:

a. Find appropriate constants a and b for the Beverton–Holt model by requiring that the growth rate is r when the population is 0 and that the growth rate is 0 when the population is at the carrying capacity, K.

b. Give the DDS for the Beverton–Holt model and implement it in Excel.

c. Find the equilibrium values for the Beverton–Holt model.

d. Compare population projections for the Beverton–Holt and Ricker models for the salmon fishery in Example 5.17.

6

BLOOD ALCOHOL CONCENTRATION AND PHARMACOKINETICS

The study of the course of a drug as it is administered, absorbed, and eliminated from the body is known as **pharmacokinetics**. In this chapter we build pharmacokinetic models for predicting the amount of drug in the body over time with a special focus on commonly used drugs such as alcohol, ibuprofen, and caffeine. We begin with the example of blood alcohol concentration (BAC) before providing a general model that can be used for many different drugs.

6.1 BLOOD ALCOHOL CONCENTRATION

Blood alcohol concentration (**BAC**) is a measure of how much alcohol, specifically ethanol, is in the body. When alcohol is ingested, it moves rapidly through the stomach to the small intestine. Since alcohol is water soluble, it is absorbed from the small intestine into the body water where it quickly becomes evenly distributed throughout the body.

For many drugs, alcohol included, the concentration of the drug in the body is more important than the total amount present because larger bodies need more of the drug in order to achieve the same effect. A 300-pound NFL lineman, for example, will feel much different after four beers than a 150-pound person would.

To calculate BAC, we proceed in stages: (i) we calculate the amount of alcohol ingested, (ii) we estimate the amount of water a person's body contains, (iii) we calculate the concentration of alcohol in the body water by dividing the amount of

Models for Life: An Introduction to Discrete Mathematical Modeling with Microsoft® Office Excel®, First Edition. Jeffrey T. Barton.
© 2016 John Wiley & Sons, Inc. Published 2016 by John Wiley & Sons, Inc.

alcohol by the amount of water, and (iv) we deduce the concentration of alcohol in the blood in light of the fact that blood is 80.6% water.

We note that there are no standard units for reporting BAC. Different countries tend to report it in different ways. In the United States, the most commonly used units are grams of ethanol per deciliter of blood, and these are the units we will use.

The question of how much body water a person has is an interesting one that depends on many factors including weight, age, and sex. The amount of body water helps explain observed differences in how males and females respond to the same dose of alcohol. Women in general have a higher percentage of body fat than men, and thus they tend to have less body water than men even when their body weight is the same. Thus a dose of alcohol will typically produce a higher BAC in a woman than in a man of the same weight. As a result, women tend to feel more intoxicated than men when consuming the same amount of alcohol. We proceed with an example of how a basic BAC calculation is done.

Example 6.1: Mark is a 180-pound male who quickly consumes two 12-oz. beers. Estimate Mark's BAC.

We assume that all of the alcohol from the two beers is quickly emptied from Mark's stomach and distributed uniformly in his total body water.

First we need to know how much alcohol, in grams, Mark consumed. A standard 12-oz. beer contains about 14 g of alcohol (as do a 5-oz. glass of wine or 1.5-oz. shot of 80-proof liquor), so our subject has approximately 28 g of alcohol in his body water.

Next we need to calculate how much body water a 180-pound male typically has. In the absence of more specific information, we use standard average values for **body water percentage**. On average males are 58% water, while females are 49% water. The lower percentage of body water for females is due primarily to their typically higher levels of body fat, which contains little water, versus muscle, which contains a lot of water. As we will see later, there are more sophisticated formulas for determining more precise estimates for body water percentage based on height, weight, sex, and age. For now though we have enough information to estimate the Mark's BAC.

1. Begin with body weight in pounds, and change the body weight to kilogram (1 kg = 2.2046 pounds):

$$180 \, \text{pounds} \times \frac{1 \, \text{kg}}{2.2046 \, \text{pounds}} = 81.65 \, \text{kg}.$$

2. Using typical sex percentages, find total body water volume (1 l of water weighs 1 kg) by multiplying body weight by body water percentage:

$$81.65 \, \text{kg} \times 58\% = 47.36 \, \text{kg} \, H_2O$$
$$= 47.36 \, l \, H_2O.$$

3. Calculate the concentration of alcohol in the body water by dividing total amount of alcohol by total body water:

$$\frac{28 \text{ g}}{47.36 \text{ l H}_2\text{O}} = 0.5912 \text{ g per l H}_2\text{O}.$$

4. Using the fact that blood is 80.6% water, calculate BAC from body water concentration:

$$\text{BAC} = 0.5912\frac{\text{g}}{1 \text{H}_2\text{O}} \times 0.806\frac{1 \text{H}_2\text{O}}{1 \text{blood}} = 0.4765 \text{ g per l blood.}$$

5. Convert our BAC into the appropriate units, which in the United States are typically grams of ethanol per deciliter of blood. There are 10 deciliters (dl) in a liter, so we have

$$\text{BAC} = 0.4765\frac{\text{g}}{1 \text{blood}} \times \frac{1 \text{l}}{10 \text{ dl}} = 0.04765 \text{ g per dl blood.}$$

Note that BAC levels are often reported as a % even though, strictly speaking, they are not true percentages. Thus we would report Mark's BAC to be 0.048%. □

We see in the previous example that our subject is well below the standard legal limit for driving, which is 0.08%. We should also regard this first estimate as conservative since we assumed that all of the alcohol from the two beers was absorbed instantaneously into the body water. In reality some of that alcohol would have already been metabolized as the drinks were being consumed.

In Excel we create a basic BAC calculator similar to ones available online by having the user input sex, body weight in pounds, and number of standard drinks consumed. The spreadsheet will then have to use the correct body water percentage and carry out the calculations necessary to report an estimate for the individual's BAC. In creating the spreadsheet, we take advantage of some features in Excel that allow us to control the type of input a user can provide.

E.18   Data Validation

In many of our spreadsheet models, we ask the user to input some or all of the parameter values we use in our calculations. Adding drop-down menus in Excel is a great way of adding some sophistication to the spreadsheet while at the same time ensuring that the user inputs are exactly what we require. For example, if we ask the user to type her or his sex into an empty cell, we may get several different responses—"Female," "female," "F," "f," etc.—and that variety can make later calculations based on sex difficult. Similarly, if we want to limit user input to whole numbers, we do not want the user to be able to enter "2.4." The way we deal with these issues in Excel is known as "Data Validation." With Data Validation we can limit the kind of input that a cell will accept, or we can create a drop-down menu for a cell so that the user must choose from a provided list of given inputs.

Example 6.2: Use Data Validation to limit a cell's input to whole numbers between 0 and 10.

Suppose we want a user to input a whole number between 0 and 10 into cell B5. We click on cell B5 and from the Data Tools group under the Data tab we select "Data Validation." Figure 6.1 shows the dialog box that appears.

FIGURE 6.1   Data Validation dialog box.

Under "Allow" we select "Whole number," under "Data" we select "between," and in the boxes for "Minimum" and "Maximum" we enter 0 and 10, respectively. After clicking "Okay" we check our work. If we enter the value "7" into cell B5, note that nothing happens. However, if we try to enter "7.5," we get the error message shown in Figure 6.2. The same message appears if we try to enter a value outside the specified range. □

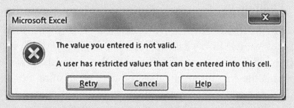

FIGURE 6.2   Data Validation error message.

The Data Validation dialog box also provides options for us to customize the feedback that the user receives when clicking on a cell that requires specific input. The "Input Message" allows us to provide guidance to the user as to what kind of input the cell requires; this message will appear whenever the cell is selected. The "Error Alert" is the message that will appear if the user attempts to enter the wrong kind of input.

If B5 is the cell where the user must enter the number of drinks consumed, our input message might be entitled "Number of Drinks" and read "Please enter a whole number between 0 and 10." In case of an invalid entry, our error alert could be set with title "Invalid Number of Drinks" and message "The entry for this cell must be a whole number between 0 and 10."

Another type of data validation involves setting up a drop-down menu for a cell so that the user must select one of the entries provided.

Example 6.3: Create a drop-down menu for cell B5 that forces the user to choose "M" for male and "F" for female.

Here we go back to Data Validation, but this time we select "Allow: List" under Settings. We have a couple choices now. In the "Source:" line we can simply list the options we want to appear, separated by commas, such as "M, F." Alternatively we can store the choices as a list elsewhere in our spreadsheet and select the range of values under Source. The second way is more work but allows for easier editing later if we want to update the list. The setup for the first option is shown in Figure 6.3.

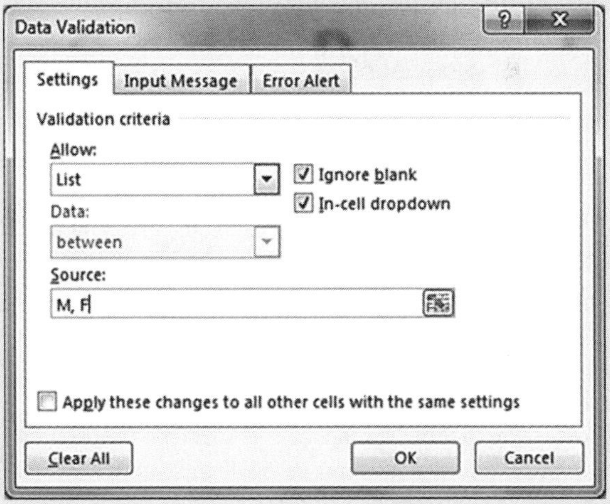

FIGURE 6.3 Creating a drop-down menu for sex.

Once entered cell B5 will now display a drop-down menu icon for the user to click whenever B5 is selected. Figure 6.4 shows the result of a user clicking on the drop-down menu for sex. □

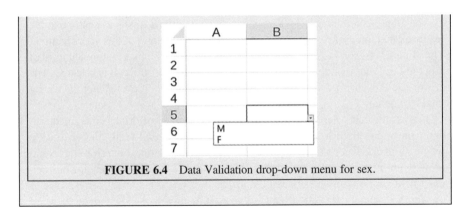

FIGURE 6.4   Data Validation drop-down menu for sex.

We are now ready to create a basic BAC calculator. We collect all of our parameters together at the top of the spreadsheet and apply whatever limits we want to impose on the inputs using Data Validation. In this case the only restriction is on sex, and we use a drop-down menu to control that as in the previous example. Our setup is shown in Figure 6.5.

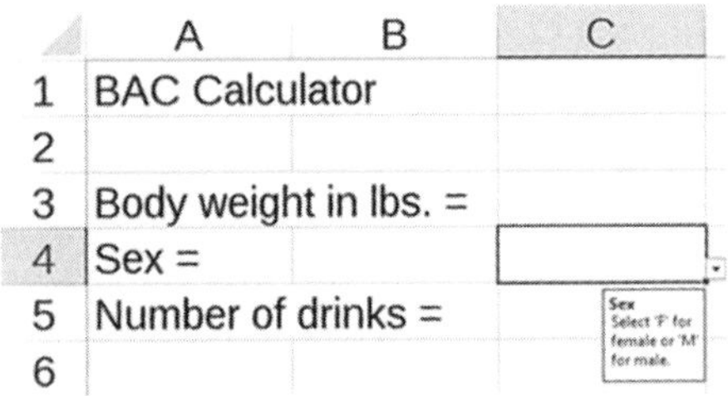

FIGURE 6.5   Basic BAC calculator setup.

Next we need to calculate the user's BAC based on the given inputs. This is a matter of following the calculations from Example 6.1 and getting Excel to carry them out. One complication is that we need to use the correct body water percentage based on the user's sex. We take care of this with an "IF" statement of the form "IF(user is male, use 0.58, else use 0.49)." With sex stored in cell C4, the exact syntax is "IF(C4 = "M",0.58,0.49)." The full formula version of the calculator is provided in Figure 6.6.

As a quick verification that the spreadsheet is working properly, we enter the parameters from Example 6.1 and note that we do in fact get BAC = 0.048. Figure 6.7 shows this result.

In Table 6.1 we present some typical effects of alcohol at various BAC levels collected from a variety of sources (see CDC, 2015a; Clemson Redfern Health Center, 2015).

In Example 6.1 we see that Mark will likely experience lowered inhibition and euphoria, but he will also exhibit some loss of coordination and impaired judgment.

In the next section we incorporate the effects of time on BAC.

	A	B	C	D	E	F
1	BAC Calculator					
2						
3	Body weight in lbs. =		180			
4	Sex =		M			
5	Number of drinks =		2			
6						
7	BAC =		=((C5*14)/((C3/2.2046)*IF(C4="M",0.58,0.49)))*0.806/10			

FIGURE 6.6   Basic BAC calculator formula.

	A	B	C
1	BAC Calculator		
2			
3	Body weight in lbs. =		180
4	Sex =		M
5	Number of drinks =		2
6			
7	BAC =	0.048	

FIGURE 6.7   BAC calculator result confirmation for Example 6.1.

TABLE 6.1   The Effects of Alcohol at Given BAC

BAC	Effects
0.02–0.03	Mild relaxation, mild euphoria, some loss of judgment, decline in ability to multitask
0.04–0.06	Lowered inhibition, feeling of euphoria, emotions intensified, impaired judgment, impaired coordination, loss of reflexes
0.08 (legal limit for driving)	Impaired balance, impaired judgment, loss of self-control, loss of concentration, poor speed control and perception while driving
0.10–0.125	Poor reaction time and control, slurred speech, poor coordination, inability to maintain lane position while driving
0.13–0.15	Further loss of muscle control and balance, possible vomiting, substantial impairment of ability to control vehicle
0.16–0.19	Possible vomiting, drinker appears very drunk
0.20	Vomiting, blackouts, inability to stand or walk without assistance, increased likelihood of injury, risk of choking on vomit
0.30–0.35	Stupor, passing out, coma possible, death possible
0.40	Coma, death likely

6.1.1 Section Exercises

1 Scott is a 200-pound man who consumes five standard glasses of wine. Estimate Scott's BAC.

2 Chris is a 120-pound woman who consumes four 12-oz. beers. Estimate her BAC assuming all of the consumed alcohol is in her body.

3 Anna is a 150-pound woman. Determine how many standard drinks she can consume and still remain below the legal driving limit for BAC.

4 Seth is a 250-pound man. Determine how many standard drinks he can consume and still remain below the legal driving limit for BAC.

5 *Extension*: Jeff is a 180-pound man who consumes two 1.5-oz. drinks of 100-proof Scotch. Estimate his BAC. You may use the fact that the mass density of alcohol is 23.3 g per fluid ounce.

6 *Extension*: Joan is a 170-pound woman who consumes three 16-oz. IPAs. If the IPA is 8% alcohol by volume, estimate Joan's BAC. You may use the fact that the mass density of alcohol is 23.3 g per fluid ounce.

7 Calculate your own BAC after having two drinks of your choosing.

8 *Extension*: Find better estimates for total body water based on parameters such as age, weight, sex, etc. and incorporate them into the basic BAC calculation spreadsheet in place of the values based on the average percentages of 58% for men and 49% for women.

9 *Extension*: Using the body water percentage estimate from Exercise 8, estimate your BAC after having the two drinks you used in Exercise 7. How much difference did the body water percentage estimate make for the BAC calculation?

10 *Extension*: Use the BAC calculator spreadsheet to compare the effects of quickly consuming 4 standard drinks for an average male US college student and an average female US college student.

6.2 THE WIDMARK MODEL

The basic calculations from the previous section provide a way for us to get a rough estimate of a person's BAC. However, these kinds of calculations suffer from being static—they only give us BAC at one moment in time. They also make use of questionable assumptions: that all consumed alcohol is present in the body, and that the alcohol is instantly distributed throughout the blood. In this section we go a step further and present our first discrete dynamical system (DDS) model for predicting BAC over time: the Widmark model.

In 1932 Widmark developed a single-compartment model for predicting BAC over time that has become the most widely used and cited BAC model due to its simplicity and its accuracy for a large percentage of the population (Heck, 2007). Here we present the Widmark model; in later sections we discuss ways the model may be improved.

As soon as alcohol is consumed, it begins to be removed from the body primarily by metabolism in the liver. A small percentage of the alcohol is excreted by passing from the body unchanged via the breath, sweat, and urine; another small percentage is metabolized in the stomach (Kent, 2012). The Widmark model does not differentiate among these different pathways; instead it treats the body as a single compartment and it treats excretion and metabolism as a single elimination process leading to an overall constant rate of decrease in BAC.

Once consumed, alcohol diffuses rapidly through the body water and hence the blood. Widmark estimated that the rate at which alcohol is then cleared from the body results in a decrease in BAC of about 0.017 each hour, or $\frac{0.017}{60} = 0.000283$ per min. This rate of elimination varies from individual to individual, and it can range from 0.010 to 0.040 per h with lower values typical for those who do not regularly consume alcohol and higher values for heavy drinkers (NHTSA, 1994). In other words, heavy drinkers tend to metabolize alcohol more quickly than others. The average value for a heavy drinker is an approximate 0.020 decrease in BAC per hour (NHTSA, 1994).

Since the Widmark model assumes the rate of change for BAC is a constant that does not depend on the amount of alcohol present, we say that the Widmark model uses **zero-order elimination** for alcohol. The flow diagram for the Widmark model is given in Figure 6.8.

FIGURE 6.8   Widmark model flow diagram.

Our DDS for BAC is thus $BAC(t) = BAC(t-1) - 0.000283$, where time is measured in minutes since the last drink. The initial BAC is calculated as in the previous section.

The fact that BAC is decreasing by a constant amount means that there will be no equilibrium values for the model. To verify this we can attempt to solve for BAC^* in the equation $BAC^* = BAC^* - 0.000283$. Subtracting BAC^* from both sides results in the false statement $0 = -0.000283$ so we have no solution.

The Excel implementation of the Widmark model is relatively straightforward once we have the basic BAC calculator from the previous section. We can modify that spreadsheet slightly so that the BAC calculation appears as our initial BAC. Then we get Excel to subtract 0.000283 from the previous BAC each minute. The formula version of the model is shown in Figure 6.9.

	A	B	C	D
1	Widmark BAC Model			
2				
3	Body weight in lbs. =		140	
4	Sex =		F	
5	Number of drinks =		5	
6				
7	*t*	BAC(t)		
8	0	=((C5*14)/((C3/2.2046)*IF(C4="M",0.58,0.49)))*0.806/10		
9	=A8+1	=B8-0.000283		
10	=A9+1	=B9-0.000283		

FIGURE 6.9   Widmark Excel model formula.

Note we have used the average elimination rate of 0.017 per h in our model. In the exercises the reader is invited to modify this rate to account for the user's drinking habits. We end this section with a computational example.

Example 6.4: Suppose Angela is a 140-pound female who has five drinks over a relatively short time. Use the Widmark model to project Angela's BAC over time and estimate how long it will be before she is legally able to drive.

Once we have the Widmark model set up, we need to input Angela's parameters and then drag the formulas down far enough for Angela's BAC to drop below 0.08. The results of our projection are shown in Figure 6.10 with most of the rows hidden.

	A	B	C
1	Widmark BAC Model		
2			
3	Body weight in lbs. =		140
4	Sex =		F
5	Number of drinks =		5
6			
7	*t*	BAC(t)	
8	0	0.181	
9	1	0.181	
367	359	0.080	
368	360	0.079	

FIGURE 6.10   Widmark model projection for Example 6.4.

The Widmark projections show Angela's BAC starting at 0.181. Note that according to Table 6.1 Angela will be very drunk at this BAC level. She will likely experience a substantial decrease in muscle control, possible vomiting, and substantial impairment in her ability to operate a vehicle. The model projects that it will take Angela's BAC 6 h from the cessation of drinking to drop below the legal limit for driving. If we project Angela's BAC far enough into the future, we end up with negative values for her BAC after about 10.5 h. This of course does not make physical sense. □

The last example highlights a major flaw in the Widmark model: if we project BAC far enough into the future, we will always end up with negative values for BAC. In the next section we present a model that addresses this issue.

6.2.1 Section Exercises

1 Give an explicit formula for the Widmark model.

2 Implement the Widmark model in Excel. Allow the user to input sex, number of drinks consumed, and body weight as parameters in the model.

3 Use Excel's Data Validation feature to have the Widmark Excel model use different elimination rates based on user-selected drinking habits. In particular, use a drop-down menu to allow the user to select light, moderate, or heavy drinker and have the BAC calculation use the corresponding elimination rate.

4 Diana is a 160-pound woman who consumes five standard glasses of wine.

   a. Use the Widmark Excel model to estimate her initial BAC.

   b. Estimate how long it will take before Diana is legally able to drive.

5 Use your own parameters and drink of choice to:

   a. Estimate your initial BAC.

   b. Estimate how long it will take before you are legally able to drive.

6 Reveal a shortcoming of the Widmark model by projecting BAC over a long period of time. Explain why the results are a problem.

7 *Extension*: Graph the Widmark model BAC projections over time for parameters of your choosing. On the Excel graph, include horizontal lines marking the BAC levels for the legal driving limit, drunkenness, coma, and death.

8 *Extension*: Make the Widmark Excel model more flexible by allowing users to input nonstandard drinks.

6.3 THE WAGNER MODEL

Though the Widmark model is still widely used today, it does have some clear problems. One issue is that the assumption of a constant decrease in BAC over time leads to projections of negative BAC concentrations, which are not physically meaningful.

In 1972 Wagner and Patel improved the Widmark model by modifying the elimination rate to behave in a more realistic way (Wagner & Patel, 1972). Rather than a constant rate of decrease in BAC, the Wagner model assumes that the rate at which alcohol is cleared depends on how much alcohol is in the body. For high BAC levels, the body's ability to eliminate alcohol is overwhelmed to the point that the elimination rate appears to be constant; we denote the body's maximum BAC elimination rate by V_{max}. However, at low BAC levels, the elimination rate will be roughly proportional to the BAC level. Unlike the Widmark model, the Wagner model takes into account the elimination rate's dependence on BAC. In this way the Wagner model avoids predicting negative BACs.

The kind of rate dependence assumed by Wagner is known as **Michaelis–Menten kinetics**, one of the most common models for enzyme kinetics. The basic formula for the rate of alcohol elimination under Michaelis–Menten (M–M) kinetics is given by

$$\frac{V_{max}}{K_m + BAC} BAC.$$

Here V_{max} is the maximum removal rate for alcohol in grams per deciliter per minute, and K_m is a parameter representing the BAC (in grams per deciliter) at which the removal rate for alcohol would be one half of V_{max}. Values for the M–M parameters, V_{max} and K_m, vary among individuals, and the Wagner model can be customized to an individual by determining their personal V_{max} and K_m based on their own BAC measurements (e.g., by using a breathalyzer). In the absence of specific individual values, we use population averages that can be found in the literature just as we do for body water percentage.

Examining the M–M equation, we use some algebra to see that it can be rewritten as

$$\frac{V_{max}}{K_m + BAC} BAC = \frac{V_{max}}{(K_m + BAC)/BAC} = \frac{V_{max}}{(K_m/BAC) + 1}$$

In this form we can see that if we have a very large BAC, then the term $\frac{K_m}{BAC}$ will be close to zero, and the rate of elimination will be close to (but not quite) V_{max}, that is, we will have near constant, or zero order, maximal elimination. We also note that for very low BAC levels, the factor $\frac{V_{max}}{K_m + BAC}$ will be approximately equal to $\frac{V_{max}}{K_m}$; hence for low BAC levels, we have roughly exponential decline at the rate of $\frac{V_{max}}{K_m}$.

Finally we note that if the BAC equals K_m, then the elimination rate is

$$\frac{V_{max}}{K_m + BAC} BAC = \frac{V_{max}}{(K_m/K_m) + 1} = \frac{V_{max}}{2}.$$

In other words, when BAC reaches K_m, the rate of elimination is one half the maximum value as the definition of K_m states.

With V_{max} and K_m left as parameters to be determined later, we have the flow diagram given in Figure 6.11.

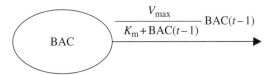

FIGURE 6.11 Wagner model flow diagram.

The corresponding DDS for the model is thus

$$BAC(t) = BAC(t-1) - \frac{V_{max}}{K_m + BAC(t-1)} BAC(t-1).$$

When we implement the Wagner model in Excel, the main change to the Widmark model is to replace the constant elimination rate, 0.000283, with the more complicated elimination rate represented by M–M kinetics. We store the parameters V_{max} and K_m in their own cells for easy experimenting later.

As noted earlier the parameters V_{max} and K_m vary among individuals, but we can use average values that will be accurate for large segments of the population. Lewis in 1986 found average values for V_{max} and K_m based on the values given in several studies (Lewis, 1986). He found $V_{max} = 0.0003693$ and $K_m = 0.00858$. Using Lewis's values for V_{max} and K_m, the Excel Wagner formula is displayed in Figure 6.12.

	A	B	C	D
1	Wagner BAC Model			
2				
3	Body weight in lbs. =		140	
4	Sex =		F	
5	Number of drinks =		5	
6	V_{max} =		0.000369	
7	K_m =		0.00858	
8				
9	t	BAC(t)		
10	0	0.181		
11		1	=B10-($C$6/($C$7+B10))*B10	

FIGURE 6.12 Wagner Excel model formula.

We end this section with a computational example.

Example 6.5: With the same parameter values as Example 6.4, use the Wagner model to project how long it will take before Angela is legally able to drive. How does the Wagner projection compare to Widmark's?

As before we need to copy the model formulas down until Angela's BAC falls below 0.08. As the output in Figure 6.13 shows, the Wagner model predicts that it will take approximately 5 h before Angela's BAC falls below 0.08.

	A	B	C
1	Wagner BAC Model		
2			
3	Body weight in lbs. =		140
4	Sex =		F
5	Number of drinks =		5
6	V_{max} =		0.000369
7	K_m =		0.00858
8			
9	t	$BAC(t)$	
10	0	0.181	
11	1	0.181	
304	294	0.080	
305	295	0.079	

FIGURE 6.13   Wagner model projection for Example 6.5.

Compared to the Widmark model, the Wagner model appears to assume a higher typical rate of alcohol elimination than the Widmark model because it takes 1 h less for Angela's BAC to fall below the legal limit. Also, unlike the Widmark model, the Wagner model will not give us negative values for Angela's BAC regardless of how far into the future we project it. □

In the next section we highlight another important difference between the Wagner and Widmark models: the presence of an equilibrium value.

6.3.1 Equilibrium Analysis

Intuitively we know that once drinking ends BAC should reach an equilibrium value of 0. The Widmark model fails on this account since it has no equilibrium values. The Wagner model, on the other hand, does exhibit an equilibrium value at 0. We have

$$BAC^* = BAC^* - \frac{V_{max}}{K_m + BAC^*} BAC^*$$

$$0 = -\frac{V_{max}}{K_m + BAC^*} BAC^*.$$

Thus either $0 = \frac{V_{max}}{K_m + BAC^*}$ or $0 = BAC^*$. Since V_{max} is a nonzero constant, our only equilibrium value occurs when $BAC^* = 0$, or when all alcohol has been eliminated from the body. Furthermore, since all initial BAC levels eventually revert to 0 once drinking has stopped, the equilibrium at 0 is a stable one.

6.3.2   Determining V_{max} and K_m from Data

Breathalyzers measure alcohol in the breath, but they are calibrated so that the output is reported in the same units as BAC, grams of ethanol per deciliter of blood. With a breathalyzer we can collect real data from which we can estimate personal values for V_{max} and K_m. We illustrate one approach to this kind of estimation in the following example using data collected by the author.

Example 6.6: The author is a 180-pound male who consumed the equivalent of one and two thirds drinks very quickly and then used a breathalyzer to measure his BAC periodically afterward. The author's BAC data is given in Table 6.2. During the period of measurement, no other food or drink was consumed. Find the parameters V_{max} and K_m that produce the best fit in the Wagner model.

TABLE 6.2   Breathalyzer BAC Data

Minutes from End of Drinking	BAC
20	0.032
30	0.029
40	0.027
85	0.015
95	0.013
115	0.010
135	0.006

Here we fit the model visually with trial and error on the two parameters V_{max} and K_m. To do so we must plot the BAC data from Table 6.2 and the Wagner model projections on the same axes. In Excel we make a third column for "BAC-Data" and input the breathalyzer values in this new column next to the appropriate times. (Note that this is the same technique we used for the Eyam plague example in Chapter 4.) Once the data is entered, we graph as usual, selecting the data column along with the model projections. The resulting graph is shown in Figure 6.14.

Note that the standard parameter values for the Wagner model already result in a good fit with the data. Through experimenting with different values, we can obtain a slightly better fit for the data if we let $V_{max} = 0.00045$ and $K_m = 0.014$. The graph with these parameter choices is shown in Figure 6.15. □

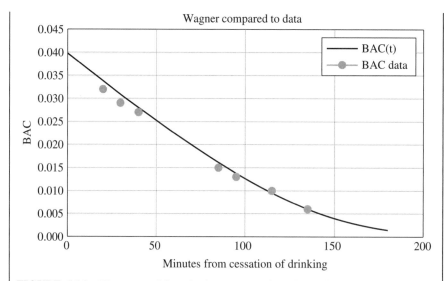

FIGURE 6.14 Wagner model projection compared to data using standard parameter values for Example 6.6.

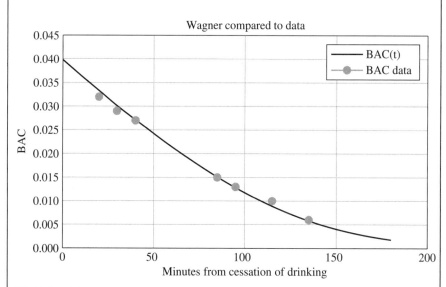

FIGURE 6.15 Wagner model projection compared to data with visually determined parameters for Example 6.6.

In the next section we improve on the Wagner model by incorporating a more realistic approach to how alcohol is typically consumed.

6.3.3   Section Exercises

1   Implement the Wagner model in Excel. Allow the user to input sex, number of drinks consumed, and body weight as parameters in the model. Store V_{max} and K_m in their own cells for easy experimentation later.

2   Diana is a 160-pound woman who consumes five standard glasses of wine.
   a. Use the Wagner Excel model to graph a projection of her BAC over time.
   b. Estimate how long it will take before Diana is legally able to drive.

3   Use the Wagner model with your own parameter choices to:
   a. Graph a projection of your BAC over time.
   b. Estimate how long it will take before you are legally able to drive.

4   *Extension*: Graph the Wagner model BAC projections over time for parameters of your choosing. On the Excel graph, include horizontal lines marking the BAC levels for the legal driving limit, drunkenness, coma, and death.

5   *Extension*: Make the Wagner Excel model more flexible by allowing users to input nonstandard drinks.

6   In a DUI court case, a defendant's BAC is measured at the scene of an accident as 0.12. The defendant is a 220-pound man who is known to have stopped drinking 2 h before being tested. Use the Wagner model to estimate how many drinks the defendant had.

7   In a DUI court case, a defendant's BAC is measured at the scene of an accident as 0.05. The defendant is a 120-pound woman who is known to have stopped drinking 3 h before being tested. At the time of the accident, she had been driving for 30 min. Use the Wagner model to determine whether she is guilty of DUI.

8   For parameter values of your choosing, compare the long-term projections for BAC for the Widmark and Wagner models.

9   Compare the average maximal elimination rate for the Wagner model with the average elimination rate for the Widmark model.

6.4   ALCOHOL CONSUMPTION PATTERNS

With both the Widmark and the Wagner models, our BAC projections are conservative in the sense that we assume all alcohol that has been consumed is in the body at once and then the BAC declines as time passes from the cessation of drinking. This will tend to overestimate BAC because the body actually starts eliminating alcohol as soon as drinking starts: having four beers in 10 min should certainly result in a higher BAC than four beers over 2 h, but with our current models, the BAC estimates would be the same.

In this section we incorporate a more realistic approach that accounts for the body metabolizing alcohol as soon as drinking starts and allows the user to control both the number of drinks and the time period over which drinking takes place.

As a first step we consider the incorporation of a constant drinking rate like "1 beer every hour" or "2 glasses of wine per hour." This rate plays a similar role to the stocking number we considered in Chapter 1. We interpret a rate like "1 beer per h" to mean that the beer is consumed uniformly over the hour as opposed to a beer being quickly consumed and then an hour elapsing before the next one.

Incorporating a drinking rate into our model requires that we are careful with our units. If the user is drinking at a rate of 1 drink per h, we cannot simply add "1" to our DDS. Instead we need to know the corresponding increase in BAC that one drink causes for the user, and then we need to convert that to a BAC increase per minute. The conversion from number of drinks per hour to increase in BAC per minute depends on the parameters for the user and involves using our basic BAC calculator. We show the details in the examples that follow.

For now we let d represent the drinking rate as an increase in BAC each minute. The flow diagram for the Wagner model with drinking rate is given in Figure 6.16.

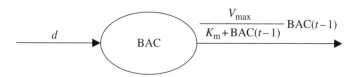

FIGURE 6.16 Flow diagram for the Wagner model with constant drinking rate.

The corresponding DDS is thus

$$BAC(t) = BAC(t-1) - \frac{V_{max}}{K_m + BAC(t-1)} BAC(t-1) + d.$$

To implement the model in Excel, we ask the user to estimate their drinking rate by inputting the number of drinks per hour they typically consume. The spreadsheet then calculates d, the increase in BAC each minute based on the user inputs. The setup with the formula for d displayed is given in Figure 6.17.

	A	B	C	D	E	F	G	H
1	Wagner BAC Model with Constant Drinking Rate							
2								
3	Body weight in lbs. =		167					
4	Sex =		F					
5	Drinks per hour =		2					
6	BAC incr. min^{-1}, d =		=((($C$5*14)/(($C$3/2.2046)*IF($C$4="M",0.58,0.49)))*0.806/10)/60					
7	V_{max} =		0.000369					
8	K_m =		0.00858					
9								
10	t	BAC(t)						
11	0	0.000						
12	1	0.001						
13	2	0.002						

FIGURE 6.17 Excel formula for converting drinking rate to increase in BAC.

To implement the DDS, we begin with a BAC of 0 and then add d to the DDS from the original Wagner model. The new formula is shown in Figure 6.18.

	A	B	C	D	E
1	Wagner BAC Model with Constant Drinking Rate				
2					
3	Body weight in lbs. =		167		
4	Sex =		F		
5	Drinks per hour =		2		
6	BAC incr. min^{-1}, d =		0.001013		
7	V_{max} =		0.000369		
8	K_m =		0.00858		
9					
10	t	$BAC(t)$			
11	0	0.000			
12	1	=B11-($C$7/($C$8+B11))*B11+$C$6			

FIGURE 6.18 Excel formula for Wagner model with constant drinking rate.

As an example we consider the following.

Example 6.7: Vincent is a 210-pound male who drinks 3 beers per h. Project Vincent's BAC at the end of 4 h using the modified Wagner model.

We need to input Vincent's parameters and copy the formulas down to $t = 240$. The result with most rows hidden is shown in Figure 6.19.

	A	B	C	D
1	Wagner BAC Model with Constant Drinking Rate			
2				
3	Body weight in lbs. =		210	
4	Sex =		M	
5	Drinks per hour =		3	
6	BAC incr. min^{-1}, d =		0.001021	
7	V_{max} =		0.000369	
8	K_m =		0.00858	
9				
10	t	$BAC(t)$		
11	0	0.000		
12	1	0.001		
250	239	0.169		
251	240	0.169		

FIGURE 6.19 Excel model BAC projection for Example 6.7.

Vincent's BAC after 4 h will be 0.169 so he will be very impaired. □

6.4.1 Equilibrium Analysis and Maintaining a Buzz

We noted in previous sections that the Widmark model has no equilibrium values, and the Wagner model has only the trivial equilibrium value of 0. The inclusion of a constant drinking rate is more interesting because it introduces a nontrivial equilibrium.

To find the equilibrium value, we need to solve the following for BAC^*:

$$BAC^* = BAC^* - \frac{V_{max}}{K_m + BAC^*} BAC^* + d.$$

Subtracting BAC^* from both sides and moving the negative term to the left-hand side give us

$$\frac{V_{max}}{K_m + BAC^*} BAC^* = d.$$

Next we multiply through by $K_m + BAC^*$ to get $V_{max} \cdot BAC^* = d(K_m + BAC^*)$. Multiplying through by d on the right and collecting like terms give us $(V_{max} - d)BAC^* = dK_m$ so that our final result is

$$BAC^* = \frac{dK_m}{V_{max} - d}.$$

The result tells us that if our drinking rate is less than the maximal elimination rate, V_{max}, then we will get a nonzero equilibrium value that depends on the drinking rate and the user-specified values for V_{max} and K_m. On the other hand, if the drinking rate exceeds the maximal elimination rate, then we do not expect a physically meaningful equilibrium value. Instead BAC will just continue to increase over time.

In the next example we show how to use the formula, and we verify it with our Excel model.

Example 6.8: Suppose Katherine is a 167-pound female who has 1 drink every 2 h, or about 0.5 drinks per h. Find Katherine's long-term BAC level assuming average values for the Wagner parameters.

First we need to determine d. We do this by plugging Katherine's parameter values into our spreadsheet and observing the value for d that corresponds to Katherine having half a drink per hour. We get $d = 0.0002533$. According to our formula for the equilibrium value, Katherine's equilibrium BAC is

$$BAC^* = \frac{dK_m}{V_{max} - d} = \frac{0.0002533 \cdot 0.00858}{0.0003693 - 0.0002533} \approx 0.019.$$

We can confirm this value by inputting 0.019 as Katherine's initial BAC and noting that her BAC would then remain constant. We can also show that with Katherine's parameter choices the equilibrium value at 0.019 is stable and thus represents her long-term BAC. To see this we start Katherine with a BAC of 0 and drag the model down far enough to see her BAC stabilize at 0.019. A graph of her projected BAC along with her equilibrium value is presented in Figure 6.20. Note that Katherine's BAC starts below the equilibrium value but tends toward it over time. ☐

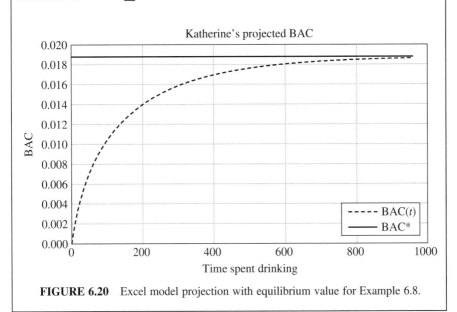

FIGURE 6.20   Excel model projection with equilibrium value for Example 6.8.

In the previous example it takes a very long time—around 14 h—for Katherine's BAC to level off at the equilibrium value, certainly far longer than we typically expect someone to be drinking. For this reason in practice, we should think of the equilibrium value as a cap, or maximum value for BAC, not necessarily a level that will be reached. This assumes an initial BAC below the equilibrium value; the situation will be different if the BAC starts above it.

In the next example we take a closer at a drinking pattern that is commonly referred to as "maintaining a buzz."

Example 6.9: Ned is a 200-pound male who wants to drink steadily in order to maintain a "buzzed" feeling without becoming drunk. Using average values for the Wagner parameters, estimate how many drinks per hour Ned should have.

   Though it is difficult to pick a specific BAC that corresponds to feeling buzzed, typically this feeling corresponds to a BAC between 0.04 and 0.06.

In this example we will use a BAC of 0.05 as the "target" for Ned. The idea is to determine d that would lead to an equilibrium value of $\mathrm{BAC}^* = 0.05$ for Ned.

First we take the time to solve the problem in the general case by isolating d in the equilibrium value formula. We get

$$\begin{aligned}
\mathrm{BAC}^* &= \frac{dK_\mathrm{m}}{V_\mathrm{max} - d}\\
(V_\mathrm{max} - d)\mathrm{BAC}^* &= dK_\mathrm{m}\\
V_\mathrm{max} \cdot \mathrm{BAC}^* &= dK_\mathrm{m} + d \cdot \mathrm{BAC}^*\\
V_\mathrm{max} \cdot \mathrm{BAC}^* &= d(K_\mathrm{m} + \mathrm{BAC}^*).
\end{aligned}$$

Thus if we know the equilibrium value, we find d by

$$d = \frac{V_\mathrm{max} \cdot \mathrm{BAC}^*}{K_\mathrm{m} + \mathrm{BAC}^*}.$$

In our current example we have

$$d = \frac{V_\mathrm{max} \cdot \mathrm{BAC}^*}{K_\mathrm{m} + \mathrm{BAC}^*} = \frac{0.0003693 \cdot 0.05}{0.00858 + 0.05} = 0.0003152.$$

Once we know that $d = 0.0003152$, we can solve for the number of drinks per hour that produces this d for Ned. We can do this algebraically, with trial and error in Excel, or by using Goal Seek. We show the Excel setup for the Goal Seek calculation in Figure 6.21.

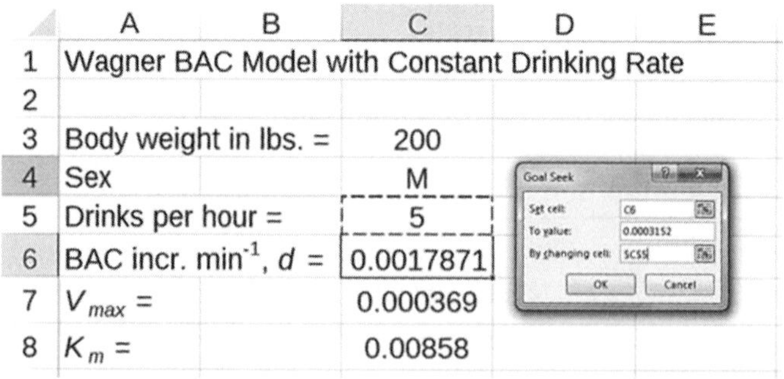
A
B
C
D
E
1
Wagner BAC Model with Constant Drinking Rate
2
3
Body weight in lbs. =
200
4
Sex
M
5
Drinks per hour =
5
6
BAC incr. min-1, d =
0.0017871
7
V max =
0.000369
8
K m =
0.00858
Goal Seek
Set cell:
C6
To value:
0.0003152
By changing cell:
$C$5
OK
Cancel

FIGURE 6.21 Goal Seek setup for determining number of drinks per hour from increase in BAC.

The results of a successful Goal Seek inform us that Ned needs to drink about 0.88 drinks per h in order for him to eventually achieve and maintain a BAC of about 0.05. □

E.19   Setting the Error Tolerance for Goal Seek

If in the previous example Goal Seek was unsuccessful, it may be due to the error tolerance setting for Goal Seek—the tolerance at which Goal Seek decides the result is "close enough." The default setting for this tolerance is 0.001, which is not small enough to give us the precision we need in the last example.

To change the setting select Options from the File drop-down, then choose Formulas. The dialog box should appear as in Figure 6.22. Under Calculation Options change the Maximum Change to 0.000001, and click "Okay." If Goal Seek is rerun, then it should find the value given in Example 6.9.

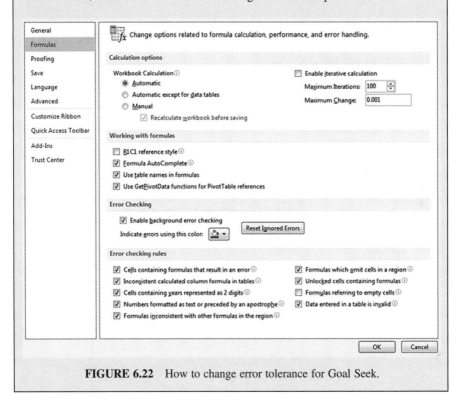

FIGURE 6.22   How to change error tolerance for Goal Seek.

In the previous example the target BAC was 0.05. While it is true that Ned's BAC will eventually rise to 0.05 if he drinks long enough, it will take much longer than is practically reasonable, in this case more than 28 h. If his goal is really to get to a BAC of 0.05 and then stay there, Ned will have to take a different approach. We explore that approach in the next example.

Example 6.10: With the parameters from the previous example, determine a drinking pattern that will start Ned at a BAC of 0.05 and keep him there.

What we need here is sometimes called a "loading dose" when discussing other drugs. The idea is to drink enough alcohol initially to get the BAC up to 0.05 quickly and then continue drinking at the rate found in Example 6.9.

To find the number of drinks required to quickly get Ned's BAC to 0.05, first we modify our model to include a "loading dose," which we will call the initial number of drinks.

We insert a row where we store the initial number of drinks; then for our initial BAC value, we calculate the BAC that results from that number of drinks. The setup with the formula for initial BAC is displayed in Figure 6.23.

	A	B	C	D	E	F	G
1	Wagner BAC Model with Constant Drinking Rate and Loading Dose						
2							
3	Body weight in lbs. =		200				
4	Sex		M				
5	Drinks per hour =		0.881866				
6	BAC incr. min^{-1}, d =		0.0003152				
7	Initial number drinks =		2				
8	V_{max} =		0.000369				
9	K_m =		0.00858				
10							
11		t	BAC(t)				
12		0	=((($C$7*14)/(($C$3/2.2046)*IF($C$4="M",0.58,0.49)))*0.806/10)				
13		1	0.043				

FIGURE 6.23 Wagner Excel model including an initial "loading dose."

Next we use Goal Seek to find the number of initial drinks that results in a BAC of 0.05 for Ned. The result of this Goal Seek is about 2.33 drinks. With an initial 2 1/3 drinks consumed quickly, Ned will raise his BAC to 0.05. Then if he continues to consume alcohol at 0.88 drinks per h, his BAC will stay at 0.05, and he will maintain his buzz. □

Before embarking on a model-designed drinking pattern, there are a couple of very important considerations to take into account. The first is that, generally speaking, one cannot "go back" when trying to maintain a buzzed feeling. What this means is that if one accidentally overshoots the BAC target and begins to experience unpleasant effects from the alcohol, then decreasing the BAC to 0.05 will not make one feel better. As a result, one should only approach BAC targets *from below*, that is, a conservative approach is best.

The second practical consideration is that this kind of maintenance drinking is very difficult to put into practice accurately, and even missing by a little bit can dramatically change the end result. We explore this fact in the next example.

Example 6.11: Suppose Ned initially consumes 2 1/3 drinks so that his BAC starts at 0.05, but then rather than consuming 0.88 drinks per h he actually consumes 1 drink per h. Examine the consequences of this small departure from the plan.

Though it may seem insignificant, consuming 1 drink per h instead of 0.88 drinks per h changes the drinking rate to $d = 0.0003574$. Thus the equilibrium value changes from $BAC^* = 0.05$ to

$$BAC^* = \frac{dK_m}{V_{max} - d} = \frac{0.0003574 \cdot 0.00858}{0.0003693 - 0.0003574} \approx 0.258.$$

Instead of maintaining a pleasant feeling of mild euphoria, Ned's BAC will continue to rise (albeit very slowly) to 0.258, a level that causes passing out, memory blackouts, stupor, and severe motor impairment. If the BAC level reaches slightly higher, 0.30, death can occur. □

The moral of this series of examples is that while it may be *possible* to maintain a BAC level that leads to pleasant side effects, it is very difficult to carry out in practice—who is going to measure out 0.88 drinks every hour? Furthermore, the consequences of drinking even slightly more than called for over time can lead to dangerous BAC levels.

The mathematical description of the phenomenon we just observed is that the equilibrium for BAC is **sensitive** to changes in the drinking rate. If we consider the formula for BAC^*, we can see why. The denominator for that formula is the term $V_{max} - d$. Thus as the drinking rate approaches V_{max}, the denominator of the BAC^* formula approaches 0, which in turn causes the BAC^* to grow. We emphasize the point with the next example.

Example 6.12: Natalie is a 130-pound female. Find the difference in the drinking rates required for her to maintain her BAC at 0.05 versus 0.10.

Using the same method as in Example 6.9, we find a drinking rate of 0.484 drinks per h for an eventual BAC level of 0.05 and a drinking rate of 0.522 drinks per h to maintain a BAC of 0.08. Thus a difference as small as $0.523 - 0.484 = 0.038$ drinks (about 1/2 oz. of beer) per hour could lead to *twice* the eventual BAC for Natalie. □

Again we emphasize that the idea of drinking just the right amount to maintain a buzz can be an attractive one; however, the Wagner model indicates that it is very difficult to accomplish in practice without accidentally overdoing it.

In the next section we recognize that people do not tend to drink indefinitely. We need a way to turn off consumption in our model.

6.4.2 Turning Off Drinking

So far we have handled the consumption part of our model in one of two ways: by assuming all drinking has already taken place and all alcohol consumed is still in the body or by assuming a constant drinking rate that continues indefinitely. Neither of these approaches is completely satisfactory, though, because they do not fully capture how most drinking takes place.

In this section we make an improvement on how we handle consumption by allowing the user to specify the number of drinks consumed, denoted by N, as well as the time period (in minutes) over which the consumption takes place. We denote the time spent drinking, or drinking time, by DT, and our assumption is that drinking occurs uniformly over the time period given.

Once the user has input the number of drinks and the time period, the drinking rate over that time is given by $\dfrac{\text{Number of drinks}}{\text{Drinking time}} = \dfrac{N}{\text{DT}}$. Then the parameter, d, will be the corresponding per-minute increase in BAC, which we get Excel to calculate as before. The difference now lies in the fact that we stop adding d to our model as soon as the drinking period is over. We illustrate the modification to our model in the next example.

Example 6.13: Natalie from Example 6.12 has five glasses of wine over a 2-h period. Project her BAC over time.

Here the time spent drinking is 120 min so we need to turn off Natalie's consumption at $t = 120$. There are several ways to go about this in Excel, including the method used in Chapter 3 when dealing with combat models. We create an "IF" statement that serves as a kind of "on/off switch." Instead of adding d at every time step, we use an IF statement that only adds d while drinking is still happening. In Excel the syntax is roughly "IF($t <$ DT, d, 0)." We show the modified Excel setup in Figure 6.24.

	A	B	C	D	E	F	G	H
1	Wagner BAC Model with Drinking Time							
2								
3	Body weight in lbs. =		130					
4	Sex =		F					
5	Drinks consumed =		5					
6	Drinking time, DT =		120					
7	BAC incr. min⁻¹, d =		=((($C$5/$C$6)*14)/(($C$3/2.2046)*IF($C$4="M",0.58,0.49)))*0.806/10					
8	V max =		0.000369					
9	K m =		0.00858					

FIGURE 6.24 Wagner Excel model setup with finite drinking time.

The new formula for BAC is given in Figure 6.25.

10		
11	*t*	*BAC(t)*
12	0	0.000
13	1	=B12-($C$8/($C$9+B12))*B12+IF(A13<$C$6,$C$7,0)
14	2	0.003

FIGURE 6.25   Wagner Excel model BAC formula with finite drinking time.

With all parameters correctly entered, we graph Natalie's BAC over 12 h. Figure 6.26 shows the result.

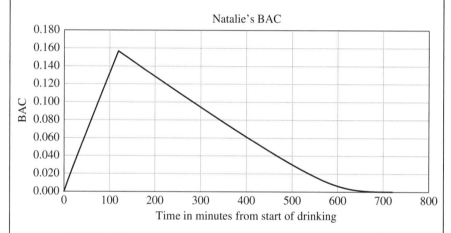

FIGURE 6.26   Natalie's projected BAC over time for Example 6.13.

Note how after 2 h Natalie's BAC stops rising and begins to fall. This indicates we have succeeded in turning off her drinking at the appropriate time. Her BAC peaks at about 0.156, and about 3 h and 45 min later it falls below the legal driving limit. It takes about 11 h for all of the alcohol to leave her system. ☐

The Wagner model is an improvement over the Widmark model in that it more accurately models the elimination of alcohol from the body. However, the Wagner model shares with the Widmark model two oversimplifications. Both models treat the body as a single reservoir from which alcohol is eliminated by a single process, and both models assume that all alcohol is instantly absorbed and distributed throughout the body water.

In reality there are several pathways by which alcohol is eliminated or metabolized, and each one of these pathways can serve as inspiration for the refinement of our model. For example, one of the largest sources of variability in BAC stems from the rate of gastric emptying, that is, the rate at which alcohol moves from the stomach to the small intestine. Factors affecting gastric emptying include the type of drink (carbonated vs. uncarbonated, high alcohol vs. low) and whether the stomach is full or empty (Kent, 2012). Furthermore, some alcohol is actually metabolized in the stomach before it ever reaches the blood. We could, therefore, include compartments for the stomach and the small intestine. We would then need to estimate parameters for the rate of gastric emptying, the rate at which alcohol is metabolized in the stomach, and the rate of absorption from the small intestine into the blood. J.E. Pieters in 1990 and D.M. Umulis (whose model also includes compartments for the liver and body muscle) in 2005 both developed such models (Pieters, Wedel, & Schaafsma, 1990; Umulis, Gurman, Singh, & Fogler, 2005). In 1998 M.D. Levitt took a different approach and proposed a two-compartment model that treats the liver as a separate compartment from the rest of the body (Levitt & Levitt, 1998).

As indicated above, the BAC models we present in this chapter only scratch the surface of what is possible. The more we come to understand about alcohol metabolism, the more factors we can incorporate into the model.

6.4.3   Section Exercises

1   Toni is a 170-pound woman who drinks 3 standard drinks per hour. Project Toni's BAC at the end of 4 h using the modified Wagner model.

2   Suppose Conrad is a 140-pound male who has 1 drink every 3 h. Find Conrad's long-term BAC level assuming average values for the Wagner parameters.

3   Using average values for the Wagner parameters, estimate how many drinks per hour you would need to have in order to maintain a BAC level of 0.045.

4   Approximately how long will it take in Exercise 3 for your BAC level to get close to the 0.045 "target?"

5   What would the loading dose need to be in order for you to quickly raise your BAC to 0.045?

6   Produce an Excel graph that shows that with the loading dose you found in Exercise 5 and the drinking rate you found in Exercise 3, your BAC would remain constant at 0.045.

7   In Exercise 3 suppose you accidentally drink 20% more per hour than your calculated rate. Find the resulting long-term BAC. Explain the significance of this result.

8   Suppose you drink six standard beers over a 5-h period and then stop.

   a. Use the Wagner model to estimate your peak BAC level.

   b. Determine how long it will be until you are legally able to drive.

   c. Determine approximately how long it will be until all alcohol has left your system.

9 *Extension*: In the same manner as we did for the Wagner model, incorporate a drinking rate and drinking time into the Widmark Excel model. Compare this modified Widmark model with the modified Wagner model for parameter values of your choosing.

6.5 MORE GENERAL DRUG ELIMINATION

As soon as a drug is ingested, the body begins to eliminate it. This can happen through **metabolism**, where enzymes break down the drug into different metabolites, or it can happen through **excretion**, where the drug is passed out of the body through the breath, sweat, or urine. In this section we will not make a distinction between these two processes, opting instead to make the simplifying assumption that treats both possibilities together as a single process that we call **elimination**. It may become necessary or expedient later to consider metabolism and excretion separately, but for now our goal is to keep our model as simple as possible.

For most drugs at usual dosages, elimination takes place at a rate that is a constant *proportion* of the amount of drug present in the body. This kind of elimination process is called **first-order elimination**. In contrast a drug that is eliminated by a constant *amount* for each time step is said to undergo **zero-order elimination**. Many common drugs, including ibuprofen and caffeine, undergo first-order elimination. Alcohol is an example of a drug that is well modeled by **zero-order** elimination (as in the Widmark model), at least for relatively high amounts of alcohol in the body. In this section we focus on first-order elimination.

Though the context is different, we should recognize first-order elimination as an exponential decay model. If we let $B(t)$ be the amount of drug in the body at time t and let r be the elimination rate, then we have the familiar flow diagram in Figure 6.27.

FIGURE 6.27 Flow diagram for first-order single-compartment drug elimination.

The corresponding DDS is then $B(t) = B(t-1) - rB(t-1)$. With $B(0)$ equal to the initial amount of drug in the body, we should have no problem now setting up an Excel model or explicit formula to model the amount of drug remaining in the body over time. The primary difficulty we face is in finding a value r for a given drug.

6.5.1 Drug Half-Life

Drug manufacturers are required to report what is known as the **half-life** of a drug, which is the time it takes the body to eliminate one half of the drug. Thus if a drug has a reported half-life of 4 h and initially 500 mg of the drug is present in the body, there will be 250 mg in the body 4 h later, 125 mg 4 h after that, and so on. We use the symbol $T_{1/2}$ to denote the half-life. Our job as modelers is to deduce the rate of elimination, r, from the half-life. We have done this sort of thing before when we deduced the growth rate for a population of deer from its doubling time. The next example shows how we can deduce the elimination rate from the half-life by using the explicit formula.

Example 6.14: The half-life for the pain reliever ibuprofen is approximately 2 h (RxList, 2015). Determine r, the approximate percentage of the drug that is eliminated from the body each minute.

We use the explicit formula for the exponential model where t is time in minutes and $B(t)$ is the amount of ibuprofen in milligrams still present in the body at time t. Our explicit formula is $B(t) = (1-r)^t B(0)$. By definition if $T_{1/2}$ is the half-life of ibuprofen, then $B(T_{1/2}) = (1/2)B(0)$ where $B(0)$ is the initial amount of ibuprofen in the body. Thus with a half-life of 120 min, we have

$$\frac{1}{2}B(0) = (1-r)^{120}B(0)$$

$$\frac{1}{2} = (1-r)^{120}$$

$$\left(\frac{1}{2}\right)^{1/120} = 1-r$$

$$1 - \left(\frac{1}{2}\right)^{1/120} = r.$$

Thus $r = 0.00576$. As a percent we have an elimination rate for ibuprofen of $r = 0.576\%$ per min. Note that the initial amount of drug present did not matter in our calculation of r. □

Generalizing the previous example we see that we can always do a similar calculation whenever we know the half-life:

$$\frac{1}{2}B(0) = (1-r)^{T_{1/2}}B(0)$$

$$\frac{1}{2} = (1-r)^{T_{1/2}}$$

$$\left(\frac{1}{2}\right)^{(1/T_{1/2})} = 1-r$$

$$1 - \left(\frac{1}{2}\right)^{(1/T_{1/2})} = r.$$

Because we will always have to perform this calculation before proceeding with our model, it makes sense to get Excel to automate the process. In the next example we set up an Excel worksheet that models a first-order elimination process. The spreadsheet will allow the user to input a drug's half-life from which it will compute the appropriate elimination rate automatically.

Example 6.15: Project the amount of ibuprofen remaining in the body over the course of 8 h if initially 500 mg is present and no new doses are administered.

E.20 Exponents

In order to get Excel to calculate an elimination rate from the half-life, we need to work with exponents. Excel uses the caret symbol "^" for exponentiation. So in order to calculate "2 raised to the 4th power," we type "=2^4" in Excel.

To find the elimination rate from the half-life, we need store the half-life in its own cell, say, cell C3, and in the cell where we want the elimination rate, we type "=1 − 0.5^(1/C3)." The setup with formulas showing is given in Figure 6.28.

	A	B	C	D
1	First Order Drug Elimination			
2				
3	Drug half-life, $T_{1/2}$ =		120	(minutes)
4	Elimination rate, r =	=1-0.5^(1/$C$3)		

FIGURE 6.28   Determining elimination rate from half-life in Excel.

As a quick check we enter the half-life for ibuprofen and verify that we get the same result: $r = 0.00576$. The result of our computation is given in Figure 6.29.

	A	B	C	D
1	First Order Drug Elimination			
2				
3	Drug half-life, $T_{1/2}$ =		120	(minutes)
4	Elimination rate, r =		0.005760	

FIGURE 6.29   Excel result for determining elimination rate from half-life.

With the elimination rate in hand, setting up the rest of the spreadsheet should be familiar. We show the setup with formula for $B(t)$ displayed in Figure 6.30.

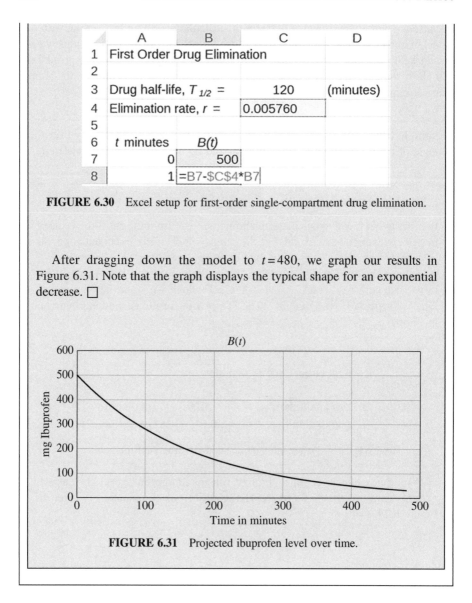

FIGURE 6.30   Excel setup for first-order single-compartment drug elimination.

After dragging down the model to $t = 480$, we graph our results in Figure 6.31. Note that the graph displays the typical shape for an exponential decrease. □

FIGURE 6.31   Projected ibuprofen level over time.

6.5.2   Variations on Half-Life

Occasionally a drug's elimination rate will be reported in a nonstandard way. In an experimental study, for example, drug levels may only be taken every 30 min. In such a case it is natural to report the percentage eliminated every 30 min rather than giving a half-life. Fortunately, as long as we know what percentage has been eliminated over a given time span, we can still find r.

Example 6.16: Suppose we know that 75% of a drug is eliminated every 10 min. Find the elimination rate, r.

If 75% of the drug is eliminated in 10 min, we know we will have 25% of the drug remaining after 10 min. Thus whatever the initial drug amount, $B(0)$, we will have $B(10) = 0.25B(0)$. Solving for r gives us

$$(1-r)^{10}B(0) = 0.25B(0)$$
$$(1-r)^{10} = 0.25$$
$$1-r = 0.25^{1/10}$$
$$r = 1 - 0.25^{1/10} \approx 0.1294.$$

Thus as a percent we have an elimination rate of $r = 12.94\%$ per min. $\square$

Automating this kind of calculation in Excel is left as an exercise.

6.5.3 Absorption

The model in the preceding section assumed the body was a single compartment, and we focused on the elimination of the drug from the body. For drugs that are administered via direct injection or intravenously, a single-compartment model makes sense because the drug is instantly present in the blood. However, many drugs, especially over-the-counter drugs, are administered orally.

Drugs taken orally do not instantly enter the bloodstream; they must be digested first. This means we need to take into account how quickly the drug is absorbed into the body from the gastrointestinal, or GI, tract. To model absorption we add the GI tract as a second compartment considered as separate from "the body," or central compartment, and we introduce a new parameter, the **absorption rate**, into the model.

Let $\text{GI}(t)$ be the amount of drug in the GI tract at time t, and let α be the absorption rate. We assume that absorption from the GI tract into the body is a first-order process so that α represents the fixed percentage of the drug being absorbed into the body at each time step. The two-compartment model is pictured in Figure 6.32.

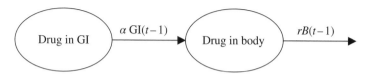

FIGURE 6.32   Two-compartment pharmacokinetic model flow diagram.

The corresponding DDS for the two-compartment model is therefore

$$\text{GI}(t) = \text{GI}(t-1) - \alpha \text{GI}(t-1)$$
$$B(t) = B(t-1) - rB(t-1) + \alpha \text{GI}(t-1).$$

The Excel spreadsheet for the two-compartment model accepts the half-life and the absorption rate as parameters, and it calculates the elimination rate automatically. The spreadsheet with formulas displayed is given in Figure 6.33.

	A	B	C
1	First Order Drug Elimina		
2			
3	Drug half-life, $T_{1/2}$ =		120
4	Elimination rate, r =		=1-0.5^(1/$C$3)
5	Absorption rate, α =		0.01
6			
7	t minutes	GI(t)	B(t)
8	0	500	0
9	=A8+1	=B8-$C$5*B8	=C8+$C$5*B8-$C$4*C8
10	=A9+1	=B9-$C$5*B9	=C9+$C$5*B9-$C$4*C9
11	=A10+1	=B10-$C$5*B10	=C10+$C$5*B10-$C$4*C10

FIGURE 6.33   Two-compartment Excel model.

Like our elimination rate the absorption rate of a drug is seldom easy to find directly in the literature. Instead we must deduce the rate from the kind of information that is available, and there are a variety of ways in which the rate can be reported. One common way is to give the drug's **GI half-life**. The GI half-life is the time it takes one half of the drug to leave the GI tract and enter the bloodstream. If a GI half-life is given in the literature, then we have already seen how to calculate α: we use the explicit formula for exponential decay and solve for the α that gives us 1/2 of the drug remaining in the GI tract at the stated half-life.

Another possibility in the literature is to find a statement such as, "research shows that 95% of the drug will enter the bloodstream within 30 min of ingestion." In these situations we still use the explicit formula.

Example 6.17: Suppose we know from research that 95% of a drug will be absorbed from the GI tract into the bloodstream within 30 min of ingestion. Find the absorption rate, α.

The explicit formula for the amount of drug remaining in the GI tract is given by $GI(t) = (1-\alpha)^t GI(0)$, where GI(0) is the initial dose of the drug. The way the absorption of the drug is reported we should have only 5% of the original dose remaining after 30 min: $GI(30) = 0.05GI(0)$. Thus we solve the following for α:

$$GI(30) = 0.05GI(0)$$
$$(1-\alpha)^{30}GI(0) = 0.05GI(0)$$
$$(1-\alpha)^{30} = 0.05$$
$$1-\alpha = 0.05^{1/30}.$$

This gives us $\alpha \approx 0.095$. $\square$

Another common way that the absorption rate is reported is by noting the time it takes for a drug to reach its **peak plasma level**. We denote this time by T_{max}, and it is the time at which the amount of drug in the plasma is at its highest following a single dose of the drug administered orally. We make the simplifying assumption that the peak plasma concentration will coincide with the peak amount in the body. Unfortunately, unlike our process for deducing r from the half-life, there is no simple formula for finding α from T_{max}. Instead we rely on our Excel model and trial and error. In other words, we set up the two-compartment model, and then we experiment with different absorption rates until we get the highest point on the $B(t)$ graph to occur at T_{max}. We illustrate the process in the next example where we deduce the absorption rate from the peak plasma level of ibuprofen.

Example 6.18: Orally ingested ibuprofen will reach its peak plasma level in about 1 h (Jannsen & Venema, 1985). Deduce the absorption rate of ibuprofen, α.

Using our two-compartment model spreadsheet and a stand-in value for α, we make a graph for $B(t)$. Then finding α is a matter of experimenting with different values for α until we get the peak of this graph to settle over $T_{max} = 60$ min. In Figure 6.34 we show the result for $\alpha = 0.10$, which turns out to be too high because the peak occurs too soon (at about 30 min).

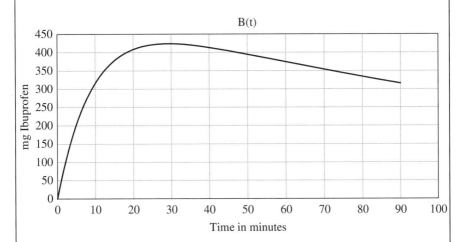

FIGURE 6.34   Determining absorption rate from peak plasma level.

After some experimenting we see that the peak occurs at the appropriate time when $\alpha = 0.036$. The graph is shown in Figure 6.35. Note the time at which the peak for $B(t)$ occurs. $\square$

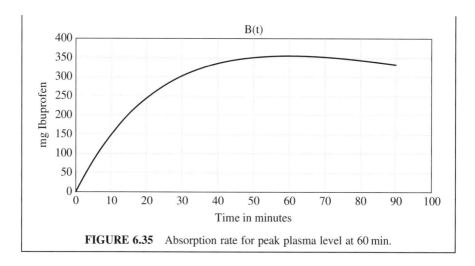

FIGURE 6.35   Absorption rate for peak plasma level at 60 min.

Up until now we have dealt only with single doses of a drug. In the next section we refine our two-compartment model to allow for multiple doses.

6.5.4 The Dosing Function

Our models up to this point—the single-compartment and two-compartment models—have been based on the assumption that we have a single dose of the drug administered either via injection (for the single-compartment model) or orally (for the two-compartment model). We know of course from experience that most drugs are not administered this way. Rather, they are generally administered on some set schedule like "one pill every 6 h." In this section we develop the mathematical and Excel tools necessary to model drug dosages in a way that is flexible enough to account for many commonly used regimens.

If we let $D(t)$ be the dosage of a drug at time t, then we call $D(t)$ the **dosing function**. The graph of a dosing function will be zero except for the times at which a dose is administered. If a 400-mg pill is administered every 6 h, the graph of the dosing function would appear as in Figure 6.36.

We include the dosing function $D(t)$ in our flow diagrams in Figures 6.37 and 6.38. The corresponding DDS is then

$$B(t) = B(t-1) - rB(t-1) + D(t)$$

for the single-compartment model (doses given via injection) and

$$G(t) = G(t-1) - \alpha G(t-1) + D(t)$$
$$B(t) = B(t-1) - rB(t-1) + \alpha G(t-1)$$

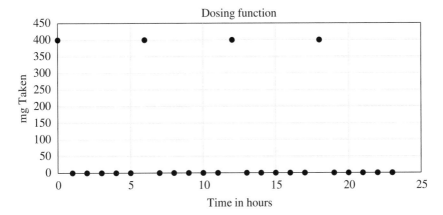

FIGURE 6.36 Example of dosing function graph.

FIGURE 6.37 Single-compartment with dosing function flow diagram.

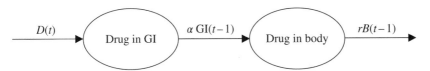

FIGURE 6.38 Two-compartment with dosing function flow diagram.

for the two-compartment model (doses given orally). The assumption in the case of injection is that the full dose is instantly available in the body. In the case of oral administration, the assumption is that the dose is immediately available in the GI tract, and from there it begins to be absorbed into the body.

To model a realistic dosing function, we need a way to set a repeating pattern of dosages. For example, one 400-mg pill every 6 h should result in a dosing function $D(t)$ that is equal to 400 whenever time is a multiple of 6 h, and 0 otherwise.

Thinking of t as representing hours from the start of medication, we can represent $D(t)$ as

$$D(t) = \begin{cases} 400 & \text{if } t \text{ is a multiple of } 6 \\ 0 & \text{otherwise} \end{cases}.$$

To make further progress on creating such a function in Excel, we first need to introduce a kind of arithmetic known as modular or clock arithmetic that has many important applications throughout mathematics including cryptography and Internet security.

6.5.5 Modular Arithmetic

The last version of $D(t)$ above had us adding 400 mg of a drug whenever t *is a multiple of 6*. Equivalently, we add a 400-mg dose whenever t *has a remainder of 0 when divided by 6*. Arithmetic that focuses on the remainder of a division is known as **modular arithmetic**, also known as **clock arithmetic**. The basic idea in modular arithmetic is that when we divide one whole number by another, we care only about the remainder. So, for example, if we divide 7 by 4 the remainder is 3, and if we divide 11 by 4 the remainder is also 3. In modular arithmetic we would say 7 and 11 are **equivalent modulo 4** because they have the same remainder when divided by 4. The numbers 3, 15, 19, and 23 are also equivalent modulo 4.

More generally, the notation $t \bmod k$ means "divide the number t by the number k and keep only the remainder." Thus $7 \bmod 4 = 3$ and $11 \bmod 4 = 3$. Similarly $11 \bmod 6 = 5$ and $61 \bmod 6 = 1$.

If we consider a divisor of 6, then dividing the first several positive integers by 6 gives us

$$0 \bmod 6 = 0$$
$$1 \bmod 6 = 1$$
$$2 \bmod 6 = 2$$
$$3 \bmod 6 = 3$$
$$4 \bmod 6 = 4$$
$$5 \bmod 6 = 5$$
$$6 \bmod 6 = 0.$$
$$7 \bmod 6 = 1$$
$$8 \bmod 6 = 2$$
$$9 \bmod 6 = 3$$
$$10 \bmod 6 = 4$$
$$11 \bmod 6 = 5$$
$$12 \bmod 6 = 0$$
$$\vdots$$

The important thing to notice about the remainders is that they form a repeating pattern; this is just the sort of pattern we need in order to create a dosing function.

Using modular arithmetic with t in hours, we write $D(t)$ as

$$D(t) = \begin{cases} 400 & \text{if } t \bmod 6 = 0 \\ 0 & \text{otherwise} \end{cases}.$$

If we prefer to think of time in minutes, we can change $D(t)$ to

$$D(t) = \begin{cases} 400 & \text{if } t \bmod 360 = 0 \\ 0 & \text{otherwise} \end{cases}.$$

Next we see how to implement such a function in Excel.

E.21   The MOD Command

The "MOD" function is what Excel uses to perform modular arithmetic. We know that if we divide 7 by 3, we get a remainder of 1. In Excel this can be computed by typing "=MOD(7,3)" into any cell. After hitting "Enter" we see the result "1." The order of the inputs is important: Excel computes the first number entered divided by the second. Had we accidentally typed "=MOD(3,7)," we would have gotten "3" as the result. In Figure 6.39 we see a few examples of the result of the "MOD" function on a column of time values for several different divisors. Notice how the pattern of remainders repeats for each choice. It is this repetition that we take advantage of when constructing a dosing function.

	A	B	C	D
1	Modular Arithmetic in Excel			
2				
3	t	t mod 3	t mod 4	t mod 9
4	0	0	0	0
5	1	1	1	1
6	2	2	2	2
7	3	0	3	3
8	4	1	0	4
9	5	2	1	5
10	6	0	2	6
11	7	1	3	7
12	8	2	0	8
13	9	0	1	0
14	10	1	2	1
15	11	2	3	2
16	12	0	0	3
17	13	1	1	4
18	14	2	2	5
19	15	0	3	6
20	16	1	0	7
21	17	2	1	8

FIGURE 6.39   Results of Excel MOD function for different divisors.

The formula version for the first three columns is given in Figure 6.40.

We are now ready to incorporate a dosing function into our two-compartment model in Excel. The modification for the single-compartment model would be similar. We start off by including a place for the user to input the required parameter values. We need the half-life for elimination from the body, $T_{1/2}$, the absorption rate from the GI tract, α, the dosage amount, d (typically in milligrams), and the dosage frequency, or time between doses, f, usually reported in hours. We also need a new column where we will keep the dosing function, $D(t)$. The Excel setup is shown in Figure 6.41.

	A	B	C
1	Modular Arithmetic in Ex		
2			
3	*t*	*t* mod 3	*t* mod 4
4	0	=MOD(A4,3)	=MOD(A4,4)
5	=A4+1	=MOD(A5,3)	=MOD(A5,4)
6	=A5+1	=MOD(A6,3)	=MOD(A6,4)
7	=A6+1	=MOD(A7,3)	=MOD(A7,4)
8	=A7+1	=MOD(A8,3)	=MOD(A8,4)

FIGURE 6.40   Excel formulas for MOD function example.

	A	B	C	D
1	Two-Compartment Drug Model			
2				
3	Drug half-life, $T_{1/2}$ =		120	(minutes)
4	Elimination rate, r =		0.005760	
5	Absorption rate, α =		0.036	
6	Dosage amount, d =		400.0	(mg)
7	Dose frequency, f =		6.0	(hours)
8				
9	t minutes	$D(t)$	$GI(t)$	$B(t)$
10	0		500	0
11	1		482.0	18.0
12	2		464.6	35.2

FIGURE 6.41   Excel setup for two-compartment model with dosing function.

Our time units will be minutes so for the dosing function we need to add a dose of d mg at every time that has a remainder of 0 when divided by $360 \times f$. An IF statement for $D(t)$ in Excel will have the following form: IF($t \bmod(60 \cdot f) = 0$, add another dose of d mg, otherwise add 0). More precisely the Excel command will look like "=IF(MOD(t, $60f$) = 0, d, 0)." The model with the formula for $D(t)$ visible is shown in Figure 6.42.

Figure 6.43 shows a graph of our completed dosing function over a period of 12 h.

Next we incorporate the dosing function into our two-compartment model by adding the column for $D(t)$ to the column for $GI(t)$. The model with the formula for $GI(t)$ showing is given in Figure 6.44.

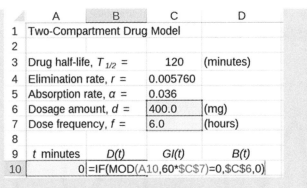

	A	B	C	D
1	Two-Compartment Drug Model			
2				
3	Drug half-life, $T_{1/2}$ =		120	(minutes)
4	Elimination rate, r =		0.005760	
5	Absorption rate, α =		0.036	
6	Dosage amount, d =		400.0	(mg)
7	Dose frequency, f =		6.0	(hours)
8				
9	t minutes	$D(t)$	$GI(t)$	$B(t)$
10		0	=IF(MOD(A10,60*$C$7)=0,$C$6,0)	

FIGURE 6.42 Excel formula for dosing function.

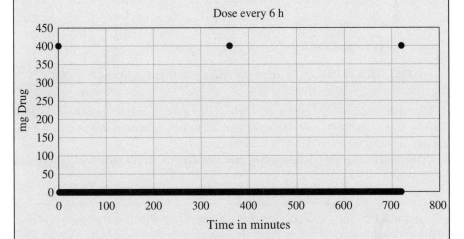

Dose every 6 h

FIGURE 6.43 Graph of Excel dosing function over 12 h.

	A	B	C	D
1	Two-Compartment Drug Model			
2				
3	Drug half-life, $T_{1/2}$ =		120	(minutes)
4	Elimination rate, r =		0.005760	
5	Absorption rate, α =		0.036	
6	Dosage amount, d =		400.0	(mg)
7	Dose frequency, f =		6.0	(hours)
8				
9	t minutes	$D(t)$	$GI(t)$	$B(t)$
10	0	400	400	0
11	1	0	=C10-$C$5*C10+B11	

FIGURE 6.44 Excel formula for GI tract with dosing function.

Finally we show how a repeated dosing schedule affects the amount of drug in the body. Figure 6.45 shows the graph for $B(t)$ over a period of 24 h given a 400-mg dose of the drug every 6 h.

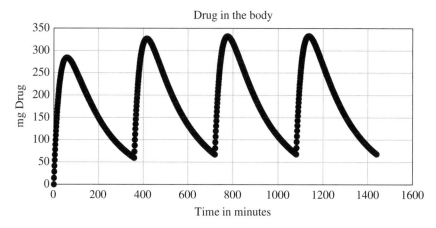

FIGURE 6.45   Model projection for drug in the body for 400 mg every 6 h.

Note that the amount of drug in the body does not stay constant. Rather it exhibits a kind of "sawtooth" pattern of a sharp uptick after every new dose and an exponential decline between doses as the drug is eliminated.

In the next example we examine how repeated doses affect the amount of ibuprofen in the body.

Example 6.19: Suppose a patient takes 200-mg doses of ibuprofen, one every 6 h. How much ibuprofen is left in the body 24 h after the first dose? Use minutes for the time unit.

	A	B	C	D
1	Two-Compartment Drug Model			
2				
3	Drug half-life, $T_{1/2}$ =		120	(minutes)
4	Elimination rate, r =		0.005760	
5	Absorption rate, a =		0.036	
6	Dosage amount, d =		200.0	(mg)
7	Dose frequency, f =		6.0	(hours)
8				
9	t minutes	$D(t)$	$GI(t)$	$B(t)$
10	0	200	200	0
11	1	0	192.8	7.2
1448	1438	0	0.0	34.4
1449	1439	0	0.0	34.2
1450	1440	200	200.0	34.0

FIGURE 6.46   Ibuprofen projections for Example 6.19.

For the basic model parameters, we know from previous examples that $\alpha = 0.036$ and $r = 0.00576$. For the dosing parameters in this example, we have $f = 6$, and $d = 200$. Once all values are input into Excel, we simply drag the model equations down until we reach $t = 24 \times 60 = 1440$ min. The result is shown in Figure 6.46 with most rows hidden. We have $B(1440) \approx 34$ mg of ibuprofen in the body. $\square$

As a further refinement to our dosing function, the next section considers the time it takes for a pill to dissolve in the stomach before it is available to be absorbed into the body.

6.5.6 Dissolution

We know when we take a pill orally that it takes some time for it to dissolve in the stomach, but at the moment our model assumes that the medicine is immediately available in the GI tract for absorption into the body. We make the model for orally administered doses more realistic by introducing a parameter that controls how long it takes for a pill to dissolve in the GI tract. We will call this parameter the **dissolution time**, and we denote it by δ.

For a wide variety of drugs, 30 min is a good approximation for how long it takes a standard (i.e., not extended release) pill to dissolve (Spitznagel, 1992). Let $\delta = 30$.

The incorporation of a dissolution time requires that rather than having the entire dose available instantly in the GI tract, we add each dose a little bit at a time over the dissolution time. If we assume a steady release for the drug, we add $\frac{1}{\delta} = \frac{1}{30}$ of the dose each minute over the $\delta = 30$ min of dissolution. In the next example we show how to construct a dosing function that takes dissolution time into account.

Example 6.20: Suppose a drug with typical dissolution time is prescribed in pill form so that the patient takes a 150-mg pill once every 4 h. Find the dosing function.

For a 150-mg pill with a typical dissolution time of 30 min, the dosing function must add $\frac{d}{\delta} = \frac{150}{30} = 5$ mg of the drug to the GI tract every minute for 30 min following ingestion, and 0 afterward until the next dose.

With $f = 4$ as the dosage frequency in hours, the dosage schedule is one pill every $60 \times f = 240$ min. Thus every 240 min a pill is taken, and the pill takes 30 min to completely dissolve into the GI tract. We arrange this in Excel by modifying the dosing function as follows:

$$D(t) = \begin{cases} 5 & \text{if } 0 \leq t \bmod 240 < 30 \\ 0 & \text{otherwise} \end{cases}.$$

This function adds 5 mg of the drug per minute for the 30 min following ingestion where the 30 min is from the minute a dose is taken, $t \bmod 240 = 0$, to 29 min later, $t \bmod 240 = 29$. $\square$

With all parameters left as unknowns, the most general form of our dosing function is given by

$$D(t) = \begin{cases} \dfrac{d}{\delta} & \text{if } 0 \le t \bmod(60f) < \delta \\ 0 & \text{otherwise} \end{cases}.$$

The effect we see in our model predictions should be that the rise in drug level in the body is more gradual than before. We include dissolution in the next example.

Example 6.21: Suppose a patient takes 200-mg caplets of ibuprofen, one every 6 h. How much ibuprofen is left in the body 24 h after the first dose if each caplet takes 30 min to dissolve? Use minutes for the time unit.

In Excel our spreadsheet only needs to be modified slightly to account for the dissolution parameter δ. We show the formula version of the modification in Figure 6.47.

	A	B	C	D	E
1	Two-Compartment Drug Model with Dosing Function and Dis				
2					
3	Drug half-life, $T_{1/2}$ =		120	(minutes)	
4	Elimination rate, r =		0.005760		
5	Absorption rate, α =		0.036		
6	Dosage amount, d =		200.0	(mg)	
7	Dose frequency, f =		6.0	(hours)	
8	Dissolution time, δ =		30.0	(minutes)	
9					
10	t minutes		$D(t)$	$GI(t)$	$B(t)$
11	0	=IF(MOD(A11,60*$C$7)<$C$8,$C$6/$C$8,0)			

FIGURE 6.47   Excel model formula with dissolution time.

For the basic model parameters, we know from previous examples that $\alpha = 0.036$ and $r = 0.00576$. For the dosing parameters in this example, we have $f = 6$, $d = 200$, and $\delta = 30$. Once all values are input into Excel, we drag the model equations down until we reach $t = 24 \times 60 = 1440$ min. We see that the model projects $B(1440) \approx 37$ mg of ibuprofen in the body. In comparison with Example 6.19, it appears that the inclusion of a dissolution time leads to a slightly increased level at the required time, 37 versus 34 mg. □

We know that for many drugs the dosing regimen does not continue indefinitely. Instead the patient takes the drug over a specified time period or until a prescription runs out. Thus we need to include a way to turn off the dosing function, just as we needed a way to turn off alcohol consumption for the BAC model.

6.5.7 Turning Off the Dosing Function

Our model so far has been set up to handle either a single dose of a drug or a dosing regimen that continues indefinitely. Of course in reality drugs are often prescribed only for a specific time frame or for a specific number of doses. For example, a patient may be given a bottle of antibiotic pills and directed to "take one pill a day until all pills have been taken." In such circumstances we employ an "on/off" switch in Excel just as we did when turning off alcohol consumption.

Here we let N be the total number of doses to be administered; N is stored in its own cell and input by the user. With time in minutes and f the time between doses in hours, we know that if we have N total doses, the entire regimen (until just before the next dose) will take $N \times 60 \times f$ min to complete. Thus if $t \geq N \times 60 \times f$, we need to turn the dosing function off. In Excel this can be done with an IF statement: $IF(t < N \cdot 60 \cdot f, 1, 0)$. If we multiply our expression for $D(t)$ by this IF statement, then $D(t)$ will be multiplied by 1 and hence unchanged until $t \geq N \times 60 \times f$, at which time $D(t)$ will be multiplied by 0 and hence will be turned off. The final version of our dosing function with formula showing is given in Figure 6.48.

	A	B	C	D	E	F	G	H	I
1	Two-Compartment Drug Model with Dosing Function, Dissolution Time, and Finite Number of Doses								
2									
3	Drug half-life, $T_{1/2}$ =		120	(minutes)					
4	Elimination rate, r =		0.005760						
5	Absorption rate, α =		0.036						
6	Dosage amount, d =		200.0	(mg)					
7	Dose frequency, f =		6.0	(hours)					
8	Dissolution time, δ =		30.0	(minutes)					
9	Total doses, N =		2						
10									
11	t minutes		$D(t)$	$GI(t)$	$B(t)$				
12			0	=IF(MOD(A12,60*$C$7)<$C$8,$C$6/$C$8,0)*IF(A12<$C$9*60*$C$7,1,0)					

FIGURE 6.48 Dosing function Excel formula with "on/off" switch included.

Our next example incorporates the "on/off" switch into our model.

Example 6.22: Using the parameters from Example 6.21, graph the amount of ibuprofen in the body over 24 h if only two doses are taken.

With only two doses the value of our new parameter is $N = 2$. All other parameters remain the same as in the previous example. We graph the results in Figure 6.49.

Note that there are now only two peaks for the two doses, after which the amount of ibuprofen in the body tends to 0. □

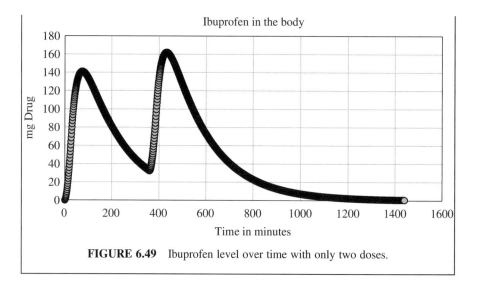

FIGURE 6.49   Ibuprofen level over time with only two doses.

In the next section we take a look at some practical considerations that affect how dose amounts and dose frequencies are set.

6.5.8   The Therapeutic Window

In the previous section we developed a very flexible dosing function and implemented it in Excel. With it we can model a wide range of dosing regimens. We have control over the dose, the frequency, the total number of doses, and the amount of time each dose takes to dissolve.

Prescriptions for different drugs come with varying directions for dosage amounts and frequency, and it is interesting to explore why this is. Why, for example, would it be better to take one 400-mg pill every 4 h rather than one 800-mg pill every 8 h? Why not 100 mg every hour?

There are lots of reasons for different dosing schedules, but one important consideration is the **therapeutic window** or **therapeutic range** for the drug. The therapeutic window is the range of drug levels in the body that produces the desired effect. If a drug level is too low, the desired effect will not be achieved, but if the drug level is too high, there could potentially be undesirable side effects. Knowing the therapeutic window helps determine how much of the drug should be administered with each dose and how often doses should be taken.

The therapeutic window is often reported in terms of plasma concentration so we need to know how to estimate the concentration of a drug in the blood plasma from the amount present in the body. In the next section we examine how to accomplish this.

6.5.9   Section Exercises

1   Create an Excel spreadsheet that will automatically calculate the elimination rate, r, from the percentage of drug eliminated over a specified period of time. The user

should be able to enter both the percentage eliminated and the time period over which it happens.

2 The half-life of the antidepressant Zoloft is approximately equal to 26 h (RxList, 2015).

   a. Determine the elimination rate, r, for Zoloft.

   b. Project the amount of Zoloft in the body over 1 week if a single 50-mg dose is administered.

3 Suppose you know that 95% of a drug is eliminated in a 24-h period. Determine the elimination rate, r, for the drug.

4 Suppose we know from research that 99% of a drug will be absorbed from the GI tract into the bloodstream within 45 min of ingestion. Find the absorption rate, α.

5 The antidepressant Zoloft reaches peak plasma concentration roughly 6 h after ingestion (RxList, 2015). Determine the absorption rate, α.

6 Suppose a patient takes 200-mg tablets of ibuprofen, two every 6 h. How much ibuprofen is left in the body 48 h after the first dose? Use minutes for the time unit and the parameter values from Example 6.19.

7 Suppose a drug with typical dissolution time is prescribed in pill form so that the patient takes a 300-mg pill once every 6 h. Find the dosing function.

8 Suppose a patient takes 100-mg caplets of ibuprofen, one every 4 h. How much ibuprofen is left in the body 24 h after the first dose if each caplet takes 30 min to dissolve? Use minutes for the time unit.

9 *Extension*: Suppose a patient takes one 100-mg pill of Zoloft every day. If each pill takes 30 min to dissolve, project the amount of Zoloft in the body over 1 week. Use hours as the time unit.

10 Using the parameters from Exercise 8, graph the amount of ibuprofen in the body over 24 h if only three doses are taken.

6.6 THE VOLUME OF DISTRIBUTION

A complication when examining the therapeutic range of a drug is that, like alcohol, it is not necessarily the *amount* of drug in the body that matters, but rather it is the *concentration* of drug in the body. A commonly used measure of drug concentration is the number of micrograms of the drug present in a milliliter of blood plasma, that is, the units of concentration are $\frac{\mu g \, drug}{ml \, plasma}$ or simply μg per ml when the context is understood. However, this measurement is equivalent to milligrams (mg) of the drug per liter (l) of plasma so we will use mg per l for our units.

   Unfortunately, estimating plasma concentration is not as simple as dividing the amount of drug in the body by the individual's plasma volume because in general drugs do not distribute only in the blood plasma. Rather they distribute throughout the body including the organs and body tissues such as muscle and fat. Thus for some drugs very little of the drug is actually found in the plasma. Still, it is much easier to measure plasma concentration with a laboratory test than, say, the concentration of drug in the liver. For this reason plasma concentration is still the preferred method of measuring drug levels in the body. From the plasma concentration the concentration at other sites throughout the body can be estimated based on the known properties of the drug.

The fact that not all of a drug will end up in the plasma introduces a complication for our model: we wish to estimate the plasma concentration, but we cannot simply divide the amount of drug in the body by the amount of plasma in the body. If we did we would get a potentially serious overestimation because most of the drug may in fact be elsewhere. What we need is a way to estimate the proportion of a drug that is actually in the plasma. Once we have that we will be able to estimate the plasma concentration with our model.

In order for us to estimate the proportion of a drug that ends up in the plasma, we introduce a fundamental pharmacological concept known as the **volume of distribution**. The volume of distribution is not an actual, physical volume in the body. Instead it refers to the theoretical volume of plasma that would be necessary in order for the *measured* plasma concentration to occur *if all of the drug were in fact in the plasma*. If only half of a drug dose ends up in the plasma, then the volume of distribution would be twice the actual plasma volume.

If we know the actual volume of plasma in a body and we also know the volume of distribution, then the proportion of the drug that ends up in the plasma can be determined by

$$\text{Proportion in plasma} = \frac{\text{Actual plasma volume}}{\text{Volume of distribution}}.$$

The good news is that we can estimate actual plasma volume based on sex and body weight, and the volume of distribution is a commonly reported property of any drug.

Example 6.23: Find the proportion of ibuprofen in the body that actually ends up in the plasma.

A variety of sources report the volume of distribution of ibuprofen as between 0.10 and 0.14 l per kg of body weight. We will use the commonly given 0.12 l per kg. We also use the fact that humans have actual plasma volume equal to approximately 0.04 l per kg body weight (Spruill et al., 2014). Thus the proportion of ibuprofen that is expected in the plasma is given by

$$\text{Proportion in plasma} = \frac{\text{Actual plasma volume}}{\text{Volume of distribution}} = \frac{0.04 \text{ l per kg}}{0.12 \text{ l per kg}} \approx 0.33.$$

We see that roughly one third of ibuprofen will actually be present in the plasma. □

We modify our Excel drug elimination model to include the volume of distribution, denoted V_D, as a user input parameter, and we have Excel calculate the proportion of drug found in the plasma, denoted by ρ. The resulting setup with the formula for ρ displayed is given in Figure 6.50.

	A	B	C	D	E	F	G	H
1	Two-Compartment Drug Model for Plasma Concentration							
2								
3	Drug half-life, $T_{1/2}$ =		120	(minutes)	Volume of dist., V_d =		0.12	(l/kg)
4	Elimination rate, r =		0.005760		Prop. in plasma, ρ =		=0.04/$G$3	
5	Absorption rate, a =		0.036					
6	Dosage amount, d =		200.0	(mg)				
7	Dose frequency, f =		6.0	(hours)				
8	Dissolution time, δ =		30.0	(minutes)				
9	Total doses, N =		2					

FIGURE 6.50   Excel calculation of plasma proportion from volume of distribution.

In the next section we continue our examination of ibuprofen.

6.6.1   Section Exercises

1 The antidepressant Zoloft (sertraline) is widely distributed throughout the body tissues and as a result has a large volume of distribution: 20 l per kg. Determine the proportion of Zoloft that is found in the plasma.

2 Find the volumes of distribution for three different drugs of your choosing. Use the volumes you find to determine the proportion of each drug that ends up in the plasma.

3 Suppose we know 25% of a particular drug ends up in the plasma. Determine this drug's volume of distribution.

4 Suppose we know 60% of a particular drug ends up in the plasma. Determine this drug's volume of distribution.

5 A patient weighing 175 pounds has a plasma concentration of 10 mg per l. Determine how much drug in total is in the patient's body if the volume of distribution for the drug is 0.16 l per kg.

6.7   COMMON DRUGS

In this final section we apply our previous work to constructing models for plasma concentration for some common drugs, in particular ibuprofen and caffeine.

6.7.1   Ibuprofen

Ibuprofen is a pain reliever, specifically a nonsteroidal anti-inflammatory drug (NSAID), that has a very wide therapeutic window: there is a lot of room between the drug level that produces pain relief and the drug level that is considered toxic.

The minimum effective level for ibuprofen is 10–50 mg per l with toxicity occurring at levels over 200 mg per l (ARUP, 2015).

In order for us to discern the concentration of ibuprofen in a specific individual for a given dose, we first need to know how much blood plasma she or he has. As we noted in the previous section, on average humans have about 0.04 l of plasma per kilogram of body weight. Thus we need to modify our Excel drug elimination spreadsheet to include body weight as a user input.

First we modify our Excel spreadsheet so that the user can enter their own body weight and have Excel calculate their plasma volume. The setup with formula for plasma volume displayed is given in Figure 6.51.

	A	B	C	D	E	F	G	H
1	Two-Compartment Drug Model for Plasma Concentration							
2								
3	Drug half-life, $T_{1/2}$ =	120	(minutes)		Volume of dist., V_d =		0.12	(l/kg)
4	Elimination rate, r =	0.005760			Prop. in plasma, ρ =		0.33	
5	Absorption rate, α =	0.036			Body weight =		200	(pounds)
6	Dosage amount, d =	200.0	(mg)		Plasma volume =		=($G$5/2.2046)*0.04	
7	Dose frequency, f =	6.0	(hours)					
8	Dissolution time, δ =	30.0	(minutes)					
9	Total doses, N =	2						

FIGURE 6.51 Calculating plasma volume from body weight.

To get Excel to calculate drug plasma concentration in addition to the total amount of drug in the body, we add a new column for plasma concentration where we multiply the amount of drug in the body by the proportion in the plasma and then divide by the calculated plasma volume. The result is shown in Figure 6.52.

	A	B	C	D	E	F	G	H
1	Two-Compartment Drug Model for Plasma Concentration							
2								
3	Drug half-life, $T_{1/2}$ =	120	(minutes)		Volume of dist., V_d =	0.12		(l/kg)
4	Elimination rate, r =	0.005760			Prop. in plasma, ρ =	0.33		
5	Absorption rate, α =	0.036			Body weight =	200		(pounds)
6	Dosage amount, d =	200.0	(mg)		Plasma volume =	3.6		(liters)
7	Dose frequency, f =	6.0	(hours)					
8	Dissolution time, δ =	30.0	(minutes)					
9	Total doses, N =	2						
10								
11	t minutes	$D(t)$		$GI(t)$		$B(t)$	$PC(t)$ (mg/l)	
12	0	6.6666667		6.7		0	=D12*$G$4/$G$6	
13	1	6.6666667		13.1		0.2	0.02	

FIGURE 6.52 Calculating plasma concentration from total amount of drug in the body.

In the next example we compare typical dosage instructions for over-the-counter ibuprofen with the stated therapeutic window.

Example 6.24: Lauren has a headache and is going to take some ibuprofen. The dosage instructions for 200-mg ibuprofen tablets are to take 1–2 tablets every 4–6 h, not to exceed 6 tablets in a 24-h period. Compare our model projections for concentration with the therapeutic window for Lauren if she weighs 130 pounds.

Recall that the low end of the therapeutic window was given as a range, 10–50 mg per l, while the upper end was given as 200 mg per l. To make these levels easy to reference, we include them on our graph as horizontal lines. In Excel we accomplish this by making an additional column for each level where every entry in the column is the stated value. In this way when the column is added to our graph, we get a horizontal line.

We examine the ibuprofen concentration over 24 h for two cases: one where Lauren takes 1 tablet every 6 h and one where she takes 2 tablets every 4 h.

Taking 1 tablet every 6 h means our parameters will be $d = 200$ and $f = 6$. Figure 6.53 shows Lauren's ibuprofen plasma concentration. Note that her ibuprofen concentration remains above the 10-mg per l mark a little over half the time she is taking the tablets. This puts Lauren just barely into the therapeutic window for a little over half the time she is taking the tablets, and this may or may not be enough to relieve her headache.

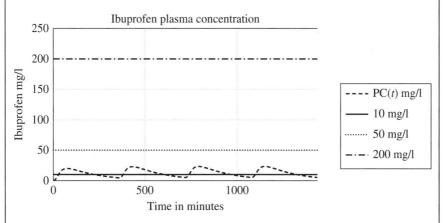

FIGURE 6.53   Ibuprofen plasma concentration over time for 1 tablet every 6 h.

Next we examine Lauren's concentration if she takes 2 tablets every 4 h. Here we have to set $d = 400$, $f = 4$, and $N = 3$. The value for N is set to avoid taking more than 6 tablets in a 24-h period. Figure 6.54 displays the results. Now Lauren's

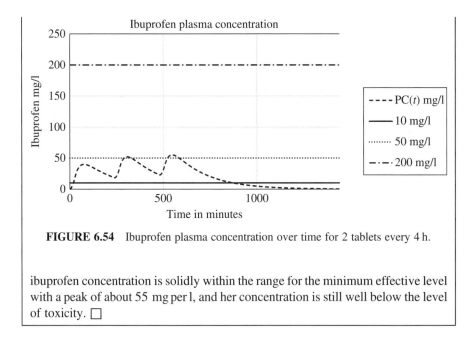

FIGURE 6.54   Ibuprofen plasma concentration over time for 2 tablets every 4 h.

ibuprofen concentration is solidly within the range for the minimum effective level with a peak of about 55 mg per l, and her concentration is still well below the level of toxicity. □

For ibuprofen it appears based on the projections of our model that the recommended doses are fairly conservative. We caution the reader against taking any dose beyond that which is recommended. For one thing we always have to remember that our model is based on many simplifying assumptions that are unlikely to hold in individual cases, and we note that there may be other reasons for the recommended dosages other than the plasma concentration level. Drug interactions, overall health of the user, and other factors all go into what is considered a safe dosage for over-the-counter use. Our model results are interesting and reasonable, but they should not lead us to ignore the sensible limits given by the FDA.

For many over-the-counter drugs, we may not know the therapeutic window, but we can use our model to deduce it from the given directions. The reader is invited to do so in the exercises.

6.7.2   Caffeine

Caffeine is a stimulant commonly found in a wide variety of foods and beverages such as coffee, tea, soft drinks, energy drinks, and chocolate; it is also found in certain medications. Though perhaps not always thought of as a drug, it is the most commonly used mood-altering drug in the world (WebMD, 2015). In low to moderate doses, its effects include increased alertness, improved concentration, and improved athletic performance (Smith, 2002; Spriet & Graham, 2015). However, an overdose of caffeine can lead to serious adverse effects such as increased blood pressure, rapid heartbeat, vomiting, and death (Healthline, 2015). It is one of few legal performance-enhancing drugs widely used by endurance athletes, though urinary concentration levels above 12 mg per l are banned by the International Olympic Committee (IOC) (Spriet, 1995).

In addition to the many foods and beverages that contain caffeine, it has also become possible in recent years to purchase pure caffeine in powdered form over the Internet. Using caffeine in this form is a bad idea. One teaspoon of powdered caffeine contains the equivalent of the caffeine found in 25 cups of coffee. Powdered caffeine has been linked to at least two deaths, and the FDA has issued a warning against its use (FDA, 2014).

When caffeine is ingested, it passes rapidly into the bloodstream through the small intestine. It is metabolized in the liver at a rate that is proportional to the amount of caffeine in the body; thus caffeine undergoes first-order elimination.

In order to use our two-compartment model to project caffeine levels, we need to find the relevant parameters for caffeine. We start with the elimination rate.

The half-life of caffeine can vary quite a bit from person to person, but it will generally be between 2.5 and 4.5 h (Fredholm B. B. et al., 1999). For smokers the half-life will be reduced by 30–50% compared to nonsmokers, and for women taking oral contraceptives the half-life can as much as double (Fredholm B. B. et al., 1999). Deducing the elimination rate from half-life is something we have done before and our Excel model now does it automatically. Using 3 h, or 180 min, for the half-life gives us an elimination rate of $r = 0.003843$. Next we find the absorption rate from the GI tract into the blood.

Within 45 min of ingestion, approximately 99% of caffeine is absorbed into the blood (Marks & Kelly, 1973). A calculation similar to finding the rate of elimination from the half-life is required to find the rate, α, at which caffeine moves from GI tract to the bloodstream. Specifically we need to solve the equation $(1-\alpha)^{45} = 0.01$, and the result is $\alpha = 0.0973$. In other words, about 10% of ingested caffeine is absorbed into the blood each minute, a very rapid rate.

Once we know α and r, we are just about ready to model caffeine. The dose depends on the source of the caffeine. A strong 8-oz. cup of coffee, for example, contains about 150 mg of caffeine. The reader will find it easy to look up similar values for other beverages such as tea or energy drinks. Since we assume that the caffeine ingested will be in beverage form, we interpret the dissolution time to be the time it takes someone to finish a drink. We assume a value of 15 min here. The frequency we understand to be the time between drinks and N will be the total number of drinks consumed in a day.

The volume of distribution of caffeine is approximately 0.40–0.75 l per kg of body weight (Abernethy, 1985; Giardinia, 2015). We use 0.50 here, which gives us $\rho = \dfrac{0.04}{0.50} = 0.08$.

Finally we note that the IOC limit for caffeine concentration is given as a urinary concentration. According to Birkett and Miners (1991), plasma concentration of caffeine tends to be roughly 1.4 times that of the urinary concentration. Thus the IOC limit of 12 mg per l for caffeine urine concentration is equivalent to a plasma concentration limit of $12 \times 1.4 = 16.8$ mg per l.

We put all of this together in the following example.

Example 6.25: Suppose Mia is a 110-pound female Olympic athlete who consumes three strong 8-oz. cups of coffee all at once, finishing the last one an hour before her competition. Will Mia's caffeine plasma concentration exceed the IOC limit?

For this example we need the amount of caffeine in a cup of strong brewed coffee. A quick Internet search reveals that this can vary quite a bit but a value of 150 mg is reasonable; thus $d = 150$. We use $\delta = 15$ min for the time it takes to drink one cup. Because she consumes all of the cups in succession, there are exactly 15 min between doses and we have $N = 3$ and $f = 0.25$. We graph Mia's plasma concentration and show the result in Figure 6.55. Note that Mia's plasma concentration remains below the IOC limit. However, if she were to have a fourth cup, she would be at some risk of exceeding the limit. Four cups of coffee in quick succession is a lot of coffee. This indicates that under normal circumstances athletes do not have to be overly concerned with exceeding the IOC limit. To do so would almost surely indicate intentional, excessive intake. $\square$

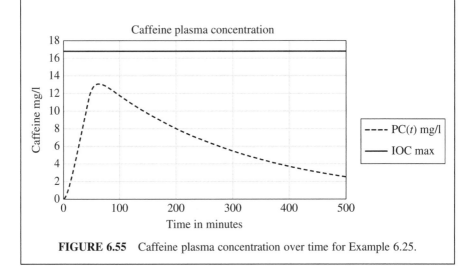

FIGURE 6.55 Caffeine plasma concentration over time for Example 6.25.

We noted earlier that caffeine does enhance athletic performance, but it does so at levels that are typically well below the IOC threshold. In fact, a dose of approximately 1.4–2.7 mg per pound of body weight is sufficient to obtain performance benefits, while higher doses confer no additional benefit (Goldstein et al., 2010). Moreover, a dose of 4–5.9 mg per pound of body weight is necessary to exceed the IOC limit (Goldstein et al., 2010), so it is in general not to an athlete's advantage to risk going over the limit. In the previous example, Mia had the equivalent of a dose of 3.5 mg per pound of body weight, more than enough to reap all of caffeine's performance benefits, and she was still below the IOC limit.

6.7.3 Exercises

1 Jessie is sore from a long run and is going to take some ibuprofen. If Jessie weighs 120 pounds and takes 2 200 mg tablets every 4 h, graph her ibuprofen plasma concentration along with horizontal lines for the therapeutic window.

2 Craig weighs 300 pounds and is going to take ibuprofen to relieve a back ache. Will taking 1 200 mg tablet every 6 h be enough for him to experience relief from the pain? Explain.

3 Suppose Steve is a 210-pound Olympic athlete who consumes four strong 8-oz. cups of coffee consecutively, finishing the last 30 min before his competition. Will Steve's caffeine plasma concentration exceed the IOC limit? Explain.

4 How many Starbucks grande lattés can Mia from Example 6.25 have before she is over the IOC limit for caffeine?

5 We based our elimination rate for caffeine on a value for the half-life that was near the middle of the given half-life range. The range was given as 2.5–4.5 h, and we used 3 h to find r. Investigate how much difference it would make in Example 6.25 if the half-life were (a) 2.5 h and (b) 4.5 h.

6 We based our proportion of caffeine in the plasma, $\rho = 0.08$, on a volume of distribution of 0.50, which is in the given range of approximately 0.40–0.75 l per kg. Investigate how much difference it would make in Example 6.25 if we used (a) 0.40 and (b) 0.75 for the volume of distribution.

7 *Extension*: Suppose Mia from Example 6.25 is taking oral contraceptives. Rework Exercise 4 taking into account this new information.

8 The NCAA competition limit on caffeine urine concentration is 15 mg per l. Determine how many of your favorite caffeinated beverage you could have 3 h before a competition and still be under the NCAA limit.

9 Use our Excel model to estimate how many 200 mg tablets of ibuprofen a 200-pound man can ingest at once before his ibuprofen plasma concentration reaches toxic levels.

10 *Extension*: Estimate the therapeutic window for Zoloft by modeling standard dosage recommendations.

11 *Extension*: Suppose a caffeine plasma concentration of 4 mg per l is enough to disturb your sleep. Estimate how late in the day you could have a grande Starbucks coffee and not disturb your sleep.

7

RANKING METHODS

In this chapter we explore how discrete dynamical systems can be used in ranking methods. We focus on a particular type of model known as a **Markov model**, named for the Russian mathematician Andrei Markov (1856–1922). In a Markov model members are divided into classes, or states, and during each time step members move from one state to another, or they may remain in the same state. Members may not "leave the model." The percentages or proportions of members moving from one state to any other state is fixed and does not depend on time. Finally, what happens in a given time step does not depend on what has happened in previous time steps: all that is important is the present state, not how that state came about.

7.1 INTRODUCTION TO MARKOV MODELS

In this section we introduce Markov models via the concrete example of a truck rental company fleet. We examine the models numerically with Excel, algebraically in finding equilibrium distributions, and graphically by examining the flow diagram. As we show in the following, we can deduce a great deal about a Markov model from its flow diagram, including whether or not the model will have a stable distribution.

7.1.1 Truck Rentals

Suppose a truck rental company has locations in Birmingham, Alabama; Columbia, South Carolina; and Dallas, Texas. The company permits one-way rentals

Models for Life: An Introduction to Discrete Mathematical Modeling with Microsoft® Office Excel®,
First Edition. Jeffrey T. Barton.
© 2016 John Wiley & Sons, Inc. Published 2016 by John Wiley & Sons, Inc.

and knows based on past experience that during a typical week the following truck movements occur:

1. Of the trucks that start in Birmingham:
   a. 60% stay in Birmingham
   b. 15% travel to Columbia
   c. 25% travel to Dallas
2. Of the trucks that start in Columbia:
   a. 45% stay in Columbia
   b. 15% travel to Birmingham
   c. 40% travel to Dallas
3. Of the trucks that start in Dallas:
   a. 70% stay in Dallas
   b. 20% travel to Birmingham
   c. 10% travel to Columbia

Example 7.1: Suppose the truck rental company has a total fleet of 150 trucks: 60 trucks in Birmingham, 40 trucks in Columbia, and 50 trucks in Dallas. Determine (i) how many trucks will end up in each city after 13 weeks, and (ii) how the company should allocate its fleet to maintain constant inventory at each location.

Our time steps will be weeks, and we let $B(t)$ be the number of trucks in Birmingham, $C(t)$ the number of trucks in Columbia, and $D(t)$ the number of trucks in Dallas after t weeks. We represent truck movements with a flow diagram as in Figure 7.1. The arrows indicate the percentage of trucks moving from one city to another each week.

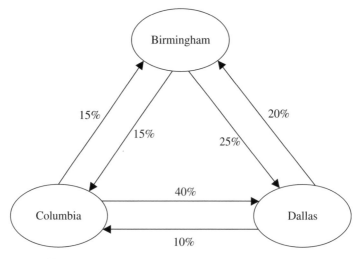

FIGURE 7.1 Percentages of trucks moving between cities.

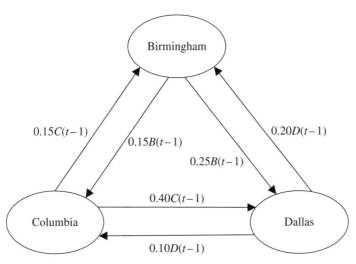

FIGURE 7.2 Numbers of trucks moving between cities.

Next we label the diagram with the numbers of trucks moving from city to city as in Figure 7.2.

We construct the corresponding DDS as usual with inward-pointing arrows representing additions and outward-pointing arrows subtractions. The truck rental DDS is given by

$$B(t) = B(t-1) - 0.25B(t-1) - 0.15B(t-1) + 0.15C(t-1) + 0.20D(t-1)$$
$$C(t) = C(t-1) - 0.40C(t-1) - 0.15C(t-1) + 0.15B(t-1) + 0.10D(t-1)$$
$$D(t) = D(t-1) - 0.20D(t-1) - 0.10D(t-1) + 0.25B(t-1) + 0.40C(t-1).$$

Implementing the model in Excel requires a column for the number of trucks in each of the cities and some care in referring to the appropriate cells. We have six parameters for the model, one for each of the arrows in the flow diagram representing truck movements between two cities. Our initial Excel setup with the formula for Birmingham displayed is given in Figure 7.3.

	A	B	C	D	E	F
1	Truck Rental Markov Model					
2						
3	From Birmingham to Columbia =				15.0%	
4	From Birmingham to Dallas =				25.0%	
5	From Columbia to Birmingham =				15.0%	
6	From Columbia to Dallas =				40.0%	
7	From Dallas to Birmingham =				20.0%	
8	From Dallas to Columbia =				10.0%	
9						
10		*t*	Birmingham	Columbia	Dallas	
11		0	60	40	50	
12		1	=B11-$E$3*B11-$E$4*B11+$E$5*C11+$E$7*D11			

FIGURE 7.3 Excel setup for truck rental example.

With 60 trucks initially in Birmingham, 50 in Dallas, and 40 in Columbia, after the first week we have 52 trucks in Birmingham, 32 in Columbia, and 66 in Dallas. If we drag the model down to week 13, we see that the number of trucks in each city stabilizes at about 48 trucks in Birmingham, 27 in Columbia, and 75 in Dallas. Figure 7.4 shows the Excel output with most rows hidden.

	A	B	C	D	E
1	Truck Rental Markov Model				
2					
3	From Birmingham to Columbia =				15.0%
4	From Birmingham to Dallas =				25.0%
5	From Columbia to Birmingham =				15.0%
6	From Columbia to Dallas =				40.0%
7	From Dallas to Birmingham =				20.0%
8	From Dallas to Columbia =				10.0%
9					
10	t	Birmingham	Columbia	Dallas	
11	0	60	40	50	
12	1	52	32	66	
23	12	48	27	75	
24	13	48	27	75	

FIGURE 7.4   Excel projections for numbers of trucks in each city.

No matter how much further we drag the model down, this long-term distribution of trucks will not change. It also answers the second question in the example. If we want to maintain a constant inventory in each city, we need to begin with 48 trucks in Birmingham, 27 in Columbia, and 75 in Dallas.

We conclude the example by noting that having a constant number of trucks in each city from week to week does not imply that we have the same trucks in each city from week to week. Trucks are still moving among the cities, it is just that the movements balance out to keep the numbers the same. □

It is interesting to note that the long-term distribution of trucks in the previous example does not depend on how we distribute our 150 trucks initially. Even if we start with all of the trucks in Columbia, for example, eventually the distribution settles down to what we found in Example 7.1.

We should expect that changing the total number of trucks in the fleet will change the final numbers in each city. What is interesting, though, is that the **distribution** of trucks does not change: we will end up with the same percentages of trucks in each city regardless of how many trucks we start with. A distribution that remains constant over time is known as an **equilibrium distribution**, and if the model tends to this distribution regardless of the initial values, it is called a **stable distribution**. When all of the percentages are strictly positive, we call a distribution **positive**. Generally speaking the requirement that the stable distribution be positive lets us know that we will get physically meaningful results from the model.

In the next example we use Excel to verify that the truck fleet has a positive stable distribution.

Example 7.2: Show that the long-term distribution of trucks in Example 7.1 remains the same regardless of how many trucks there are initially.

As it stands now our Excel model keeps track of how *many* trucks are in each city but not what *percentage* of trucks is in each city. Thus to work this example, we must set up additional columns in our Excel spreadsheet where we keep track of the percentage of trucks in each location. For example, to find the percentage of trucks that are in Dallas at time t, we need to calculate the number of trucks in Dallas divided by the total number of trucks: $\dfrac{D(t)}{B(t)+C(t)+D(t)}$. Figure 7.5 shows the Excel setup with the formula for Birmingham displayed.

	A	B	C	D	E	F	G
1	Truck Rental Markov Model						
2							
3	From Birmingham to Columbia =				15.0%		
4	From Birmingham to Dallas =				25.0%		
5	From Columbia to Birmingham =				15.0%		
6	From Columbia to Dallas =				40.0%		
7	From Dallas to Birmingham =				20.0%		
8	From Dallas to Columbia =				10.0%		
9							
10	t	Birmingham	Columbia	Dallas	Birm. %	Col. %	Dal. %
11	0	60	40	50	=B11/(B11+C11+D11)		

FIGURE 7.5   Excel setup for calculating distribution of trucks.

With our new columns formatted as percentages, we copy the columns down to find the long-term stable distribution. We end up with about 32% of the trucks in Birmingham, 18% in Columbia, and 50% in Dallas. All percentages are positive so we have a positive stable distribution. By experimenting with different initial numbers of trucks, we can confirm that these percentages do not change. □

The stable distribution can provide useful information for planning purposes as we see in the next example.

Example 7.3: Suppose the truck rental company is considering expanding its fleet to 250 total trucks. Determine the company's storage needs at each city's facility.

Because the fleet has a stable distribution, the rental company knows that in the long term it can expect about $250 \times 32\% = 80$ trucks to end up in Birmingham, $250 \times 18\% = 45$ to end up in Columbia, and $250 \times 50\% = 125$ trucks to end up in Dallas. Based on these requirements the company knows approximately how many trucks their facility in each city must house. Furthermore the company can assess whether or not their current facilities are sufficient or if there is a need to build additional capacity. □

Not all Markov models have a positive stable distribution. In the next section we show how to determine when a model will have such a distribution.

7.1.2   Existence of a Positive Stable Distribution

The existence of a positive stable distribution is a common feature of Markov models, though it is not guaranteed. Two characteristics are sufficient in order to guarantee that a Markov model will have a positive stable distribution. The first is that the flow diagram for the model must be **irreducible**. A flow diagram is irreducible if it is possible to travel between any two states (not necessarily in one step). The second is trickier to define, and it is that the flow diagram must be **aperiodic**.

To define what we mean by an aperiodic flow diagram, we first must define what we mean by an aperiodic state. The **period** of a state in a flow diagram is the greatest common divisor of the numbers of steps required for all possible return trips to the state. A state in a flow diagram is said to be **periodic** if its period is any number other than 1. If the period of a state is 1, the state is said to be **aperiodic**. We then define an aperiodic flow diagram as one for which all states are aperiodic. (For a more rigorous treatment of the definition of periodic and aperiodic, see any introductory graph theory text.) Below we present examples of flow diagrams to help clarify the definitions.

Example 7.4: Figure 7.6 shows a flow diagram that is irreducible but not aperiodic.

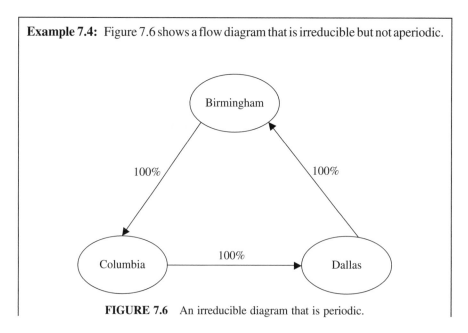

FIGURE 7.6   An irreducible diagram that is periodic.

It is irreducible because it is possible to travel from any city to any other city. The diagram is not aperiodic because in order to return to any starting state, it is necessary to travel around a forced loop of three states: any trip beginning and ending at the same state must consist of a number of steps that is a multiple of 3. Thus every state is periodic with period equal to 3. □

A nice way to tell if the state is aperiodic is if it is possible to simply remain in the state. If it is possible to remain in a state, then the period for that state is automatically 1. Thus if we have a flow diagram where it is possible to remain in every state, the diagram will be aperiodic.

Example 7.5: The flow diagrams in Figure 7.7 and in Figure 7.8 are not irreducible.

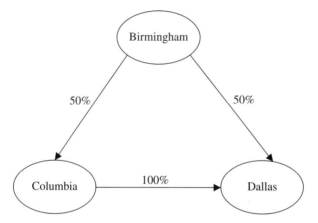

FIGURE 7.7　A diagram that is not irreducible.

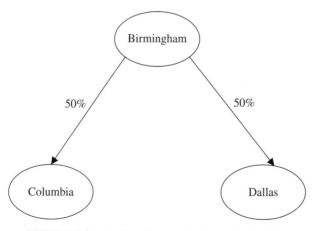

FIGURE 7.8　Another diagram that is not irreducible.

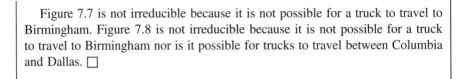

Figure 7.7 is not irreducible because it is not possible for a truck to travel to Birmingham. Figure 7.8 is not irreducible because it is not possible for a truck to travel to Birmingham nor is it possible for trucks to travel between Columbia and Dallas. □

Before getting to our main result, we state without proof the following fact about irreducible flow diagrams: for any irreducible flow diagram, if one state isaperiodic, then all states are aperiodic. Thus to determine whether or not an *irreducible* flow diagram is aperiodic, we need only find *one* state that we know is aperiodic.

The following theorem will be our main result in this chapter. Its proof, which we omit here, can be found in any introductory text on Markov models. By a **finite** Markov model, we mean one with a finite number of states.

Theorem 7.1   If a finite Markov model's flow diagram is both irreducible and aperiodic, then the model will have a positive stable distribution.

Note that in Figure 7.1 for the truck rental example, it is possible to travel from any of the three cities to any of the others. It is also possible for a truck to remain in any city. Thus for the truck rental example, we have an irreducible, aperiodic flow diagram, and we are thus guaranteed a positive stable distribution, that is, one that does not depend on the initial location or number of trucks.

The model represented by Figure 7.6 will not have a stable distribution because all trucks continually make a forced loop through the three cities. Unless we start with equal proportions of trucks in each city, the proportions in each city change every time step.

The model represented by Figure 7.7 will have a stable distribution because all trucks end up in Dallas regardless of where they start. However, the distribution is not positive because 0% of the trucks end up in Birmingham and Columbia.

The model represented by Figure 7.8 will not have a stable distribution because the final percentages in each city depend on the initial distribution of trucks. If all trucks start in Birmingham, then we end up with 50% in Columbia and 50% in Dallas. However, if all trucks start in Columbia, we end up with 100% in Columbia. The distribution is also not positive because we always end up with 0% of the trucks in Birmingham.

Next we show how to determine equilibrium distributions algebraically.

7.1.3 Equilibrium Distributions

In Chapter 1 if we knew we had a stable equilibrium, we could find it in one of two ways. We could look at the long-term values projected by Excel, or we could solve for the equilibrium value algebraically. With a stable equilibrium both methods must produce the same result.

The same is true when dealing with Markov models and long-term distributions. If we know that we will get a stable positive distribution, then we can find it with Excel or we can find it algebraically.

Any finite Markov model with an irreducible flow diagram will have an equilibrium distribution that can be determined algebraically. It is when the flow diagram is also aperiodic that we know the equilibrium distribution is in fact the positive stable distribution guaranteed by Theorem 7.1. We illustrate this fact in the following example.

Example 7.6: Find the equilibrium distribution for the model in Example 7.1.

We first must find the equilibrium point (B^*, C^*, D^*) at which the number of trucks in each city does not change. For the truck rental fleet in the previous section, this means solving the following system for (B^*, C^*, D^*):

$$B^* = B^* - 0.25B^* - 0.15B^* + 0.15C^* + 0.20D^*$$
$$C^* = C^* - 0.40C^* - 0.15C^* + 0.15B^* + 0.10D^*$$
$$D^* = D^* - 0.20D^* - 0.10D^* + 0.25B^* + 0.40C^*.$$

Our method is to get all of the variables in terms of one of the others and then to find the value for that one variable. Here we work to write B^* and C^* in terms of D^*. First we collect like terms to get

$$0 = -0.40B^* + 0.15C^* + 0.20D^*$$
$$0 = -0.55C^* + 0.15B^* + 0.10D^*$$
$$0 = -0.30D^* + 0.25B^* + 0.40C^*.$$

The first equation allows us to write B^* in terms of C^* and D^*:

$$0.40B^* = 0.15C^* + 0.20D^*$$
$$B^* = 0.375C^* + 0.50D^*.$$

Next we substitute the expression for B^* into the second equation and simplify. This gives

$$0 = -0.55C^* + 0.15B^* + 0.10D^*$$
$$= -0.55C^* + 0.15(0.375C^* + 0.5D^*) + 0.10D^*$$
$$= -0.49375C^* + 0.175D^*.$$

Thus we can write C^* in terms of D^*: $C^* = \dfrac{0.175}{0.49375}D^* = 0.3544D^*$. This in turn allows us to write B^* in terms of D^* as well:

$$B^* = 0.375C^* + 0.50D^* = 0.375(0.3544D^*) + 0.50D^* = 0.6329D^*.$$

We have found that our equilibrium values for the number of trucks in each city are related by

$$(B^*, C^*, D^*) = (0.6329D^*, 0.3544D^*, D^*).$$

The last step is to solve for D^*, which we do by recalling that the total number of trucks in our fleet has to be 150. Thus at equilibrium we must have

$$B^* + C^* + D^* = 150$$
$$0.6329D^* + 0.3544D^* + D^* = 150$$
$$1.9873D^* = 150$$
$$D^* = 75.48.$$

The result is that we should end up with about 75 trucks in Dallas long term, and we note that this is the same value we found before with Excel.

To finish up we need to calculate the number of trucks in Birmingham and Columbia and note that these match up with our previous work as well:

$$B^* = 0.6329D^* = 0.6329 \cdot 75.48 = 47.77$$
$$C^* = 0.3544D^* = 0.3544 \cdot 75.48 = 26.75.$$

Once we have the equilibrium point, we can find the equilibrium distribution. For the percentage of trucks that end up in Dallas, we calculate

$$\frac{D^*}{B^* + C^* + D^*} = \frac{D^*}{0.6329D^* + 0.3544D^* + D^*}.$$

Note that we can cancel the D^* in the numerator and denominator to get

$$\frac{D^*}{B^* + C^* + D^*} = \frac{1}{0.6329 + 0.3544 + 1} = \frac{1}{1.9873} = 0.5032 \approx 50\%.$$

As shown by the cancellation of D^*, we did not need the number of trucks to get this result: it is independent of how many trucks we have and where they start. The computations for Birmingham and Columbia are similar, and all agree with our previous Excel result. $\square$

In this example the equilibrium distribution of trucks found from the equilibrium point turned out to be the same as the long-term distribution of trucks from our Excel work. Once again, this is because the flow diagram in Example 7.1 is both irreducible and aperiodic: Theorem 7.1 guarantees a positive stable distribution.

In the case of an unstable equilibrium in Chapter 1, the equilibrium value we found algebraically did not turn out to be the long-term value for the model. The same phenomenon can occur with an equilibrium distribution. The next example shows that it is possible for the equilibrium distribution to not be the long-term distribution of a Markov model.

Example 7.7: Determine the equilibrium distribution for the model represented by the flow diagram in Figure 7.6.

Recall that the diagram in Figure 7.6 in Example 7.4 is irreducible but not aperiodic, and it will not have a stable distribution. As we see in the following, it does still have an equilibrium distribution.

The DDS for the model represented in Figure 7.6 is given by

$$B(t) = B(t-1) - B(t-1) + D(t-1)$$
$$C(t) = C(t-1) - C(t-1) + B(t-1)$$
$$D(t) = D(t-1) - D(t-1) + C(t-1).$$

This simplifies to give

$$B(t) = D(t-1)$$
$$C(t) = B(t-1)$$
$$D(t) = C(t-1).$$

Thus we need to solve the following for (B^*, C^*, D^*):

$$B^* = D^*$$
$$C^* = B^*$$
$$D^* = C^*.$$

The last equation implies that the numbers of trucks in each city must all be equal, which means the equilibrium distribution must occur when $\frac{1}{3}$ of the trucks are in each city. Thus if we begin with $\frac{1}{3}$ of the trucks in each city, the proportion in each city will remain at $\frac{1}{3}$. Otherwise the proportion in each city will change with every time step. □

In the next section we give an alternative way of interpreting Markov flow diagrams.

7.1.4 Interpreting Markov Flow Diagrams

A useful way to interpret a Markov flow diagram is to change our point of view from a global one to an individual one. In our truck rental example, the global view is that every time step 20% of trucks in Dallas will move on to Birmingham. However, we do not know which trucks will move from Dallas to Birmingham. An equivalent interpretation would be that from an individual truck's perspective if it is in Dallas, then it has a 20% chance of heading to Birmingham during the next time step. Similarly, rather than taking the global view that 40% of trucks in Columbia head to Dallas during the next time step, we could say that an individual truck in Columbia has a 40% chance of being taken to Dallas during the next time step. Thus if we imagine ourselves as a truck in the rental fleet, then from week to week we move randomly among the three cities with our movements governed by the chances of traveling from one city to the next given in our flow diagram. This kind of interpretation of a Markov flow diagram is known as a **random walk**.

Generally speaking viewing a Markov model as a random walk means experiencing the model on an individual level. At each oval the outward-pointing arrows show us the possible moves we can make, and we can interpret the attached percentages as the percent chance that we will take that path. Thus we can view our every step as a random one governed by the chances that we will move on to another oval or remain in the current one. Once at the next oval the individual is faced with another random movement in the subsequent step. The individual is "randomly walking" from oval to oval in the flow diagram.

With this interpretation we also view our stable distribution differently. From a global perspective, the stable distribution represents the percentages of the fleet that end up in each city. We need to remember though that a stable distribution does not mean that trucks have quit moving. Trucks are still moving among cities every week, it is just that the movements balance out to keep the percentages the same. From a random walk perspective, we view the stable distribution as the relative amount of time an individual truck will spend in each city. For Example 7.2 we expect any particular truck to spend an average of 50% of its weeks in Dallas, 32% in Birmingham, and 18% in Columbia.

In the remainder of this chapter, we use Markov models in the context of ranking systems, first for sports teams and then for web search results.

7.1.5   Section Exercises

1   Suppose a truck rental company has locations in Birmingham, Alabama; Columbia, South Carolina; and Dallas, Texas. The company permits one-way rentals and knows based on past experience that during a typical week the following truck movements occur:

Of the trucks that start in Birmingham:

- 50% stay in Birmingham
- 25% travel to Columbia
- 25% travel to Dallas
  Of the trucks that start in Columbia:
- 40% stay in Columbia
- 20% travel to Birmingham
- 40% travel to Dallas
  Of the trucks that start in Dallas:
- 80% stay in Dallas

- 15% travel to Birmingham
- 5% travel to Columbia
- **a.** Give a flow diagram for the model.
- **b.** Implement the model in Excel.
- **c.** If 200 trucks begin in each city, use Excel to determine the long-term numbers of trucks in each city.
- **d.** Use Excel to determine the long-term distribution of trucks in each city.

2 For the situation in Exercise 1, suppose the facilities in each city can hold 100 trucks each. Determine the largest number of trucks the company can have in its fleet.

3 Suppose a truck rental company has locations in Birmingham, Alabama; Columbia, South Carolina; Dallas, Texas; and Evansville, Indiana. The company permits one-way rentals and knows based on past experience that during a typical week the following truck movements occur:

    Of the trucks that start in Birmingham:
- 50% stay in Birmingham
- 25% travel to Columbia
- 15% travel to Dallas
- 10% travel to Evansville
    Of the trucks that start in Columbia:
- 40% stay in Columbia
- 14% travel to Birmingham
- 34% travel to Dallas
- 12% travel to Evansville
    Of the trucks that start in Dallas:
- 75% stay in Dallas
- 15% travel to Birmingham
- 5% travel to Columbia
- 5% travel to Evansville
    Of the trucks that start in Evansville:
- 65% stay in Evansville
- 10% travel to Birmingham
- 10% travel to Columbia
- 15% travel to Dallas
- **a.** Give a flow diagram for the model.
- **b.** Implement the model in Excel.
- **c.** If 200 trucks begin in each city, use Excel to determine the long-term numbers of trucks in each city.
- **d.** Use Excel to determine the long-term distribution of trucks in each city.

4 For the situation in Exercise 3, suppose the facilities in each city can hold 100 trucks each. Determine the largest number of trucks the company can have in its fleet.

5 For the situation in Exercise 1, find the largest fleet the company can accommodate if it can house 100 trucks in Birmingham, 50 trucks in Columbia, and 150 trucks in Dallas.

6 For the situation in Exercise 3, find the largest fleet the company can accommodate if it can house 60 trucks in Birmingham, 70 trucks in Columbia, 80 trucks in Dallas, and 60 trucks in Evansville.

7 Considering the flow diagram in Exercise 1, how do we know that we will end up with a stable distribution?

8 Considering the flow diagram in Exercise 3, how do we know that we will end up with a stable distribution?

9 Give an example of a four-city flow diagram for which there will be no positive stable distribution. What condition(s) of Theorem 7.1 does the diagram violate?

10 Algebraically determine the long-term distribution for the truck fleet in Exercise 1.

11 Interpret the flow diagram for Exercise 1 from a random walk point of view.

12 *Extension*: Suppose the truck rental company from Exercise 1 wishes to expand its fleet. It does so by adding two new trucks to the Dallas facility each week:

a. Give the flow diagram for the new situation.

b. Implement the change in Excel.

c. Compare and contrast the long-term model behavior with the behavior from Exercise 1.

13 Give an example of a flow diagram with six states that is not irreducible.

7.2 RANKING SPORTS TEAMS

A popular activity among sports analysts and fans is to produce rankings of who is the best team. In contexts where teams or individuals play only a small percentage of possible opponents such as in professional chess, Scrabble™, and tennis, mathematical ranking systems are especially important. Until recently college football relied on a composite of several different ranking systems, some of which use advanced mathematics, to decide which 2 teams should play in the national championship game. There are many different mathematical methods for producing rankings beyond those discussed here, and an excellent survey of many of them can be found in the book *Who's #1?* By Langville and Meyer (2012).

Ranking sports teams simply based on wins and losses or winning percentage can be problematic. In Major League Baseball where teams play a large number of games against a relatively high percentage of all possible opponents each season, using straight wins and losses makes some sense. For college football, though, where teams only play a small percentage of available teams and teams vary dramatically in their abilities, using wins and losses is insufficient. Strength of schedule, or the difficulty level of the teams played, must play a prominent role. A team that is undefeated against teams that all have losing records should not be given as much credit as a team that is undefeated against teams with winning records. The competition faced by the second team is much stiffer and so the second team should get more credit.

One method of ranking teams is to turn the idea of "getting more credit for good wins" into a mathematical process using a Markov model that tracks the "distribution of credit" among all teams. We begin with the assumption that each time a team defeats another, the losing team gives some proportion of its credit to the winning team. We set up the model so that a team with only one loss will give a relatively high proportion of its credit to the team that beat it, but a team with many losses will give a reduced proportion of credit to each of the teams that defeated it. In this way the winning team rightfully receives more credit from good wins (against teams with few losses) than from unimpressive wins (against teams with many losses). We allow teams to keep some of their credit (the equivalent of trucks remaining in a city), but bad teams get to keep less credit than good teams. Thus good teams will have to give away less of their total credit than bad teams.

Once we decide how to assign credit from losing teams to winning teams, our model will represent the "flow of credit" among the teams. Since credit must either be passed to a winning team or kept, the total amount of credit will not change and we will have a Markov model. By dragging the model down far enough, we will get a stable distribution of credit, and on the basis of that distribution, we will rank our teams according to who has the most credit: the team with the most credit will be number 1, and so on.

The passing of credit from 1 team to another is certainly more abstract than the idea of trucks moving from one city to another, but once we decide how exactly to assign credit, the mathematics is the same. We illustrate the entire process by considering a 5-team league where each team plays every other team once.

Consider 5 teams: the Bears, the Cardinals, the Dolphins, the Eagles, and the Falcons. Wins and losses from the entire season are organized in Table 7.1. As we move across a row, we record victories for the row team over the column team with a "1."

Looking at the first row, we see that the Bears defeated the Cardinals and the Dolphins but lost to the Eagles and the Falcons. Similarly the Falcons defeated the Bears, the Cardinals, and the Dolphins but lost to the Eagles. From the point of view of going down a column, the 1's represent losses by the column team to the row team. We compile each team's record in Table 7.2.

Ordering the teams by wins (or winning percentage) gives us a way to check the reasonableness of our eventual ranking: the Falcons should likely be number 1, and

TABLE 7.1   Complete Wins and Losses for a 5-Team League

Teams	Bears	Cardinals	Dolphins	Eagles	Falcons
Bears	0	1	1	0	0
Cardinals	0	0	1	1	0
Dolphins	0	0	0	1	0
Eagles	1	0	0	0	1
Falcons	1	1	1	0	0

TABLE 7.2   Team Records for a 5-Team League

Team	Wins	Losses
Bears	2	2
Cardinals	2	2
Dolphins	1	3
Eagles	2	2
Falcons	3	1

the Dolphins should likely be last. However, the middle 3 teams all have identical records so we cannot just use records to generate a definitive ranking.

The next step is to decide how much of its credit a losing team should give to a winning team. In keeping with our belief that victories over good teams should earn more credit than victories over poor teams, we adopt the following two conventions:

1. The proportion of the losing team's credit that a winning team earns is $\dfrac{1}{L+2}$ where L is the number of losses that the losing team has suffered.
2. Every team keeps any credit that it does not give to teams that defeated it.

For example, the Cardinals lost two games: one to the Bears and one to the Falcons. Thus for the Cardinals $L=2$, and the Cardinals will give $\dfrac{1}{L+2}=\dfrac{1}{4}$ of its credit to the Bears, $\dfrac{1}{L+2}=\dfrac{1}{4}$ of its credit to the Falcons, and the remaining $1-\dfrac{1}{4}-\dfrac{1}{4}=\dfrac{1}{2}$ they will keep for themselves. Similarly the Falcons lost only to the Eagles, so the Eagles will receive $\dfrac{1}{L+2}=\dfrac{1}{1+2}=\dfrac{1}{3}$ of the Falcons' credit while the Falcons keep the remaining $\dfrac{2}{3}$ for themselves. Continuing in this way we transform Table 7.1, which records only wins and losses, into Table 7.3, which records the proportion of the losing team's credit the winning team receives.

TABLE 7.3   Proportions of Losing Team's Credit Given to Winning Team

Teams	Bears	Cardinals	Dolphins	Eagles	Falcons
Bears	0	1/4	1/5	0	0
Cardinals	0	0	1/5	1/4	0
Dolphins	0	0	0	1/4	0
Eagles	1/4	0	0	0	1/3
Falcons	1/4	1/4	1/5	0	0

Moving across the first row, we see that the Bears receive $\frac{1}{4}$ of the Cardinals' credit and $\frac{1}{5}$ of the Dolphins' credit. Note the Dolphins lost more games than the Cardinals, so the Bears receive less credit for beating the Dolphins. We can also view the table by moving down a column. Moving down the column for the Bears, for example, we see that the Bears give $\frac{1}{4}$ of their credit to each of the 2 teams that beat them: the Eagles and the Falcons.

The choice of $\frac{1}{L+2}$ as the proportion of credit distributed by a losing team is in a sense an arbitrary one. We want the proportion to be less than $\frac{1}{L}$, for example, since otherwise a team would distribute *all* of its credit to other teams. But why use $\frac{1}{L+2}$ rather than $\frac{1}{L+1}$ or $\frac{1}{L+10}$? There is really no definite answer to that question; we are trying to strike a balance between losing teams distributing enough, but not too much, credit to winning teams. Different choices will result in different rankings, and it is up to the modeler to decide what choice seems to result in the most appropriate rankings. The reader is invited to investigate alternative choices in the exercises. Next we show how to generate a ranking with our current choice.

Example 7.8: Rank the 5 teams using a Markov model for the flow of credit among them.

Following our usual practice, we first create a flow diagram, then we find the corresponding DDS, and finally we implement the model in Excel.

Creating a table like Table 7.3 makes it easier for us to create the flow diagram for the flow of credit among the teams. Each team is represented by an oval, and the proportion of credit given by 1 team to another is represented by an arrow connecting the two. The flow diagram for our 5-team league is given in Figure 7.9. To make the diagram easier to read, we only include the proportions on our arrow labels.

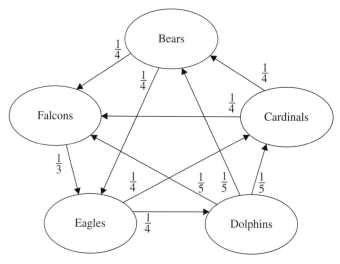

FIGURE 7.9 Flow of credit among 5 teams.

Note in the flow diagram that both conditions of Theorem 7.1 are met: it is possible for credit to flow from any team to any other team so it is irreducible, and credit may remain with a team so it is aperiodic. Thus our model is guaranteed to have a positive stable distribution. In a ranking context having all stable distribution percentages be positive means we will get a useful ranking that applies to all teams.

For the DDS we use only the first letter of each team's name to keep our notation as simple as possible. With t representing steps in the model rather than time, the resulting DDS is

$$B(t) = B(t-1) - \frac{1}{2}B(t-1) + \frac{1}{4}C(t-1) + \frac{1}{5}D(t-1)$$

$$C(t) = C(t-1) - \frac{1}{2}C(t-1) + \frac{1}{5}D(t-1) + \frac{1}{4}E(t-1)$$

$$D(t) = D(t-1) - \frac{3}{5}D(t-1) + \frac{1}{4}E(t-1)$$

$$E(t) = E(t-1) - \frac{1}{2}E(t-1) + \frac{1}{4}B(t-1) + \frac{1}{3}F(t-1)$$

$$F(t) = F(t-1) - \frac{1}{3}F(t-1) + \frac{1}{4}B(t-1) + \frac{1}{4}C(t-1) + \frac{1}{5}D(t-1).$$

Implementing this DDS in Excel is no more difficult than it was for the truck rental example, though we do have five columns now rather than three. Setting up a general model is more difficult because we have a large number of parameters that all depend on the wins and losses during the season. For now we content ourselves with implementing this particular example. Since we are interested in the stable distribution, we have columns for percentages for each team as well. The Excel setup with the formula for the Bears showing is given in Figure 7.10.

	A	B	C	D	E	F
1	Football Team Rankings					
2						
3	t	Bears	Cardinals	Dolphins	Eagles	Falcons
4	0	1	1	1	1	1
5	1	=B4-(1/2)*B4+(1/4)*C4+(1/5)*D4			1.08	1.37

FIGURE 7.10   Excel setup for a 5-team ranking.

Unlike the truck rental example, with our ranking model we do not have initial values to plug into Excel. However, we know from Theorem 7.1 that we are going to get a positive stable distribution so the initial assignment of credit should not matter—we will still end up with the same percentages of credit for each team. For this reason it will be our custom to just assign initial credit equal to 1 for each team.

In Figure 7.11 we show the output of the Excel model with most rows hidden. By experimenting with different initial amounts of credit, we can verify that the distribution of credit does not change. This stable distribution is the basis of our ranking of the teams. We report the ranking based on the stable distribution in Table 7.4.

	A	B	C	D	E	F	G	H	I	J	K
1	Football Team Rankings										
2											
3	*t*	Bears	Cardinals	Dolphins	Eagles	Falcons	B%	C%	D%	E%	F%
4	0	1	1	1	1	1	20.00%	20.00%	20.00%	20.00%	20.00%
5	1	0.95	0.95	0.65	1.08	1.37	19.00%	19.00%	13.00%	21.67%	27.33%
24	20	0.67	0.90	0.56	1.35	1.52	13.48%	17.98%	11.24%	26.97%	30.34%
25	21	0.67	0.90	0.56	1.35	1.52	13.48%	17.98%	11.24%	26.97%	30.34%
26	22	0.67	0.90	0.56	1.35	1.52	13.48%	17.98%	11.24%	26.97%	30.34%

FIGURE 7.11 Excel output for a 5-team ranking.

TABLE 7.4 Markov Method Rankings for a 5-Team League

Team	Credit (%)
Falcons	30.34
Eagles	26.97
Cardinals	17.98
Bears	13.48
Dolphins	11.24

Note how the rankings agree with our intuition that the Falcons should be the highest ranked team and the Dolphins the lowest. Note also that among the 3 teams with a 2-2 record, our system has ranked the Eagles, then the Cardinals, then the Bears. The Eagles had the "best wins"—one over the Falcons (ranked number 1) and one over the Bears (ranked number 4). The Cardinals had wins over the Eagles (ranked 2) and the Dolphins (ranked 5), and the Bears had wins over the Cardinals (ranked 3) and the Dolphins (ranked 5). Our Markov system has rewarded the teams who have victories over better opponents and produced a sensible ranking. □

In the next section we provide an interpretation of the flow diagram for credit much like the one given in Section 7.1.4 for the flow diagram for truck rentals.

7.2.1 Interpreting the Flow Diagram

The interpretation of a flow diagram in the context of ranking football teams takes some getting used to. As we move from one step to the next, we are tracing the abstract "flow of credit" among teams. From a global point of view, we say that from one step to the next $\frac{1}{3}$ of the Falcons' credit is given to the Eagles. If we stretch our abstraction a little farther, we can consider the point of view of a single "unit of credit" as we did for a single truck in the rental company example. We imagine the unit of credit flowing from 1 team to any of the teams that defeated it. If our unit of credit starts with the Falcons, then it has a $\frac{1}{3}$ chance of flowing to the Eagles in the next step and a $\frac{2}{3}$ chance of staying with the Falcons. If the unit does flow to the Eagles, then on the next step it faces a $\frac{1}{4}$ chance of flowing to the Bears, a $\frac{1}{4}$ chance of flowing to the Cardinals, and a $\frac{1}{2}$ chance of staying with the Eagles. Each successive step, or iteration, of our model traces the path of the unit of credit as it randomly flows among the teams. The unit of credit is taking a random walk among the teams.

From a global point of view, the eventual stable distribution shows us how all credit is distributed among teams. From the point of view of an individual unit, the stable distribution tells us the percentage of time the unit spends visiting each team. The higher the percentage, the more credit a team deserves.

7.2.2 Undefeated and Winless Teams

The presence of undefeated teams causes a problem for our current ranking system. To see why, we consider the effect of an undefeated team on our flow diagram. A team with no losses will give no credit to any other team, so in the flow diagram the oval for the undefeated team will have inward-pointing arrows but no outward-pointing arrows. From a random walk perspective, any unit of credit that travels to the undefeated team will stay there permanently. Ultimately what we will see is that *all* credit will end up with the undefeated team. In terms of Theorem 7.1, we will not get a positive stable distribution because every team but the undefeated team ends up with zero credit. This happens because the presence of an undefeated team prevents the flow diagram from being irreducible: it is no longer possible to travel between any two states.

We illustrate the problem in the next example.

Example 7.9: Consider our 5-team conference from before, only this time we assume the Falcons go undefeated. Rank the 5 teams.

The only game that changes is the one between the Eagles and the Falcons, which is now a victory for the Falcons. The season's wins and losses are summarized in Table 7.5.

TABLE 7.5 Season Wins and Losses with Undefeated Team

Teams	Bears	Cardinals	Dolphins	Eagles	Falcons
Bears	0	1	1	0	0
Cardinals	0	0	1	1	0
Dolphins	0	0	0	1	0
Eagles	1	0	0	0	0
Falcons	1	1	1	1	0

As before we use the season wins and losses to compile the records for each team as shown in Table 7.6.

TABLE 7.6 Season W–L Records with Undefeated Team

Team	Wins	Losses
Bears	2	2
Cardinals	2	2
Dolphins	1	3
Eagles	1	3
Falcons	4	0

The proportions of credit that are transferred change because the Eagles now have three losses and no win over the Falcons. These are shown in Table 7.7. From Table 7.7 we construct the flow diagram given in Figure 7.12.

TABLE 7.7 Credit Given by Losing Teams to Winning Teams for Example 7.9

Teams	Bears	Cardinals	Dolphins	Eagles	Falcons
Bears	0	1/4	1/5	0	0
Cardinals	0	0	1/5	1/5	0
Dolphins	0	0	0	1/5	0
Eagles	1/4	0	0	0	0
Falcons	1/4	1/4	1/5	1/5	0

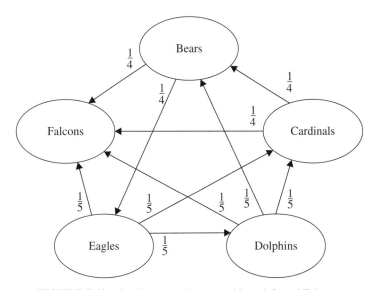

FIGURE 7.12   Credit among 5 teams with undefeated Falcons.

The corresponding DDS is now

$$B(t) = B(t-1) - \frac{1}{2}B(t-1) + \frac{1}{4}C(t-1) + \frac{1}{5}D(t-1)$$

$$C(t) = C(t-1) - \frac{1}{2}C(t-1) + \frac{1}{5}D(t-1) + \frac{1}{5}E(t-1)$$

$$D(t) = D(t-1) - \frac{3}{5}D(t-1) + \frac{1}{5}E(t-1)$$

$$E(t) = E(t-1) - \frac{3}{5}E(t-1) + \frac{1}{4}B(t-1)$$

$$F(t) = F(t-1) + \frac{1}{4}B(t-1) + \frac{1}{4}C(t-1) + \frac{1}{5}D(t-1) + \frac{1}{5}E(t-1).$$

Implementing this model in Excel requires only minor modifications to the original. Once implemented in Excel and with formulas dragged down far enough, we get the results shown in Figure 7.13 with most rows hidden.

	A	B	C	D	E	F	G	H	I	J	K
1	Football Team Rankings										
2											
3	*t*	Bears	Cardinals	Dolphins	Eagles	Falcons	*B%*	*C%*	*D%*	*E%*	*F%*
4	0	1	1	1	1	1	20.00%	20.00%	20.00%	20.00%	20.00%
5	1	0.95	0.90	0.60	0.65	1.90	19.00%	18.00%	12.00%	13.00%	38.00%
41	37	0.00	0.00	0.00	0.00	5.00	0.00%	0.00%	0.00%	0.00%	100.00%
42	38	0.00	0.00	0.00	0.00	5.00	0.00%	0.00%	0.00%	0.00%	100.00%
43	39	0.00	0.00	0.00	0.00	5.00	0.00%	0.00%	0.00%	0.00%	100.00%

FIGURE 7.13   Excel results for 5 teams with undefeated Falcons.

Our ranking is based on the long-term distribution of credit shown in Table 7.8. The method has correctly identified the Falcons as the best team, but it has broken down for the rest of the rankings. Because the Falcons have no losses, no credit flows from them to any other team. Thus all credit eventually flows to the Falcons where it stops. When this happens we say the team is an **absorbing state**, and we note that the presence of an absorbing state automatically prevents the diagram from being irreducible. We can spot the absorbing state on the flow diagram since it is the only oval with no outgoing arrows.

TABLE 7.8   Stable Distribution of Credit when Falcons are Undefeated

Team	Credit (%)
Falcons	100
Bears	0
Cardinals	0
Dolphins	0
Eagles	0

A winless team presents a similar, though less problematic, issue. The oval for a winless team will have only outward-pointing arrows. Thus a winless team will also prevent a diagram from being irreducible since no credit can flow to it. The result is that winless teams end up with zero credit, and so we do not get a positive stable distribution. However, the presence of a winless team, unlike the presence of an undefeated team, does not prevent the model from ranking the rest of the teams sensibly. The good news is that the fix we employ in the following to deal with undefeated teams has the effect of fixing the issue of winless teams as well.

When all teams in a conference play each other exactly once, we can have at most one undefeated team (and at most one winless team). Since we know that team to be the best, one way to proceed is to simply remove it from our ranking system at the beginning and then just rank the remaining 4 teams.

First we adjust the season's win/loss results to get Table 7.9.

TABLE 7.9   Wins and Losses with Undefeated Falcons Removed

Teams	Bears	Cardinals	Dolphins	Eagles
Bears	0	1	1	0
Cardinals	0	0	1	1
Dolphins	0	0	0	1
Eagles	1	0	0	0

The corresponding table of credit is given in Table 7.10.

TABLE 7.10   Credit Proportions with Undefeated Falcons Removed

Teams	Bears	Cardinals	Dolphins	Eagles
Bears	0	1/3	1/4	0
Cardinals	0	0	1/4	1/4
Dolphins	0	0	0	1/4
Eagles	1/3	0	0	0

The DDS is then

$$B(t) = B(t-1) - \frac{1}{3}B(t-1) + \frac{1}{3}C(t-1) + \frac{1}{4}D(t-1)$$

$$C(t) = C(t-1) - \frac{1}{3}C(t-1) + \frac{1}{4}D(t-1) + \frac{1}{4}E(t-1)$$

$$D(t) = D(t-1) - \frac{1}{2}D(t-1) + \frac{1}{4}E(t-1)$$

$$E(t) = E(t-1) - \frac{1}{2}E(t-1) + \frac{1}{3}B(t-1).$$

After modifying our Excel model, we get the stable credit distribution shown in Table 7.11.

TABLE 7.11   Ranking Results with Undefeated Falcons Removed

Team	Credit (%)
Bears	36.36
Cardinals	27.27
Eagles	24.24
Dolphins	12.12

Thus our final ranking using the method of removing the undefeated team is

1. Falcons
2. Bears
3. Cardinals
4. Eagles
5. Dolphins

Note that this result does not agree with the rankings we had when the Falcons were included as a 1-loss team. This makes sense because with the Falcons undefeated, the Eagles have one fewer wins and no longer receive credit for being the only team to defeat the Falcons. Now the Bears have the "best wins" and are rewarded accordingly. $\square$

There are two problems with the approach in the previous example. One is that it is tedious to have to change the model. Two is that in many sports such as college football, teams do not all play each other so it is possible to have multiple undefeated teams. How do we decide among those? What we would like is a method that will work even if undefeated teams are included, and we describe one way to accomplish that with a slight adjustment to our original system.

The problem with including undefeated teams in our ranking method is that their inclusion in a flow diagram prevents the diagram from being irreducible. In other words, as soon as we include an undefeated team, we no longer satisfy the requirements of Theorem 7.1. Thus whatever adjustment we make to our model for undefeated teams will necessarily involve the undefeated team giving some credit to another team; that is the only way to keep all of the credit from getting "stuck" at an undefeated team.

Our approach can be thought of as a "good sport" approach. Here we require that all losing teams receive a small amount of credit from the winner just for playing. In this way we guarantee that undefeated teams no longer create an absorbing state in our model because credit must flow out of every team to every opponent. We also guarantee that winless teams end up with some credit for participating. There is something appealing about this approach: it seems right that a team who at least plays should receive more credit than one who sits idle.

We can experiment with different values for the proportion of credit to assign to a losing team just for playing, but we should at least be careful that the proportion is never more than what a team would receive for winning.

The least proportion of credit that a winning team can possibly receive is for defeating a winless team. If there are N teams in the league, then the worst possible record for a team would be 0 wins and $N-1$ losses. Thus the least proportion of credit a winning team can receive is $\frac{1}{L+2} = \frac{1}{N-1+2} = \frac{1}{N+1}$. One way to guarantee that credit to a losing team never exceeds this is to make the "just-for-playing" proportion of credit that winning teams distribute to losing teams equal to $\frac{1}{N^2}$. As long as $N \geq 2$ we will have $\frac{1}{N^2} < \frac{1}{N+1}$. The reader is invited to verify this fact in the exercises.

In the next example we examine the effect of the "just-for-playing" adjustment on our model.

Example 7.10: Consider again our 5-team conference where the Falcons are undefeated. Rank the 5 teams by including a "just-for-playing" credit for losing teams.

Recall that the season results are given in Table 7.5 and that Table 7.6 gives the season won/loss records for all teams.

The table of proportions of credit must change to reflect credit given to losing teams just for playing, in this case $\frac{1}{N^2} = \frac{1}{25}$. We see the result in Table 7.12.

TABLE 7.12 Credit Given to Opponents Including Just-for-Playing Adjustment

Teams	Bears	Cardinals	Dolphins	Eagles	Falcons
Bears	0	1/4	1/5	1/25	1/25
Cardinals	1/25	0	1/5	1/5	1/25
Dolphins	1/25	1/25	0	1/5	1/25
Eagles	1/4	1/25	1/25	0	1/25
Falcons	1/4	1/4	1/5	1/5	0

The inclusion of just-for-playing credit causes dramatic changes in the flow diagram given in Figure 7.14. As we see in Figure 7.14, we no longer have an absorbing state at the Falcons. Credit can now flow from any team to any other team. The diagram now satisfies the requirements of Theorem 7.1, so we are guaranteed that the method will produce a proper ranking of all 5 teams.

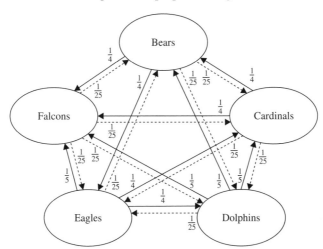

FIGURE 7.14 Credit among 5 teams where credit is given just for playing.

The DDS for the model is now

$$B(t) = B(t-1) - \frac{29}{50}B(t-1) + \frac{1}{4}C(t-1) + \frac{1}{5}D(t-1) + \frac{1}{25}E(t-1) + \frac{1}{25}F(t-1)$$

$$C(t) = C(t-1) - \frac{29}{50}C(t-1) + \frac{1}{25}B(t-1) + \frac{1}{5}D(t-1) + \frac{1}{5}E(t-1) + \frac{1}{25}F(t-1)$$

$$D(t) = D(t-1) - \frac{16}{25}D(t-1) + \frac{1}{25}B(t-1) + \frac{1}{25}C(t-1) + \frac{1}{5}E(t-1) + \frac{1}{25}F(t-1)$$

$$E(t) = E(t-1) - \frac{16}{25}E(t-1) + \frac{1}{4}B(t-1) + \frac{1}{25}C(t-1) + \frac{1}{25}D(t-1) + \frac{1}{25}F(t-1)$$

$$F(t) = F(t-1) - \frac{4}{25}F(t-1) + \frac{1}{4}B(t-1) + \frac{1}{4}C(t-1) + \frac{1}{5}D(t-1) + \frac{1}{5}E(t-1).$$

	A	B	C	D	E	F
1	Football Team Rankings					
2						
3	t	Bears	Cardinals	Dolphins	Eagles	Falcons
4	0	1	1	1	1	1
5	1	=B4-(29/50)*B4+(1/4)*C4+(1/5)*D4+(1/25)*E4+(1/25)*F4				

FIGURE 7.15   Excel setup for ranking with just-for-playing credit.

TABLE 7.13   Ranking with Just-for-Playing Credit

Team	Credit (%)
Falcons	58.80
Bears	12.30
Cardinals	11.06
Eagles	9.68
Dolphins	8.16

In Figure 7.15 we show the Excel model with the formula for the Bears displayed.

After modifying our Excel model and finding the stable distribution, we get the ranking shown in Table 7.13. This ranking agrees with the ranking we produced after running our original model with the Falcons removed. □

Our choice of the proportion of credit to assign just for playing accounted for the fact that it should not be more than any winning team can receive for winning. We also must make sure that the total proportion of credit distributed by a team never exceeds 1, that is, that a team never distributes more than 100% of its credit to other teams. In fact our use of $\frac{1}{N^2}$ guarantees this, and we refer the reader to Appendix F for a verification of this fact.

7.2.3   Equilibrium Distribution

As we know from previous sections, we can find the equilibrium distribution for our Markov models by employing some algebra. Provided our flow diagram is irreducible and aperiodic, this allows us to find the positive stable distribution of credit, and hence our rankings, without having to resort to Excel. Unfortunately this approach is difficult to carry out by hand for even a modest number of teams, though inputting the Excel formulas can also take some time. For our 5-team conference we would need to solve a system of five equations in five unknowns.

We illustrate the procedure with an example using 4 teams to keep the algebra manageable.

Example 7.11: Consider the situation where we removed the Falcons and then ranked the remaining 4 teams with our Markov method that did not include the just-for-playing adjustment. The DDS for this situation was given by

$$B(t) = B(t-1) - \frac{1}{3}B(t-1) + \frac{1}{3}C(t-1) + \frac{1}{4}D(t-1)$$

$$C(t) = C(t-1) - \frac{1}{3}C(t-1) + \frac{1}{4}D(t-1) + \frac{1}{4}E(t-1)$$

$$D(t) = D(t-1) - \frac{1}{2}D(t-1) + \frac{1}{4}E(t-1)$$

$$E(t) = E(t-1) - \frac{1}{2}E(t-1) + \frac{1}{3}B(t-1).$$

Letting B^*, C^*, D^*, and E^* represent the equilibrium values for each team, we know that these are the values such that if we plug them into the right side of the system we get them out on the left side as well. Thus we need to solve

$$B^* = B^* - \frac{1}{3}B^* + \frac{1}{3}C^* + \frac{1}{4}D^*$$

$$C^* = C^* - \frac{1}{3}C^* + \frac{1}{4}D^* + \frac{1}{4}E^*$$

$$D^* = D^* - \frac{1}{2}D^* + \frac{1}{4}E^*$$

$$E^* = E^* - \frac{1}{2}E^* + \frac{1}{3}B^*.$$

As in the truck rental example, our method is to write each of the first three variables in terms of the fourth. Initial simplifying gives

$$0 = -\frac{1}{3}B^* + \frac{1}{3}C^* + \frac{1}{4}D^*$$

$$0 = -\frac{1}{3}C^* + \frac{1}{4}D^* + \frac{1}{4}E^*$$

$$0 = -\frac{1}{2}D^* + \frac{1}{4}E^*$$

$$0 = -\frac{1}{2}E^* + \frac{1}{3}B^*.$$

We note that the system has a trivial solution where all variables are equal to 0, but it is not a solution of practical interest. Further simplifying allows us to get the variables D^* and B^* in terms of E^*:

$$B^* = C^* + \frac{3}{4}D^*$$

$$C^* = \frac{3}{4}D^* + \frac{3}{4}E^*$$

$$D^* = \frac{1}{2}E^*$$

$$B^* = \frac{3}{2}E^*.$$

Next we substitute the value $D^* = \frac{1}{2}E^*$ into the second equation to get C^* in terms of E^*: $C^* = \frac{3}{4}D^* + \frac{3}{4}E^* = \frac{3}{4} \cdot \frac{1}{2}E^* + \frac{3}{4}E^* = \frac{9}{8}E^*$. We now have all of our variables in terms of E^*:

$$B^* = \frac{3}{2}E^*$$

$$C^* = \frac{9}{8}E^*$$

$$D^* = \frac{1}{2}E^*.$$

To get the equilibrium distribution, we must finish by computing the percentages for each equilibrium value. For B^* we calculate

$$\frac{B^*}{B^* + C^* + D^* + E^*} = \frac{\frac{3}{2}E^*}{\frac{3}{2}E^* + \frac{9}{8}E^* + \frac{1}{2}E^* + E^*} = \frac{\frac{3}{2}}{\frac{3}{2} + \frac{9}{8} + \frac{1}{2} + 1} = 0.3636 = 36.36\%.$$

Note that this value agrees with the one found by Excel in Example 7.9. The computations for the other values are similar and also agree with our Excel work. □

Though we can solve these kinds of systems of equations by hand for a small number of teams, we can also see that the algebraic method soon becomes laborious. Linear algebra is a branch of mathematics that provides efficient ways of solving such systems using matrices and vectors. These methods are beyond the scope of this text, but a quick Internet search on Markov methods will show that the techniques of linear algebra are the preferred ways of analyzing such systems.

Even with the tools of linear algebra, solving a system with a large number of variables still requires the use of software to solve in a reasonable amount of time. Some commercial software packages that will solve such systems are Maple, MATLAB, and Mathematica. The creators of Mathematica have put a feely available system for doing complex mathematical computations on the web called Wolfram Alpha (Wolfram, 2015). If you are careful with your syntax, Wolfram Alpha will solve systems like the previous one in seconds.

7.2.4   SEC Football

To conclude this section we rank the teams from the 2014 Southeastern Conference football season. With 14 teams each playing eight conference games (we do not include the championship here), the DDS becomes pretty unwieldy, as does the Excel implementation. Thus we will only outline the setup and present the results.

The 2014 season results are presented in Table 7.14.

TABLE 7.14 Complete Wins and Losses for 2014 SEC Football

Team	Alabama	Arkansas	Auburn	Florida	Georgia	Kentucky	LSU	Miss. St.	Missouri	Ole Miss	South Carolina	Tennessee	Texas A&M	Vanderbilt
Alabama	0	1	1	1	0	0	1	1	0	0	0	1	1	0
Arkansas	0	0	0	0	0	0	1	0	0	1	0	0	0	0
Auburn	0	1	0	0	0	0	1	0	0	1	1	0	1	0
Florida	0	0	0	0	1	1	0	0	0	0	0	1	0	1
Georgia	0	1	1	0	0	1	0	0	1	0	0	1	0	1
Kentucky	0	0	0	0	0	0	0	0	0	0	1	0	0	1
LSU	0	0	0	1	0	1	0	0	0	1	0	0	1	0
Miss. St.	0	1	1	0	0	1	1	0	0	0	0	0	1	1
Missouri	0	1	0	1	0	1	0	0	0	0	1	1	1	1
Ole Miss	1	0	0	0	0	0	0	1	0	0	0	1	1	1
South Carolina	0	0	0	1	1	0	0	0	0	0	0	0	0	1
Tennessee	0	0	0	0	0	1	0	0	0	0	1	0	0	1
Texas A&M	0	1	0	0	0	0	0	0	0	0	1	0	0	0
Vanderbilt	0	0	0	0	0	0	0	0	0	0	0	0	0	0

TABLE 7.15   W–L Records for 2014 SEC Football

Team	Wins	Losses
Alabama	7	1
Arkansas	2	6
Auburn	4	4
Florida	4	4
Georgia	6	2
Kentucky	2	6
LSU	4	4
Miss. St.	6	2
Missouri	7	1
Ole Miss	5	3
South Carolina	3	5
Tennessee	3	5
Texas A&M	3	5
Vanderbilt	0	8

TABLE 7.16   Ranking for 2014 SEC Football

Team	Credit (%)
Alabama	19.02
Ole Miss	14.94
Georgia	10.57
Missouri	9.00
Auburn	8.43
Miss. St.	8.05
LSU	6.93
Arkansas	5.85
South Carolina	4.93
Florida	4.76
Texas A&M	4.30
Tennessee	1.63
Kentucky	1.25
Vanderbilt	0.35

The records of each team are given in Table 7.15.

After carrying out our ranking method with just-for-playing credit, we get the ranking presented in Table 7.16.

Our method has produced some interesting results. First we note that Vanderbilt as the only winless team has rightly been ranked lowest. Similarly, Alabama, as 1 of 2 teams with only one loss has reasonably been ranked first. The fact that Alabama's wins are of much better quality than Missouri's gives Alabama the edge between the two. One surprising result is that Ole Miss, a team with a 5-3 record, is ranked above Georgia, a 6-2 team, and Missouri, a 7-1 team. It appears that Ole Miss has gotten a lot of credit for being the only team to defeat Alabama. Similarly, Georgia appears to have been given a lot of credit for being the only team to defeat Missouri. While it does appear that 7-1 Missouri is being short changed by the method, it is also true that Missouri's wins are overall unimpressive. In fact, if we take the total number of wins for the teams that Ole Miss, Georgia, and Missouri defeated, we get 19, 18, and 17, respectively. Even though Missouri has *more* wins, they are not *good* wins. Their best win is over 4-4 Florida, the only team with a nonlosing record that they defeated.

We see a similar result in the bottom half of the rankings where 2-6 Arkansas is ranked ahead of several teams with better records. Arkansas appears to have received a lot of credit for quality wins over 6-2 Mississippi State and 4-4 LSU.

Even though we can make an argument that the ranking in Table 7.16 is reasonable, it does go against our intuition in a few instances. The point of playing is to win as many games as possible, and even taking into account strength of schedule, it is difficult to justify ranking Ole Miss ahead of Missouri or Arkansas ahead of Florida. Our method seems to be *overvaluing* good wins and undervaluing just winning. In the exercises the reader is invited to modify the model to lessen this effect.

In the next section we present an application of Markov ranking methods to Internet searches.

7.2.5 Section Exercises

1 Table 7.17 gives the results of a season of games with no undefeated teams. Rank the teams using the Markov method from Example 7.8. Discuss any rankings that differ from your expectations.

TABLE 7.17 Complete Wins and Losses for a 6-Team League

Teams	Bears	Cardinals	Dolphins	Eagles	Falcons	Giants
Bears	0	1	1	0	1	1
Cardinals	0	0	1	1	0	1
Dolphins	0	0	0	1	1	1
Eagles	1	0	0	0	1	0
Falcons	0	1	0	0	0	0
Giants	0	0	0	1	1	0

2 Table 7.18 gives the results of a season of games with one undefeated team. Use the Markov method including the just-for-playing credit to rank the teams. Discuss any rankings that differ from your expectations.

TABLE 7.18 Complete Wins and Losses for a 6-Team League

Teams	Bears	Cardinals	Dolphins	Eagles	Falcons	Giants
Bears	0	1	1	1	1	1
Cardinals	0	0	1	1	0	1
Dolphins	0	0	0	1	1	1
Eagles	0	0	0	0	1	0
Falcons	0	1	0	0	0	0
Giants	0	0	0	1	1	0

3 If teams do not all play each other, it is possible to have more than one undefeated team, and when we have more than one undefeated team, there is more than one way we can proceed with a ranking. One way is to use the just-for-playing credit, but here we look at another option similar to Example 7.9: instead of just-for-playing credit, we can first use our original ranking method on all teams, then we remove the undefeated teams and rerun the method. Table 7.19 gives the results of a season of games with 2 undefeated teams:

   a. Rank the teams without the just-for-playing credit.

   b. Interpret the results of the ranking.

   c. After removing the undefeated teams, rerun the method.

   d. Compare the final ranking of this method with the ranking produced by the just-for-playing credit.

TABLE 7.19   Complete Wins and Losses for a 6-Team League

Teams	Bears	Cardinals	Dolphins	Eagles	Falcons	Giants
Bears	0	1	1	1	0	0
Cardinals	0	0	0	1	1	0
Dolphins	0	0	0	0	1	0
Eagles	0	0	0	0	0	0
Falcons	0	0	0	0	0	0
Giants	0	0	1	1	1	0

4 Table 7.20 gives the 2014 NCAA football season results for the SEC West. Only games between Western division teams are included. Rank the teams. Discuss any ranking that differs from your expectations.

TABLE 7.20   Complete Wins and Losses for 2014 SEC West

Teams	Alabama	Arkansas	Auburn	LSU	Miss. St.	Ole Miss	Texas A&M
Alabama	0	1	1	1	1	0	1
Arkansas	0	0	0	1	0	1	0
Auburn	0	1	0	1	0	1	0
LSU	0	0	0	0	0	1	1
Miss. St.	0	1	1	1	0	0	1
Ole Miss	1	0	0	0	1	0	1
Texas A&M	0	1	1	0	0	0	0

5 Table 7.21 gives the 2014 NCAA football season results for the SEC East. Only games between Eastern division teams are included. Rank the teams. Discuss any ranking that differs from your expectations.

6 *Extension*: Many sports have teams play each other more than once during a season. All major professional team sports in the United States, for example, have this feature. Discuss whether or not our Markov method with just-for-playing credit will still work in this situation.

TABLE 7.21   Complete Wins and Losses for 2014 SEC East

Teams	Florida	Georgia	Kentucky	Missouri	South Carolina	Tennessee	Vanderbilt
Florida	0	1	1	0	0	1	1
Georgia	0	0	1	1	0	1	1
Kentucky	0	0	0	0	1	0	1
Missouri	1	0	1	0	1	1	1
South Carolina	1	1	0	0	0	0	1
Tennessee	0	0	1	0	1	0	1
Vanderbilt	0	0	0	0	0	0	0

7 *Extension*: For a sport and league or conference of your choosing, rank the teams with the Markov just-for-playing method.

8 Prove $\frac{1}{N^2} < \frac{1}{N+1}$ for all $N \geq 2$.

9 *Extension*: Discuss how to use a Markov model to rank the importance of words in the English language.

7.3   GOOGLE PAGERANK

Google was cofounded by Sergey Brin and Lawrence Page who met while studying computer science at Stanford University in 1995. Having started Google and seen it grow to a company worth about $400 billion, Brin and Page each currently have a net worth of about $30 billion (Forbes, 2015). Together the pair developed what is now known as the Google PageRank algorithm for ranking the importance of webpages, and this algorithm is the basis of the Google search engine (Brin & Page, 1998). When a search term is entered into Google's search engine, the results of the search are relevant webpages that are ranked from most to least important, but how to rank webpages in this way is not obvious.

At the heart of the Google PageRank algorithm is a Markov model based on a simple idea: the importance of a webpage should be determined by how many webpages link to it and how important *those* pages are. As we will see, the idea is very similar to the idea of sports teams giving "credit" to teams they play, only in the context of the worldwide web, each webpage is a state, and arrows connecting pages represent hyperlinks from one webpage to another. A webpage that is linked by many webpages will be assigned a higher importance than one that has few links from other pages. The notion of "web importance" is, like credit given to a winning sports team, an abstract idea meant to be a relative, not absolute, measure.

We illustrate how the Google PageRank algorithm works by considering a very limited Internet consisting of only six webpages: *B*, *C*, *D*, *E*, *F*, and *G*. To represent the web as a flow diagram, we let each page be an oval, and we let each hyperlink from one page to another be represented by an arrow pointing to the destination page. The link structure for our miniature Internet is shown in the diagram given in Figure 7.16.

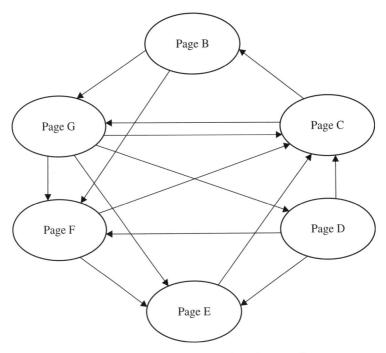

FIGURE 7.16   Link structure for a 6-page web.

Webpages that have links to many others are often referred to as **hubs**, while pages that many other pages link to are called **authorities**. In this example we would call Page G a hub and Page C an authority.

In the example that follows, we show how to take the link structure of our 6-page Internet and turn it into a ranking of the pages' importance.

Example 7.12: Use a Markov model to rank the 6 pages, B, C, D, E, F, and G, in order of importance.

First we need to assign labels to the hyperlink diagram in Figure 7.16. To do so we use an idea very similar to our setup for ranking sports teams: the starting webpage distributes its importance in equal proportions to each destination page it links to. For example, a webpage that links to four others would distribute 1/4 of its importance to each destination page. Figure 7.17 shows the flow diagram for our miniature web.

The flow diagram depicts the possible paths a web surfer can take by clicking on links. Note that because each webpage distributes all of its importance to destination pages, the surfer must leave the current page at every step, t. The only way a web surfer can stay on a page from one step to the next is if there are no outgoing links from that page.

We construct a table that reflects the link structure in our diagram as well as how much importance is assigned from 1 page to another. (This table is very similar to

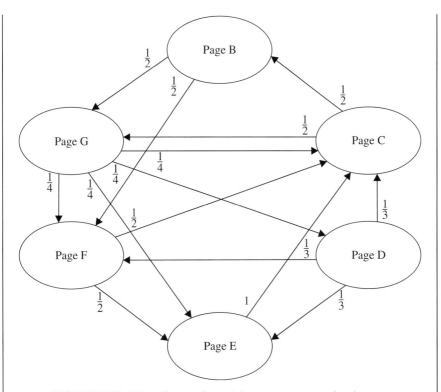

FIGURE 7.17 Flow diagram for web importance among 6 webpages.

TABLE 7.22 Hyperlink Table for a 6-Page Internet

HT	B	C	D	E	F	G
B	0	1/2	0	0	0	0
C	0	0	1/3	1	1/2	1/4
D	0	0	0	0	0	1/4
E	0	0	1/3	0	1/2	1/4
F	1/2	0	1/3	0	0	1/4
G	1/2	1/2	0	0	0	0

the table of how credit is distributed among sports teams.) We call this the **hyperlink table**, HT, and it is given in Table 7.22.

Columns represent outgoing links/importance, while rows represent incoming links/importance. Moving across the row for Webpage C, for example, we see that C receives links and hence importance from pages D, E, F, and G. Similarly, moving down the column for D shows that page D links to pages C, E, and F, and so it distributes 1/3 of its importance to each of those pages.

Corresponding to the flow diagram and Table 7.22, the DDS for the system is given by

$$B(t) = B(t-1) - \frac{1}{2}B(t-1) - \frac{1}{2}B(t-1) + \frac{1}{2}C(t-1)$$

$$C(t) = C(t-1) - \frac{1}{2}C(t-1) - \frac{1}{2}C(t-1) + \frac{1}{3}D(t-1) + E(t-1) + \frac{1}{2}F(t-1) + \frac{1}{4}G(t-1)$$

$$D(t) = D(t-1) - \frac{1}{3}D(t-1) - \frac{1}{3}D(t-1) - \frac{1}{3}D(t-1) + \frac{1}{4}G(t-1)$$

$$E(t) = E(t-1) - E(t-1) + \frac{1}{3}D(t-1) + \frac{1}{2}F(t-1) + \frac{1}{4}G(t-1)$$

$$F(t) = F(t-1) - \frac{1}{2}F(t-1) - \frac{1}{2}F(t-1) + \frac{1}{2}B(t-1) + \frac{1}{3}D(t-1) + \frac{1}{4}G(t-1)$$

$$G(t) = G(t-1) - \frac{1}{4}G(t-1) - \frac{1}{4}G(t-1) - \frac{1}{4}G(t-1) - \frac{1}{4}G(t-1) + \frac{1}{2}B(t-1) + \frac{1}{2}C(t-1).$$

Because each webpage must distribute all of its importance at each time step, the DDS simplifies quite a bit. We get

$$B(t) = \frac{1}{2}C(t-1)$$

$$C(t) = \frac{1}{3}D(t-1) + E(t-1) + \frac{1}{2}F(t-1) + \frac{1}{4}G(t-1)$$

$$D(t) = \frac{1}{4}G(t-1)$$

$$E(t) = \frac{1}{3}D(t-1) + \frac{1}{2}F(t-1) + \frac{1}{4}G(t-1)$$

$$F(t) = \frac{1}{2}B(t-1) + \frac{1}{3}D(t-1) + \frac{1}{4}G(t-1)$$

$$G(t) = \frac{1}{2}B(t-1) + \frac{1}{2}C(t-1).$$

From this point forward, it will be our custom to begin with this simplified form for the DDS. As in the case of sports team rankings, we think of t as representing the number of steps or iterations of our model, not time. Each iteration or step indicates a visit to the next webpage by a web surfer.

Entering this system into Excel is straightforward if a bit time consuming, and it is very similar to the work required for the sports team ranking examples. We include columns for each webpage and also columns for the distribution of importance for each webpage. The setup with the formula for Page E displayed is given in Figure 7.18.

We arbitrarily let all webpages start with a value of 1 for importance. We drag our formulas down until we see the emergence of a stable distribution. The Excel output with most rows hidden is shown in Figure 7.19. By experimenting with different initial values of importance, we can gain some confidence that the example in fact has a positive stable distribution.

Once we have the distribution, we rank the pages by their relative importance as shown in Table 7.23. It should not be a surprise that C is ranked highest since it

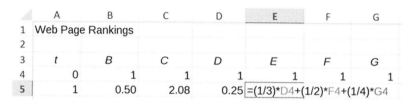

FIGURE 7.18   Excel setup for web importance for a 6-page web.

	A	B	C	D	E	F	G	H	I	J	K	L	M
1	Web Page Rankings												
2													
3	t	B	C	D	E	F	G	B%	C%	D%	E%	F%	G%
4	0	1	1	1	1	1	1	16.67%	16.67%	16.67%	16.67%	16.67%	16.67%
5	1	0.50	2.08	0.25	1.08	1.08	1.00	8.33%	34.72%	4.17%	18.06%	18.06%	16.67%
6	2	1.04	1.96	0.25	0.88	0.58	1.29	17.36%	32.64%	4.17%	14.58%	9.72%	21.53%
34	30	0.87	1.75	0.33	0.87	0.87	1.31	14.55%	29.09%	5.45%	14.55%	14.55%	21.82%
35	31	0.87	1.75	0.33	0.87	0.87	1.31	14.55%	29.09%	5.45%	14.55%	14.55%	21.82%

FIGURE 7.19   Excel output for web importance for a 6-page web.

TABLE 7.23   **Webpages Ranked by Their Relative Importance**

Page	Importance (%)
C	29.09
G	21.82
B	14.55
E	14.55
F	14.55
D	5.45

receives links from 4 of the other pages. What may be surprising are the next 2 pages on the list, *G* and *B*. Both of these pages benefit a lot from the fact that they are the only 2 pages receiving links from the most important page, *C*. □

In the next section we consider our Markov model for web importance from the point of view of a web surfer.

7.3.1   Interpreting the Flow Diagram

From a global perspective the flow diagram for our web shows the proportion of importance that is given by a link's origination page to its destination page. However, we can also view the flow diagram from an individual perspective.

If we consider the point of view of a web surfer, then when visiting a webpage with outgoing links to others, the proportions represent the probability of the web surfer choosing to click on a particular link. For a webpage with outgoing links to 2 pages, for example, there is probability, 1/2, that the web surfer will visit either page next. Thus every click for the web surfer brings them to a new webpage from which they make another random decision about which link to follow. Continuing in this way the web surfer engages in a random walk among the webpages. The stable distribution of importance can be thought of as the percentage of time the web surfer spends at each webpage. The more important the page, the more often the web surfer will visit it and the more time on average the web surfer will spend on it.

In the next two sections, we illustrate two potential problems for our Markov method for ranking webpages, and we show how Brin and Page overcame them.

7.3.2   Dangling Nodes

Pages with no links to others are called **dangling nodes**. They are problematic in the same way our undefeated sports teams were: with no outward-pointing arrows, all incoming web importance gets stuck at such a page. Eventually *all* web importance will end up at such a page and our ranking system will fail to produce a useful ranking. This is a serious problem for PageRank because there are many real webpages that do not contain links to others. Brin and Page employed a fix for dangling nodes similar to our just-for-playing credit for sports teams.

Our assumption will be that a web surfer who arrives at a page with no outward-pointing links will not actually remain there indefinitely. Instead such a surfer may just type a new URL into the browser and head to a completely unrelated page. Since we cannot predict where such a surfer will go, we assume that the next destination will be a random selection with every webpage having the same chance of being selected.

In the next examples we illustrate both the problem and the solution for dangling nodes.

Example 7.13: Consider our mini-web from the previous example, only this time assume there is no outward link from page E to page C. Rank the webpages in order of importance.

After eliminating the link from E to C, our hyperlink table changes. The result is Table 7.24.

TABLE 7.24   Hyperlink Table with E as a Dangling Node

HT	B	C	D	E	F	G
B	0	1/2	0	0	0	0
C	0	0	1/3	0	1/2	1/4
D	0	0	0	0	0	1/4
E	0	0	1/3	0	1/2	1/4
F	1/2	0	1/3	0	0	1/4
G	1/2	1/2	0	0	0	0

The new table brings about a corresponding change in the simplified DDS as well:

$$B(t) = \frac{1}{2}C(t-1)$$

$$C(t) = \frac{1}{3}D(t-1) + \frac{1}{2}F(t-1) + \frac{1}{4}G(t-1)$$

$$D(t) = \frac{1}{4}G(t-1)$$

$$E(t) = E(t-1) + \frac{1}{3}D(t-1) + \frac{1}{2}F(t-1) + \frac{1}{4}G(t-1)$$

$$F(t) = \frac{1}{2}B(t-1) + \frac{1}{3}D(t-1) + \frac{1}{4}G(t-1)$$

$$G(t) = \frac{1}{2}B(t-1) + \frac{1}{2}C(t-1).$$

After the required modifications, if we run our Excel model for this DDS, we end up with Table 7.25 for rank by importance. Note that this is not a useful ranking. Our dangling node at page E has caused all of the importance in the entire web to be collected there.

TABLE 7.25   Rank of Importance with E as a Dangling Node

Page	Importance (%)
E	100
B	0
C	0
D	0
F	0
G	0

This is the same issue we ran into the first time we considered an undefeated sports team, and it keeps our method from producing a useful ranking. We get a flow diagram that is not irreducible, and the model fails to have a positive stable distribution. ☐

Next we show how to implement our random page selection idea as a fix for the dangling node problem. We assume that anyone visiting page E will afterwards have an equal chance of jumping to any of the 6 webpages by typing in a new URL or remaining on the current page. This eliminates the dangling node problem and results in the flow diagram given in Figure 7.20.

Note that the flow diagram is now both irreducible and aperiodic so we know we will get a positive stable distribution. Along with the new flow diagram, we have a new table of how importance is distributed to destination pages that we call the **modified hyperlink table**, HT′, shown in Table 7.26.

Now we are ready to produce a useful ranking.

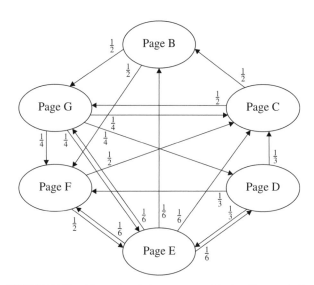

FIGURE 7.20   Flow of web importance with dangling node fix.

TABLE 7.26   Modified Hyperlink Table After Eliminating Dangling Node

HT′	B	C	D	E	F	G
B	0	1/2	0	1/6	0	0
C	0	0	1/3	1/6	1/2	1/4
D	0	0	0	1/6	0	1/4
E	0	0	1/3	1/6	1/2	1/4
F	1/2	0	1/3	1/6	0	1/4
G	1/2	1/2	0	1/6	0	0

Example 7.14: Rerank the webpages using the modified hyperlink table given in Table 7.26.

First we have the corresponding simplified DDS:

$$B(t) = \frac{1}{2}C(t-1) + \frac{1}{6}E(t-1)$$

$$C(t) = \frac{1}{3}D(t-1) + \frac{1}{6}E(t-1) + \frac{1}{2}F(t-1) + \frac{1}{4}G(t-1)$$

$$D(t) = \frac{1}{6}E(t-1) + \frac{1}{4}G(t-1)$$

$$E(t) = \frac{1}{3}D(t-1) + \frac{1}{6}E(t-1) + \frac{1}{2}F(t-1) + \frac{1}{4}G(t-1)$$

$$F(t) = \frac{1}{2}B(t-1) + \frac{1}{3}D(t-1) + \frac{1}{6}E(t-1) + \frac{1}{4}G(t-1)$$

$$G(t) = \frac{1}{2}B(t-1) + \frac{1}{2}C(t-1) + \frac{1}{6}E(t-1).$$

TABLE 7.27   Ranking After Dangling Node
has been Eliminated

Page	Importance (%)
C	20.11
E	20.11
G	20.11
F	17.88
B	13.41
D	8.38

Now implementing the model in Excel produces the ranking given in Table 7.27. This is clearly a more useful ranking than what we produced with E as a dangling node. □

In the next section we examine another complication for the Markov ranking method that is closely related to the dangling node issue.

7.3.3   The Presence of Subwebs

In Section 7.3.2 we saw that a dangling node caused our Markov page ranking method to fail because of the lack of irreducibility. We fixed the dangling node issue by forcing dangling nodes to distribute their importance equally among all webpages.

In this section we consider a related problem: the possibility of a **subweb**. If our web turns out to have a subset of webpages, possibly with links among each other, for which it is not possible to move to a webpage outside of the subset, we have what we call a subweb. The presence of a subweb means that the flow diagram will not be irreducible since it will not be possible to move between any pair of webpages: once we get to the subweb, we will be stuck there.

If we encounter a subweb, our ranking system to this point can give strange results. We examine one such possibility in the following example.

Example 7.15: Rank the importance of webpages in an 8-page web given the link structure represented in Figure 7.21.

With the links pictured in Figure 7.21, the corresponding importance labels are as in Figure 7.22. Thus the hyperlink table is given by Table 7.28. There are no dangling nodes, so we do not need to modify the hyperlink table. The corresponding simplified DDS is given by

$$B(t) = \frac{1}{2}C(t-1) + \frac{1}{4}D(t-1)$$

$$C(t) = \frac{1}{4}E(t-1)$$

$$D(t) = B(t-1) + \frac{1}{4}E(t-1)$$

$$E(t) = \frac{1}{2}C(t-1) + \frac{1}{4}D(t-1)$$

$$F(t) = \frac{1}{4}D(t-1) + \frac{1}{4}E(t-1) + \frac{1}{2}G(t-1) + \frac{1}{3}H(t-1) + \frac{1}{2}I(t-1)$$

$$G(t) = \frac{1}{4}D(t-1) + \frac{1}{4}E(t-1) + \frac{1}{2}F(t-1) + \frac{1}{3}H(t-1)$$

$$H(t) = \frac{1}{2}G(t-1) + \frac{1}{2}I(t-1)$$

$$I(t) = \frac{1}{2}F(t-1) + \frac{1}{3}H(t-1).$$

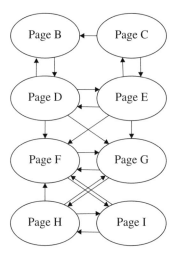

FIGURE 7.21   Link structure for an 8-page web.

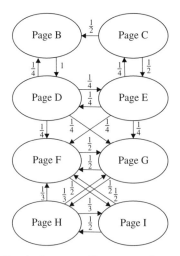

FIGURE 7.22   Assignment of importance for an 8-page web.

TABLE 7.28   Hyperlink Table for an 8-Page Web

HT	B	C	D	E	F	G	H	I
B	0	1/2	1/4	0	0	0	0	0
C	0	0	0	1/4	0	0	0	0
D	1	0	0	1/4	0	0	0	0
E	0	1/2	1/4	0	0	0	0	0
F	0	0	1/4	1/4	0	1/2	1/3	1/2
G	0	0	1/4	1/4	1/2	0	1/3	0
H	0	0	0	0	0	1/2	0	1/2
I	0	0	0	0	1/2	0	1/3	0

Implementing the DDS in Excel with initial values of importance all set to 1, we get the rankings presented in Table 7.29.

TABLE 7.29   Rankings by Importance for an 8-Page Web with Subweb

Page	Importance (%)
F	30.77
G	23.08
H	23.08
I	23.08
B	0
C	0
D	0
E	0

This is a strange result. Pages B, C, D, and E all have incoming links, so they receive importance from other webpages, but they still end up with 0 importance. The reason is that pages F, G, H, and I form a subweb that contains links among themselves, has incoming links from B, C, D, and E but has no outgoing links to B, C, D, and E. This gives us another example of a flow diagram that is not irreducible. We cannot travel between every pair of pages because there is no way to get from F, G, H, or I to pages B, C, D, and E. All importance eventually ends up in the subweb so that those pages are the only ones that end up with a useful ranking. We get a stable distribution just not a positive one. □

The last step in completing the Google PageRank algorithm is to eliminate the possibility of subwebs. We do this by following Brin and Page in modeling more realistic behavior in web surfers. People who are surfing the web do not necessarily follow links from 1 page to another: just as they would from a page with no links, surfers might at any time type in a URL to jump to any page on the web. In accounting for this possibility, we connect every webpage with every other webpage and thus guarantee that the web is irreducible and aperiodic. It will therefore have a positive stable distribution.

We start by creating a table that models the situation where a web surfer can go from any webpage to any other webpage with equal likelihood, including staying on the current page. The table is not too difficult to create: if we let N be the total number of webpages in our web, then we will have N rows, N columns, and we let every entry be $\frac{1}{N}$. For our 8-page web from Example 7.15, this gives us Table 7.30.

TABLE 7.30   Teleportation Table for an 8-Page Web

T	B	C	D	E	F	G	H	I
B	1/8	1/8	1/8	1/8	1/8	1/8	1/8	1/8
C	1/8	1/8	1/8	1/8	1/8	1/8	1/8	1/8
D	1/8	1/8	1/8	1/8	1/8	1/8	1/8	1/8
E	1/8	1/8	1/8	1/8	1/8	1/8	1/8	1/8
F	1/8	1/8	1/8	1/8	1/8	1/8	1/8	1/8
G	1/8	1/8	1/8	1/8	1/8	1/8	1/8	1/8
H	1/8	1/8	1/8	1/8	1/8	1/8	1/8	1/8
I	1/8	1/8	1/8	1/8	1/8	1/8	1/8	1/8

Because this table represents a surfer's ability to go *to* any webpage *from* any webpage, we refer to it as the **teleportation table**, T. We interpret this table in a way that is similar to the just-for-playing credit in the context of sports teams. Just by virtue of being a page on the web, all pages receive some importance from every other page.

The next step is to somehow combine this table with the modified hyperlink table, HT', in a way that keeps the sum of entries in each column equal to one. If we were to simply add the two tables together, we would get columns whose entries sum to more than one and that would result in pages transferring more than 100% of their importance to others.

The way we avoid this potential excess of importance is to introduce a new parameter. The idea is that at every time step, a web surfer has two options: either follow one of the hyperlinks on the current page or "teleport" to another by typing in a randomly selected URL. The new parameter represents the probability that a web surfer will elect to follow a link versus teleporting to another page. We call this parameter the **Google parameter**, α; it is the probability that a web surfer will follow one of the provided hyperlinks from the current page. This makes $(1-\alpha)$ the probability that the surfer will choose to teleport instead. For example, if $\alpha = 0.85$, then at every step there is an 85% chance that a web surfer will follow a hyperlink and a 15% chance that the surfer will type in a new URL and teleport to another page.

We use α to combine the modified hyperlink table, HT', with the teleportation table, T. The final table will be the sum of α times the modified hyperlink table plus

$(1-\alpha)$ times the teleportation table. We can experiment with different values for α, but for now, we use a common estimate for the value Google uses: $\alpha = 0.85$. We call the final table the **Google table**, GT, where

$$GT = \alpha \times HT' + (1-\alpha) \times T.$$

We illustrate how the Google table is constructed with the 8-page web from Example 7.15. In that example there were no dangling nodes so the modified hyperlink table is the same as the original, that is, $HT' = HT$.

We calculate α times the modified hyperlink table, HT', by multiplying each entry of HT' by α. Because rounding errors can cause problems for us later in Excel, we use fractions rather than decimals so $\alpha = \dfrac{17}{20}$. The result is given in Table 7.31.

TABLE 7.31   Modified Hyperlink Table Times α

$\alpha \times HT'$	B	C	D	E	F	G	H	I
B	0	17/40	17/80	0	0	0	0	0
C	0	0	0	17/80	0	0	0	0
D	17/20	0	0	17/80	0	0	0	0
E	0	17/40	17/80	0	0	0	0	0
F	0	0	17/80	17/80	0	17/40	17/60	17/40
G	0	0	17/80	17/80	17/40	0	17/60	0
H	0	0	0	0	0	17/40	0	17/40
I	0	0	0	0	17/40	0	17/60	0

We also need to form $(1-\alpha) \times T$ by multiplying each entry of the teleportation table by $(1-\alpha) = \dfrac{3}{20}$. The resulting table is given in Table 7.32.

TABLE 7.32 $(1-\alpha) \times T$ for an 8-Page Web

$(1-\alpha) \times T$	B	C	D	E	F	G	H	I
B	3/160	3/160	3/160	3/160	3/160	3/160	3/160	3/160
C	3/160	3/160	3/160	3/160	3/160	3/160	3/160	3/160
D	3/160	3/160	3/160	3/160	3/160	3/160	3/160	3/160
E	3/160	3/160	3/160	3/160	3/160	3/160	3/160	3/160
F	3/160	3/160	3/160	3/160	3/160	3/160	3/160	3/160
G	3/160	3/160	3/160	3/160	3/160	3/160	3/160	3/160
H	3/160	3/160	3/160	3/160	3/160	3/160	3/160	3/160
I	3/160	3/160	3/160	3/160	3/160	3/160	3/160	3/160

Finally we form the Google table,

$$GT = \frac{17}{20} \times HT' + \frac{3}{20} \times T,$$

by adding the corresponding entries from Tables 7.31 and 7.32 together. The result is presented in Table 7.33. Note that if we sum the proportions of importance given out by a webpage by adding entries down the appropriate column, we get exactly one as required.

In our next example we examine the effect of using the Google table to rank the webpages from Example 7.15 where we had an 8-page web that included a subweb.

Example 7.16: Use the Google table in Table 7.33 to rerank the webpages from Example 7.15 in order of importance.

TABLE 7.33   The Google Table for an 8-Page Web with $\alpha = 0.85$

GT	B	C	D	E	F	G	H	I
B	3/160	71/160	37/160	3/160	3/160	3/160	3/160	3/160
C	3/160	3/160	3/160	37/160	3/160	3/160	3/160	3/160
D	139/160	3/160	3/160	37/160	3/160	3/160	3/160	3/160
E	3/160	71/160	37/160	3/160	3/160	3/160	3/160	3/160
F	3/160	3/160	37/160	37/160	3/160	71/160	29/96	71/160
G	3/160	3/160	37/160	37/160	71/160	3/160	29/96	3/160
H	3/160	3/160	3/160	3/160	3/160	71/160	3/160	71/160
I	3/160	3/160	3/160	3/160	71/160	3/160	29/96	3/160

Using the Google table instead of the hyperlink table to produce our ranking means we have to reconstruct our DDS. As with the modified hyperlink table, the use of the Google table means it is possible for a web surfer to remain on the same webpage for more than one time step. We can view the entries along the main diagonal as a webpage distributing importance to itself or as just the remaining importance that is not distributed to other pages.

In the corresponding simplified DDS below, we suppress the "$t-1$" on the right to make the equations easier to read.

$$B(t) = \frac{3}{160}B + \frac{71}{160}C + \frac{37}{160}D + \frac{3}{160}E + \frac{3}{160}F + \frac{3}{160}G + \frac{3}{160}H + \frac{3}{160}I$$

$$C(t) = \frac{3}{160}B + \frac{3}{160}C + \frac{3}{160}D + \frac{37}{160}E + \frac{3}{160}F + \frac{3}{160}G + \frac{3}{160}H + \frac{3}{160}I$$

$$D(t) = \frac{139}{160}B + \frac{3}{160}C + \frac{3}{160}D + \frac{37}{160}E + \frac{3}{160}F + \frac{3}{160}G + \frac{3}{160}H + \frac{3}{160}I$$

$$E(t) = \frac{3}{160}B + \frac{71}{160}C + \frac{37}{160}D + \frac{3}{160}E + \frac{3}{160}F + \frac{3}{160}G + \frac{3}{160}H + \frac{3}{160}I$$

$$F(t) = \frac{3}{160}B + \frac{3}{160}C + \frac{37}{160}D + \frac{37}{160}E + \frac{3}{160}F + \frac{71}{160}G + \frac{29}{96}H + \frac{71}{160}I$$

$$G(t) = \frac{3}{160}B + \frac{3}{160}C + \frac{37}{160}D + \frac{37}{160}E + \frac{71}{160}F + \frac{3}{160}G + \frac{29}{96}H + \frac{3}{160}I$$

$$H(t) = \frac{3}{160}B + \frac{3}{160}C + \frac{3}{160}D + \frac{3}{160}E + \frac{3}{160}F + \frac{71}{160}G + \frac{3}{160}H + \frac{71}{160}I$$

$$I(t) = \frac{3}{160}B + \frac{3}{160}C + \frac{3}{160}D + \frac{3}{160}E + \frac{71}{160}F + \frac{3}{160}G + \frac{29}{96}H + \frac{3}{160}I.$$

TABLE 7.34 Ranking of Importance by Google Table for an 8-Page Web

Webpage	Importance (%)
F	25.50
G	20.19
H	18.03
I	17.82
D	6.64
B	4.49
E	4.49
C	2.83

After implementing the DDS in Excel and assigning initial values of importance, we drag the formulas down until we reach the positive stable distribution, which we know from Theorem 7.1 will exist. The result is given in Table 7.34.

Every webpage is now assigned a positive percentage of importance, and the Google method has ranked all webpages in a sensible way. Furthermore the Google method has broken what was a 3-way tie for second in Example 7.15. □

In the next section we indicate how to derive the equilibrium distribution for a web algebraically.

7.3.4 Equilibrium Points

As we mentioned in Section 7.2.3, we can find the equilibrium distribution for the systems in this chapter by hand. For the previous example we would find points $(B^*, C^*, D^*, E^*, F^*, G^*, H^*, I^*)$ such that

$$B^* = \frac{3}{160}B^* + \frac{71}{160}C^* + \frac{37}{160}D^* + \frac{3}{160}E^* + \frac{3}{160}F^* + \frac{3}{160}G^* + \frac{3}{160}H^* + \frac{3}{160}I^*$$

$$C^* = \frac{3}{160}B^* + \frac{3}{160}C^* + \frac{3}{160}D^* + \frac{37}{160}E^* + \frac{3}{160}F^* + \frac{3}{160}G^* + \frac{3}{160}H^* + \frac{3}{160}I^*$$

$$D^* = \frac{139}{160}B^* + \frac{3}{160}C^* + \frac{3}{160}D^* + \frac{37}{160}E^* + \frac{3}{160}F^* + \frac{3}{160}G^* + \frac{3}{160}H^* + \frac{3}{160}I^*$$

$$E^* = \frac{3}{160}B^* + \frac{71}{160}C^* + \frac{37}{160}D^* + \frac{3}{160}E^* + \frac{3}{160}F^* + \frac{3}{160}G^* + \frac{3}{160}H^* + \frac{3}{160}I^*$$

$$F^* = \frac{3}{160}B^* + \frac{3}{160}C^* + \frac{37}{160}D^* + \frac{37}{160}E^* + \frac{3}{160}F^* + \frac{71}{160}G^* + \frac{29}{96}H^* + \frac{71}{160}I^*$$

$$G^* = \frac{3}{160}B^* + \frac{3}{160}C^* + \frac{37}{160}D^* + \frac{37}{160}E^* + \frac{71}{160}F^* + \frac{3}{160}G^* + \frac{29}{96}H^* + \frac{3}{160}I^*$$

$$H^* = \frac{3}{160}B^* + \frac{3}{160}C^* + \frac{3}{160}D^* + \frac{3}{160}E^* + \frac{3}{160}F^* + \frac{71}{160}G^* + \frac{3}{160}H^* + \frac{71}{160}I^*$$

$$I^* = \frac{3}{160}B^* + \frac{3}{160}C^* + \frac{3}{160}D^* + \frac{3}{160}E^* + \frac{71}{160}F^* + \frac{3}{160}G^* + \frac{29}{96}H^* + \frac{3}{160}I^*.$$

Though we could solve this system by hand, it would certainly take a long time, and with so much algebra to do there would be a high likelihood of making a mistake along the way. This is the kind of system we would instead solve using a computer algebra system such as Maple, Mathematica, or MATLAB.

Once we have the equilibrium point $(B^*, C^*, D^*, E^*, F^*, G^*, H^*, I^*)$, we can determine the equilibrium distribution. Page B, for example, will end up with proportion of web importance equal to

$$\frac{B^*}{B^* + C^* + D^* + E^* + F^* + G^* + H^* + I^*}.$$

Because the Google PageRank method imposes structure on our flow diagram that guarantees both irreducibility and aperiodicity, our algebraic solution for the equilibrium distribution is certain to be the positive stable distribution for the model. Hence we can use it rather than Excel to produce the webpage importance rankings.

7.3.5 PageRank Summary

We have shown how to rank webpages by importance using the link structure of the web. The steps of the method are outlined below:

1. Create a hyperlink table that shows how much importance a webpage distributes to those it links to. The fraction of importance given to each destination page is the reciprocal of the number of outgoing links.

2. Fix any dangling nodes by creating the modified hyperlink table. This is done by forcing any dangling nodes to distribute their importance equally to every page on the web.

3. Fix any subwebs by forming the Google table. This is done by first creating the teleportation table that represents a web surfer's ability to visit any page on the web at any time by typing in a URL. Next we combine the modified hyperlink table with the teleportation table using the Google parameter.

4. From the Google table we form a DDS, implement the DDS in Excel, find the long-term distribution of importance, which Theorem 7.1 now guarantees, and finally rank the webpages in order of importance.

With approximately 1 billion active websites currently on the Internet (Internet Live Stats, 2015), ranking webpages in the way we have shown clearly involves practical issues beyond the basic idea of the method. Our tables, for example, would require about 1 billion rows and 1 billion columns each. Furthermore, we would have to map the entire link structure of the web. Information on how to deal with these practical difficulties can be found in other sources, including (Brin & Page, 1998; Langville & Meyer, 2006).

7.3.6 Section Exercises

1 Determine whether the web represented in Figure 7.23 contains any dangling nodes or subwebs.

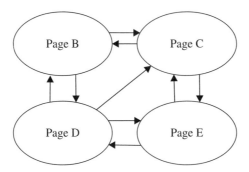

FIGURE 7.23 Web structure for Exercise 1.

2 Determine whether the web represented in Figure 7.24 contains any dangling nodes or subwebs.

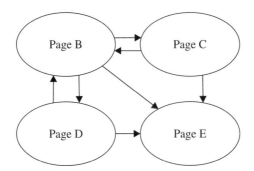

FIGURE 7.24 Web structure for Exercise 2.

3 Determine whether the web represented in Figure 7.25 contains any dangling nodes or subwebs.

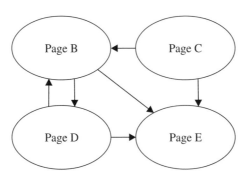

FIGURE 7.25 Web structure for Exercise 3.

4 Determine whether the web represented in Figure 7.26 contains any dangling nodes or subwebs.

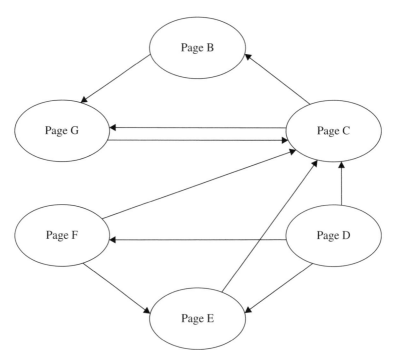

FIGURE 7.26   Web structure for Exercise 4.

5 Consider the web represented in Figure 7.23:
   a. Label the arrows with the proportions of importance that each origination page distributes to the destination page.
   b. Create a hyperlink table from the flow diagram.
   c. Fix any dangling nodes by creating the modified hyperlink table.
   d. Form the teleportation table.
   e. Fix any subwebs by forming the Google table.
   f. Rank the webpages in order of importance.

6 Consider the web represented in Figure 7.24:
   a. Label the arrows with the proportions of importance that each origination page distributes to the destination page.
   b. Create a hyperlink table from the flow diagram.
   c. Fix any dangling nodes by creating the modified hyperlink table.
   d. Form the teleportation table.
   e. Fix any subwebs by forming the Google table.
   f. Rank the webpages in order of importance.

7 Consider the web represented in Figure 7.25:
   a. Label the arrows with the proportions of importance that each origination page distributes to the destination page.

b. Create a hyperlink table from the flow diagram.

c. Fix any dangling nodes by creating the modified hyperlink table.

d. Form the teleportation table.

e. Fix any subwebs by forming the Google table.

f. Rank the webpages in order of importance.

8 Consider the web represented in Figure 7.26.

    a. Label the arrows with the proportions of importance that each origination page distributes to the destination page.

    b. Create a hyperlink table from the flow diagram.

    c. Fix any dangling nodes by creating the modified hyperlink table.

    d. Form the teleportation table.

    e. Fix any subwebs by forming the Google table.

    f. Rank the webpages in order of importance.

8

BODY WEIGHT AND BODY COMPOSITION

In this chapter we present some models for projecting weight and body composition changes brought about by changes in diet or activity level. However, it is important to understand that weight management is a complex issue that involves many behavioral and societal factors beyond the scope of this book. Models like the ones we develop can predict weight changes based on assumed eating habits, but they cannot explain *why* individuals may have the eating habits they do, nor should the models be used to recommend a particular diet.

Furthermore, the models in this chapter deal almost entirely with "calories in" and "calories out." With the exception of the Weight Watchers™ model in Section 8.6, they make no distinction between calories that come from soda and calories that come from spinach. Clearly one is healthier than the other, but by and large our models will not address this issue.

What our models can do is predict weight change based on a given caloric intake and activity level, and these predictions can be tailored to the individual user. Along with models for predicting weight change, we present some common ways of determining a healthy weight or body composition for a given individual.

Though they are certainly related, there is a distinction between weight and health, and our models deal primarily with weight. We should keep in mind that even under established guidelines for what is considered a healthy weight, there are no absolutes. An overweight person may be metabolically healthy, while a thin person or person with normal body weight may be very unhealthy (Ortega et al., 2013). Obsessing over body weight is definitely not healthy and can be a sign of a serious underlying eating

Models for Life: An Introduction to Discrete Mathematical Modeling with Microsoft® Office Excel®, First Edition. Jeffrey T. Barton.
© 2016 John Wiley & Sons, Inc. Published 2016 by John Wiley & Sons, Inc.

disorder. Ultimately our focus should be on health and well-being and not body weight. We take a step in this direction later in the chapter when we shift our modeling focus from body weight to body composition.

8.1 CONSTANT CALORIE EXPENDITURE

Mathematical models for weight change are based on the **energy balance principle**, which is an application of the first law of thermodynamics (Thomas et al., 2009). In the context of body weight, the energy balance principle states that if more calories are consumed than burned, then weight increases (i.e., the excess energy must be stored in the body), and if more calories are burned than consumed, then weight decreases (i.e., energy stored in the body must be used to make up the difference). Thus, according to the energy balance principle, we can predict changes in weight by keeping careful track of "calories in" and "calories out."

As a first attempt at modeling weight changes, we make two simplifying assumptions: (1) that 1 pound of body weight contains 3500 calories of energy and (2) that an individual burns a constant number of calories per day. Hence in order to lose 1 pound of weight, an individual must create a net calorie deficit of 3500 calories. We note that in the context of diet and weight loss when the term **calorie** is used, it actually refers to **kilocalories (kcal)**. In keeping with standard usage, we will use *calorie* and *kcal* interchangeably.

Let t represent time in days and $W(t)$ body weight in pounds after t days. We illustrate the most basic model for $W(t)$ with the example below.

Example 8.1: Set up a model for weight change in an individual who burns 2000 calories per day and follows a diet consisting of 1900 calories per day.

By burning more calories than are being consumed, the individual should lose weight. (In fact under our current assumptions, the individual should lose 1 pound every 35 days.) Figure 8.1 shows the flow diagram for the situation after converting calories to pounds. Using the standard 3500 calories per pound, we have an increase in body weight of $\frac{1900}{3500} = 0.543$ pounds per day and a decrease in body weight of $\frac{2000}{3500} = 0.571$ pounds per day.

FIGURE 8.1   Constant calorie flow diagram for Example 8.1.

The corresponding DDS for the model is given by $W(t) = W(t-1) + 0.543 - 0.571$ or $W(t) = W(t-1) - 0.028$. According to the model, each day this individual will lose 0.028 pounds. □

More generally if we let the daily calorie intake be the constant, I_0, and the daily calorie expenditure be the constant, E_0, our flow diagram will be as shown in Figure 8.2.

FIGURE 8.2   General constant calorie expenditure flow diagram.

Correspondingly, our more general DDS is given by

$$W(t) = W(t-1) + \frac{I_0}{3500} - \frac{E_0}{3500}.$$

If we store our initial weight, daily intake, and daily expenditure as parameters in Excel, our spreadsheet setup with body weight formula showing will appear as in Figure 8.3.

	A	B	C	D
1	Constant Calorie Expenditure Model			
2				
3	Current Body Weight =		215	(pounds)
4	Daily Intake, I_0 =		1750	(calories)
5	Daily Expenditure, E_0 =		2200	(calories)
6				
7	t	$W(t)$		
8		0	215	
9		1	=B8+$C$4/3500-$C$5/3500	

FIGURE 8.3   Excel setup for constant calorie expenditure model.

Note that as usual we use absolute addressing when referring to the intake and expenditure parameters. The following example provides a quick check of our work.

Example 8.2: John currently weighs 215 pounds. Assuming he consumes 1750 calories per day and burns 2200 calories per day, predict John's weight 2 weeks from today.

This is a straightforward application of the model. We type in the correct parameter values, copy our formulas down, and then note the weight after day 14. The results are shown in Figure 8.4. The model predicts that John will weigh 213.2 pounds in 2 weeks. $\square$

	A	B	C	D
1	Constant Calorie Expenditure Model			
2				
3	Current Body Weight =		215	(pounds)
4	Daily Intake, I_0 =		1750	(calories)
5	Daily Expenditure, E_0 =		2200	(calories)
6				
7	t	$W(t)$		
8	0	215		
9	1	214.9		
21	13	213.3		
22	14	213.2		

FIGURE 8.4   Excel results for Example 8.2.

Next we take a look at equilibrium values for the model.

8.1.1   Equilibrium Analysis

Finding equilibrium values for the constant calorie model requires that we find values for W^* such that $W^* = W^* + \dfrac{I_0}{3500} - \dfrac{E_0}{3500}$. After canceling the W^* on both sides of the equation, we are left with $0 = \dfrac{I_0}{3500} - \dfrac{E_0}{3500}$. This implies that the only way for us to have an equilibrium value is if $I_0 = E_0$, and if this is the case, we will have an equilibrium regardless of the actual body weight. This result should appeal to our intuition: if we burn the same number of calories that we consume, then our weight should not change.

While this constant calorie model has the benefit of simplicity, it also has serious limitations. First, the model's long-term predictions are unreasonable: in the long run it predicts an infinite weight if $I_0 > E_0$ and a negative weight if $I_0 < E_0$. Second, the model assumes that both I_0 and E_0 are constants. We can always arrange for I_0 to be constant by carefully tracking our diet and counting calories, but our daily calorie expenditure E_0 is more problematic. In general E_0 is not a constant but instead depends on factors such as body weight, sex, and body composition. In subsequent sections we explore ways of finding estimates for E_0.

8.1.2   Section Exercises

1   Implement the constant calorie expenditure model with Excel using your own body weight and calorie parameters. Note what the model predicts for your body weight over 1 week, 1 month, 6 months, 1 year, and 5 years. At what point does the model start to become unreasonable?

2   Suppose your daily intake and expenditure are equal so that your body weight is in an equilibrium state. Use Excel to show that the state is unstable with regard to changes in intake or expenditure.

3 Suppose someone consumes 100 calories per day more than they burn. How much weight will the person gain over a 2-week period?

4 What would be the required daily calorie deficit for someone who wants to lose 10 pounds in 30 days?

5 How long will it take to lose 10 pounds with a calorie deficit of 1 calorie per day?

6 *Extension*: As we age we typically burn fewer calories, largely due to a decrease in lean muscle mass. A good estimate for the age effect on calorie expenditure is that we burn 5 fewer calories per day for each year we age.

 a. Modify the constant expenditure model to reflect this fact.

 b. Use the new model and your parameters from Exercise 1 to project your body weight over 1 week, 1 month, 6 months, 1 year, and 5 years.

 c. Has the prediction changed significantly from the original?

8.2 VARIABLE CALORIE EXPENDITURE

It is difficult to know with certainty how many calories an individual will burn on a daily basis. It can also be medically important to know when a physician needs to carefully control a patient's caloric intake. An off-the-cuff guess as in the previous section will not suffice if we hope to create a realistic model.

 A first step in estimating calorie needs is to find an individual's **resting energy expenditure** (REE). REE is the number of calories an individual burns at rest, and it serves as a baseline for calories burned before taking into account daily activities or exercise. We should intuitively expect that larger individuals burn more calories than smaller ones, and we will see that this is generally true.

 The most reliable method for actually measuring (as opposed to estimating) REE is a kind of breath test known as **indirect calorimetry.** With this test energy expenditure is calculated by measuring the amount of oxygen taken in and the amount of carbon dioxide exhaled in the breath (Haugen, Chan, & Li, 2007). Indirect calorimetry is not a common test for healthy individuals, but its increasing availability and decreasing cost have made it feasible for an interested individual to have the test done. Still, we cannot assume that an individual will have the means, access, or desire to actually measure their REE, so for the purposes of our model, we estimate REE using equations developed by other researchers for this purpose.

8.2.1 Mifflin–St. Jeor Equations for REE

A quick web search on "resting energy expenditure" reveals that many different equations for estimating REE have been proposed over the years. Among the most commonly used are the Harris–Benedict, Katch-McArdle, Livingston–Kohlstadt, and Mifflin–St. Jeor (MSJ) equations. A review article examining several such equations found that the MSJ equations were most likely to be accurate (Frankenfield, Roth-Yousey, & Compher, 2005). In this section we use the MSJ equations, which were developed in 1990 by Mifflin et al. (1990).

Mifflin and St. Jeor found that the most important factors determining an individual's REE were weight, height, sex, and age. Similar to the sex difference in total body water (see Chapter 6), the sex difference in how much energy is expended at rest is largely due to the different amounts of lean body mass typically carried by men and women; however, some researchers have found evidence that this alone does not explain all of the observed difference (Arciero, Goran, & Poehlman, 1993). We give the MSJ equations for males and for females below:

$$\text{Males: REE} = 4.536 \cdot W + 15.875 \cdot H - 5 \cdot A + 5$$
$$\text{Females: REE} = 4.536 \cdot W + 15.875 \cdot H - 5 \cdot A - 161.$$

Here REE is given in calories burned per day, W is weight in pounds, H is height in inches, and A is age in years. More compactly, if we let $1 = $ male and $0 = $ female, we can use the single equation

$$\text{REE} = 4.536 \cdot W + 15.875 \cdot H - 5 \cdot A + 166 \cdot S - 161.$$

where S is sex (Mifflin et al., 1990).

In developing our model we treat height, age, and sex as constants so that body weight is the only factor in REE that varies. Certainly height and sex will remain constant, but this treatment of age will introduce some error into our model. However, unless we are projecting weight over a very long time period, it should not have a dramatic effect. In the following we provide an example of how to use the MSJ formula to estimate REE.

Example 8.3: Susan is a 135-pound woman who is 20 years old and 5'4″ tall; estimate her resting energy expenditure.

Here we simply plug all of our variables into the MSJ equation for females, being careful to convert 5'4″ to 64″:

$$\text{REE} = 4.536 \cdot 135 + 15.875 \cdot 64 - 5 \cdot 20 - 161 = 1367.36.$$

Susan's REE would be about 1367 calories per day. □

Once we have an estimate for REE the next step in determining total calorie needs is to account for daily activities.

8.2.2 Activity Level

It is important to realize that the MSJ equations estimate one's calorie needs *at rest*. However, most people do not rest all day; rather, they go to school or work, run errands, play sports, work in their gardens, etc. To account for the increased calorie needs brought about by our daily activities, we multiply REE by a number called the **activity level**. The more active we are in our daily lives, the higher our activity level will be. Once we multiply REE by this activity level, we have a true estimate for our daily calorie expenditure, E_0. We denote the activity level by a lowercase Greek lambda, λ_0. Table 8.1 summarizes guidelines for assigning the activity level based on a report by the Food and Agriculture Organization (FAO) of the United Nations (FAO, 2001).

TABLE 8.1 Values Assigned to λ_0 and Examples for a Variety of Activity Levels

Activity Level	Description	λ_0
Sedentary	Desk job, little physical activity	1.2
Light activity	Jobs involving some standing such as retail sales, some walking as exercise, and light housework	1.375
Moderate activity	Mason, construction worker, or sedentary occupation with daily hour of moderate intensity exercise	1.55
High activity	Strenuous work or exercise for several hours daily, hard manual labor such as nonmechanized farming, or nonsedentary occupation with 2 h of moderate to intense exercise daily	1.725
Extreme activity	Multiple bouts of long and intense exercise daily such as for serious athletes in season or a strenuous occupation with additional leisure exercise	1.9

For example, we might assign a computer programmer who spends most of her day sitting at a desk an activity level of 1.2, while a construction worker who engages in regular intense exercise might be assigned a level of 1.725.

Example 8.4: For Example 8.3, estimate Susan's daily energy expenditure, E_0, if she works as a bike courier.

We already have an estimate for Susan's REE, namely, 1367 calories. Because of her job, we assign Susan an activity level of 1.725 for high activity; this results in a total daily calorie expenditure of $E_0 = 1.725 \cdot 1367 = 2358$ calories per day. □

We are now ready to modify our constant calorie model by incorporating our new knowledge of how energy expenditure varies from person to person. In particular we note that the MSJ equations produce different estimates for REE for different body weights. This adds a complication to our model because as our body weight changes, so do our calorie needs.

First, we update our notation for calorie expenditure to indicate that it will no longer be constant but will instead depend on time. Hence we use $E(t-1)$ instead of E_0, and our new DDS for body weight over time is

$$W(t) = W(t-1) + \frac{I_0}{3500} - \frac{E(t-1)}{3500}.$$

Next we find the formula for $E(t-1)$ by multiplying the MSJ equation by the activity level, λ_0. We get

$$E(t-1) = \lambda_0 \cdot (4.536 \cdot W(t-1) + 15.875 \cdot H - 5 \cdot A + 166 \cdot S - 161).$$

It is important to notice that in the MSJ equation we have plugged in $W(t-1)$ for weight. This is what makes our new model an improvement over the old—it updates our daily calorie expenditure as our weight changes. This change is in agreement with what we observe in weight loss studies: if we lose weight, our calorie needs decrease, and it subsequently becomes a harder to lose more weight.

Substituting the new expression for $E(t-1)$ into the DDS, we get the final version of the variable calorie expenditure model:

$$W(t) = W(t-1) + \frac{I_0}{3500} - \frac{\lambda_0}{3500}(4.536 \cdot W(t-1) + 15.875 \cdot H - 5 \cdot A + 166 \cdot S - 161).$$

At first glance this new model may appear to be very complicated; however, it is just another example of an affine model, albeit with more complicated constants. We demonstrate this fact when we perform an equilibrium analysis on the model in the next section. For now we just need to be careful when entering the formula into Excel.

The daily calorie intake (I_0), height (H), age (A), sex (S), and activity level (λ_0) are all user-supplied parameters. To implement the model in Excel, we set aside space at the top of our worksheet for the user to input the required parameters, and to make the spreadsheet more user friendly, we make use of Excel's Data Validation feature for sex and activity level (see Section 6.1). Figure 8.5 shows a screenshot of our setup before the equations are entered.

	A	B	C	D
1	Variable Calorie Expenditure Model			
2				
3	Sex, S =		M	
4	Current Body Weight =		215	(pounds)
5	Age, A =		25	(years)
6	Height, H =		70	(inches)
7	Daily Intake, I_0 =		1750	(calories)
8	Activity Level, λ_0 =		1.55	
9			Enter activity level: 1.2 = sedentary 1.375 = light 1.55 = moderate 1.725 = high 1.9 = extreme	
10	t	W(t)		
11	0	215		

FIGURE 8.5   Excel setup for variable calorie expenditure model.

Typing the DDS in correctly will take some care since there are so many inputs. Figure 8.6 shows the final product with the equation for body weight displayed.

	A	B	C	D	E	F	G	H	I	J
1	Variable Calorie Expenditure Model									
2										
3	Sex, $S =$		M							
4	Current Body Weight =		215	(pounds)						
5	Age, $A =$		25	(years)						
6	Height, $H =$		70	(inches)						
7	Daily Intake, $I_0 =$		1750	(calories)						
8	Activity Level, $\lambda_0 =$		1.55							
9										
10	t		$W(t)$							
11	0		215							
12	1	=B11+$C$7/3500-($C$8/3500)*(4.536*B11+15.875*$C$6-5*$C$5+166*IF($C$3="M",1,0)-161)								
13	2		214.3							

FIGURE 8.6   Excel setup with body weight formula.

Note that we have referred to all parameters with absolute addressing and that we have used the single MSJ equation with an IF statement for sex. We try out our new model with an example.

Example 8.5: Suppose Karen is a 25-year-old computer programmer who currently weighs 165 pounds and is 5'8" tall. To help offset the sedentary nature of her job, she rides her bike to work and walks a few blocks at lunch every day. Predict Karen's weight 1 month from now and her long-term weight if she eats approximately 1900 kcal per day.

   Though it is a matter of judgment, we assign Karen an activity level of $\lambda_0 = 1.375$ for light activity. The rest of the problem is a matter of entering the parameters in the correct units and dragging our equations down to the appropriate time.

   Our spreadsheet predicts Karen's weight after 31 days to be about 163.1 pounds. Though the model predicts it will take several years, it appears Karen's long-term weight will stabilize at about 129.8 pounds. Figure 8.7 shows the results

	A	B	C	D
1	Variable Calorie Expenditure Model			
2				
3	Sex, $S =$		F	
4	Current Body Weight =		165	(pounds)
5	Age, $A =$		25	(years)
6	Height, $H =$		68	(inches)
7	Daily Intake, $I_0 =$		1900	(calories)
8	Activity Level, $\lambda_0 =$		1.375	
9				
10	t	$W(t)$		
11	0	165		
12	1	164.9		
41	30	163.2		
42	31	163.1		
3662	3651	129.8		
3663	3652	129.8		

FIGURE 8.7   Excel results for Example 8.5.

with unnecessary rows hidden. Intuitively this is already an improvement over the constant expenditure model. We have Karen's weight stabilizing at some reasonable value rather than tending to zero or growing to infinity. $\square$

As the last example shows, the variable calorie expenditure model predicts that if we consume the same number of calories every day, eventually our calorie needs change to exactly balance the intake. We examine this feature of the model further by examining its equilibrium value.

8.2.3 Equilibrium Analysis

Finding the equilibrium value for our model is no different than for previous models; however, we will benefit from simplifying the model with some algebra first. Our first step is to reorganize the DDS to get

$$W(t) = W(t-1) - \frac{4.536 \cdot \lambda_0}{3500} W(t-1) + \left\{ \frac{I_0}{3500} - \frac{\lambda_0}{3500} (15.875 \cdot H - 5 \cdot A + 166 \cdot S - 161) \right\}.$$

This allows us to recognize the model as a complicated-looking affine DDS, $W(t) = W(t-1) + rW(t-1) + a$, where $r = -\dfrac{4.536 \cdot \lambda_0}{3500}$ and

$$a = \left\{ \frac{I_0}{3500} - \frac{\lambda_0}{3500} (15.875 \cdot H - 5 \cdot A + 166 \cdot S - 161) \right\}.$$

From our previous work with affine systems, we know that the equilibrium value is given by

$$W^* = -\frac{a}{r} = -\frac{\dfrac{I_0}{3500} - \dfrac{\lambda_0}{3500} (15.875 \cdot H - 5 \cdot A + 166 \cdot S - 161)}{-\dfrac{4.536 \cdot \lambda_0}{3500}}.$$

After some simplifying we get

$$W^* = \frac{I_0}{4.536\lambda_0} - \frac{15.875}{4.536} H + \frac{5}{4.536} A - \frac{166}{4.536} S + \frac{161}{4.536}.$$

Writing coefficients as decimals gives

$$W^* = 0.2205 \frac{I_0}{\lambda_0} - 3.5 \cdot H + 1.102 \cdot A - 36.596 \cdot S + 35.494.$$

As a quick check that our algebra is correct, in the next example we confirm the Excel result for Karen's long-term weight from Example 8.5.

Example 8.6: Use the formula for W^* to calculate Karen's equilibrium weight from our last example.

This is just a matter of plugging in all of the parameters for Karen into the equation for W^*. We get $W^* = 0.2205\dfrac{1900}{1.375} - 3.5 \cdot 68 + 1.102 \cdot 25 - 36.596 \cdot 0 + 35.494 \approx 129.73$.

The resulting 129.73 pounds is the same long-term weight (within rounding error) we predicted using Excel. □

From our previous work on affine systems in the context of populations, we know that the equilibrium for weight will be stable. Thus we can view W^* as our long-term weight for any reasonable constant daily intake of I_0 calories. By examining the formula for W^*, we make two general observations that should agree with our intuition about how body weight works:

1. If we increase our daily intake, I_0, our long-term weight will increase.
2. If we become more active, that is, if λ_0 increases, our long-term weight will decrease.

The equilibrium value for our model does more than just predict long-term weight. It actually provides us with a reasonable way to set our daily caloric intake, I_0, for a long-term weight goal. We illustrate how to do so in the next example.

Example 8.7: Allen is a UPS delivery driver who is 35 years old and 6 feet tall. His current weight is 250 pounds, but he would eventually like to weigh 200 pounds. Find Allen's target daily calorie consumption in order for him to reach his goal.

In terms of the equilibrium for our model, we need to arrange for $W^* = 200$ by finding the required I_0. We have all other needed parameters except for Allen's activity level, λ_0. Driving a delivery truck for UPS is a physically demanding job, so we assign Allen a high activity level of $\lambda_0 = 1.725$. Now we are ready to plug all known parameters into the equation for W^*:

$$200 = 0.2205\frac{I_0}{1.725} - 3.5 \cdot 72 + 1.102 \cdot 35 - 36.596 \cdot 1 + 35.494.$$

As we can see the only remaining unknown is I_0, so we can solve for it. We get

$$200 = 0.2205\frac{I_0}{1.725} - 3.5 \cdot 72 + 1.102 \cdot 35 - 36.596 \cdot 1 + 35.494$$

$$200 = 0.1278 I_0 - 214.532$$

$$414.532 = 0.1278 I_0.$$

Thus $I_0 \approx 3244$. Allen should consume about 3244 calories per day to eventually get his weight down to 200 pounds. □

We note that the equilibrium value W^* does not depend on the initial body weight: this model would predict the same long-term weight for Karen if she started at 100 or 300 pounds. Similarly, the model predicts the same long-term weight for a given diet and activity level for *anyone* who has the same height, age, and sex, regardless of body weight. This turns out to be at odds with studies of long-term weight (Hall et al., 2011). In fact a person's initial weight does influence their long-term weight. Not only that, but body composition is also an important factor: people who weigh the same can have very different calorie needs depending on how much of that weight is fat versus muscle. Later in the chapter we develop a more sophisticated model that takes body composition into account.

Finally, we note that in our discussion of long-term weight we have continued to treat age as a constant. The longer the time period in question, the more error this assumption will introduce into the model. Since we generally require fewer calories as we age, the long-term weight values projected here will tend to be underestimates.

8.2.4 Section Exercises

1 Implement the variable calorie expenditure model with Excel using your own parameters for initial weight, height, age, activity level, and sex, and select a daily calorie intake that would result in weight gain.

   a. Note what the model predicts for your body weight over 1 week, 1 month, 6 months, 1 year, and 5 years.

   b. Use Excel to determine your long-term stable weight.

2 Implement the variable calorie expenditure model with Excel using your own parameters for initial weight, height, age, activity level, and sex, and select a daily calorie intake that would result in weight loss.

   a. Note what the model predicts for your body weight over 1 week, 1 month, 6 months, 1 year, and 5 years.

   b. Use Excel to determine your long-term stable weight.

3 For the parameters you used in Exercise 1, determine the long-term weight by finding the equilibrium value algebraically.

4 For the parameters you used in Exercise 2, determine the long-term weight by finding the equilibrium value algebraically.

5 For parameters of your choosing, set a long-term goal weight.

   a. Determine the daily calorie intake required to reach your long-term goal weight.

   b. Confirm your result in part (a) with Excel.

6 According to the Mifflin–St. Jeor equation, determine which of the following changes would result in the largest change in resting energy expenditure: an additional 3 inches in height, an additional 10 pounds of weight, or being 10 years younger. Explain how you arrived at your conclusion.

7 Using your own parameter values and choice of daily intake, compare the long-term predictions of the constant calorie and variable calorie expenditure models. Explain in a complete sentence or two the reasons for any difference.

8 *Extension*: Our variable calorie expenditure model was based on the Mifflin–St. Jeor equations for resting energy expenditure. Construct a variable calorie expenditure model based on the Harris–Benedict equations for resting energy expenditure instead.

   a. Construct an Excel spreadsheet for the model.

   b. Compare the new model's long-term projection for your own weight with the MSJ-based model. Which seems more reasonable?

   c. Do a general equilibrium analysis for males or females.

9 *Extension*: Our variable calorie expenditure model was based on the Mifflin–St. Jeor equations for resting energy expenditure. Construct a variable calorie expenditure model based on the Livingston–Kohlstadt equations for resting energy expenditure instead.

   a. Construct an Excel spreadsheet for the model.

   b. Compare the new model's long-term projection for your own weight with the MSJ-based model. Which seems more reasonable?

   c. Do a general equilibrium analysis for males or females and compare the result to MSJ.

10 *Extension*: As we age we typically burn fewer calories, largely due to a decrease in lean muscle mass. This effect of aging is incorporated in the Mifflin–St. Jeor equations, but we elected to treat age as a constant to keep the model as simple as possible. In this problem we include age effects in the model.

   a. Modify the variable calorie expenditure model to include age as a variable. It may help to include a separate column for tracking changes in age.

   b. Use the new model and your parameters from Exercise 2 to project your body weight over 1 week, 1 month, 6 months, 1 year, and 5 years.

   c. Has the prediction changed significantly from the original?

   d. What effect does varying age have on the equilibrium value?

11 Most college students have heard about the "Freshman 15," a term first introduced in an article in *Seventeen* magazine in 1989 for the purported gaining of 15 pounds experienced by students during their first year of college (Karasu, 2013). Many studies have found that the Freshman 15 is a myth

(Karasu, 2013). In fact many college students actually lose weight during their first year, and those who do gain weight typically gain somewhere around 3–5 pounds, certainly nowhere near the assumed 15 (Karasu, 2013). Moreover, the weinght that is gained could simply be due to natural growth and development into adulthood (Posterli, n.d.). Though many factors such as cafeteria-style eating, stress, and irregular sleep habits conspire to make it difficult to eat a healthy diet when starting college, the change most associated with weight gain during this time is an excess of alcohol consumption (Zagorsky & Smith, 2011).

a. Use the variable calorie expenditure model to estimate the required daily calorie intake in order for you to gain 15 pounds in 9 months.

b. Use the variable calorie expenditure model to estimate the required daily calorie intake in order for you to gain 3 pounds in 9 months.

c. Use the variable calorie expenditure model to estimate the amount of weight you would gain in 9 months if you increased your alcohol consumption by 2 drinks per day and changed nothing else.

8.3   HEALTH METRICS

So far the models we have created have focused exclusively on predicting body weight over time. While this is certainly important, body weight is generally not the best metric for gauging an individual's health or disease risk. Other body measures, such as **body mass index** (BMI) or **waist-to-height ratio** (WHR), provide physicians with much more useful information than body weight alone. We discuss these measures in the following sections.

8.3.1   BMI

While weight is an important health metric, it should not be our sole focus. We know intuitively that a $6'6''$ male weighing 200 pounds is likely at a healthier weight than a $5'4''$ male weighing the same. One attempt to capture this intuitive idea is a measurement that takes into account both weight and height called the **BMI**. This is a very popular metric used by health professionals in part because it is very easy to determine the necessary parameters in an office setting and it is straightforward to calculate. The formula for calculating BMI is

$$\text{BMI} = 703.07 \times \frac{W}{H^2},$$

where W is a person's weight in pounds and H is a person's height in inches. In our next example we compare BMIs for two men of the same weight who have different heights.

Example 8.8: John is $6'6''$ tall and Steve is $5'4''$ tall. Both weigh 200 pounds. Calculate and compare their BMIs.

John is 78 inches tall, so his BMI is $703.07 \times \frac{200}{78^2} = 23.11$. Steve is 64 inches tall, so his BMI is $703.07 \times \frac{200}{64^2} = 34.33$. Steve's BMI is much higher than John's because of his shorter stature. $\square$

The numbers we just calculated do not mean very much without some context about what is considered a healthy BMI and what is not. According to the National Institutes of Health (NIH), the BMI guidelines presented in Table 8.2 provide ranges appropriate for use in adults of both sexes (Obesity Education Initiative, 1998).

Clinicians often use BMI as an important indicator of disease risk. Individuals whose BMI falls in the overweight or obese categories are at increased risk for a variety of diseases including hypertension, diabetes, and cardiovascular disease (Obesity Education Initiative, 1998). According to Table 8.2, John from Example 8.8 has a BMI in the normal range, while Steve is obese. This provides us with an important distinction that considering weight alone would not.

TABLE 8.2   NIH Ranges for Interpreting BMI Values

BMI	Health Status Category
Below 18.5	Underweight
Between 18.5 and 24.9	Normal
Between 25.0 and 29.9	Overweight
Above 29.9	Obese

Though it has its limitations, BMI has the advantage over many other measurements in that it is very easy to calculate using easily obtainable data. In the next section we look at another such measurement.

8.3.2   WHR

One of the limitations of BMI as a health indicator is that it does not distinguish between healthy, desirable body weight like that due to lean muscle and unhealthy, undesirable body weight like that due to excess fat. Moreover, some kinds of fat, particularly abdominal fat, are known to be worse for our health than others (CDC, 2011).

The WHR is a simple metric that makes a distinction between healthy kinds of weight and unhealthy kinds of weight by providing a rough measure of abdominal fatness. Current research indicates that the WHR is actually a better predictor of disease risk than BMI (Ashwell, Gunn, & Gibson, 2012). Furthermore, WHR is appropriate to use across different ethnicities and ages (Ashwell & Hsieh, 2005).

The ratio is easy to compute. All we need is a person's height and waist circumference. Then we calculate

$$WHR = \frac{\text{Waist circumference}}{\text{Height}}.$$

The units we use do not matter as long as the same unit is used for each measurement. Waist circumference should be measured with a measuring tape wrapped around the bare waist at the navel. The tape should be snug but should not compress the belly.

The next example shows the basic calculation.

Example 8.9: Stella is 5′2″ tall and has a 30.5″ waist. Find Stella's WHR.

All we need to do here is convert Stella's height to inches and then plug our two values into the formula. We get $\text{WHR} = \dfrac{30.5}{62} \approx 0.49.$ □

Like the BMI the calculation for WHR needs some context to make it meaningful. Table 8.3 is similar to Table 8.2 and gives ranges for interpreting the WHR for both men and women based on the work of Ashwell and Hsieh (2005).

TABLE 8.3   Ranges for Interpreting WHR Values

Waist to Height Ratio	Health Status Category
Below 0.40	Underweight
Between 0.40 and 0.50	Healthy
Between 0.50 and 0.60	Overweight
Above 0.60	Obese

We see that based on Table 8.3, Stella from our previous example would be classified as healthy, though she is on the border of being overweight.

Assuming that height remains constant, our WHR only changes when our waist circumference changes. If we knew how changes in body weight are related to changes in waist size, we could use our body weight models to predict WHR. One study found that in Japanese men, waist size changed by 1 inch for every 4.87 pound change in body weight. In Japanese women, the association was 1 inch for every 5.94 pound change in body weight (Miyatake et al., 2007). While this is only one study and while it is unclear if the results extend to non-Japanese groups, these figures might allow us to use our body weight projections to make predictions about the more important WHR. The reader is invited to do so in the exercises.

8.3.3   Section Exercises

1   Find and assess the BMI for a 5′3″ person who weighs 140 pounds.

2   Find and assess the WHR for a 5′10″ person who has a 40″ waist.

3   How tall would a 250-pound person have to be in order to have a healthy BMI?

4   How tall would a person with a 45″ waist have to be in order to have a healthy WHR?

5   Use your own height and the BMI guidelines in Table 8.2 to determine a healthy weight range for yourself.

6 *Extension*: Have Excel model BMI over time by including a new column for BMI in the variable calorie expenditure model for body weight. Include a graph of BMI over time that includes horizontal lines that represent the BMI levels for underweight, normal, overweight, and obese.

7 For parameter values of your choosing, find the projected long-term BMI by using the equilibrium analysis from Section 8.2.3.

8 Use Excel, your own height, and the WHR guidelines in Table 8.3 to determine a healthy waist circumference range for yourself.

9 *Extension*: Assume that the results of the study referenced at the end of this section hold across all ethnicities. That is, assume that for men, waist size changes by $1''$ for every 4.87 pound change in body weight. For women, assume that waist size changes by $1''$ for every 5.94 pound change in body weight (Miyatake et al., 2007). For the variable calorie expenditure model, include a new parameter for waist circumference in inches.

    a. Have the worksheet compute an initial value for waist-to-height ratio.

    b. Have the worksheet model WHR over time by including new columns for waist circumference and WHR next to body weight in the model.

8.4 BODY COMPOSITION

The concept of BMI is an improvement overly merely focusing on weight because it provides an indication of body fatness, which is more correlated with disease risk than weight alone. This is indeed a better way of determining healthy weight, but it still lacks precision. The formula for BMI makes no distinction between healthy weight like muscle and unhealthy weight like fat. A bodybuilder, for example, might score poorly on a BMI chart because she may weigh a lot for her height. But because she is very lean, she may still be very healthy. Or we can look at it this way: a very fit 180-pound, $6'$ tall male will have the same BMI as a very out of shape 180-pound, $6'$ tall male, though the former is clearly healthier in general than the latter. The fit person will tend to have much less fat than one who is out of shape.

    A better determination of healthfulness is WHR because it will distinguish between a muscular, fit person and a fat, sedentary person who are the same weight and height due to differences in body shape. The muscular, fit person will have a much smaller waist and therefore score healthier on the WHR scale. Still, though, WHR is an indirect measure of body fat, and we can potentially improve our model for body weight by explicitly distinguishing between lean body mass and fat.

    While the estimates for REE given by the MSJ equation and others are useful and accurate for a large segment of the population, research indicates that most such equations are missing an important point: for a given body weight, individual calorie needs can be very different based on **body composition**, that is, the percentage of body weight that is fat (F) or lean body mass (L) (Nelson, Weinsier, Long, & Schutz, 1992). By lean body mass we mean anything that is not fat; this includes muscle but also bones and other body tissues. Though both fat and lean body mass burn calories, lean body mass burns many more calories per pound than fat; thus lean, muscular people have higher calorie needs than people with high body fat even if they weigh the same. For example, a woman who weighs 130 pounds and carries 39 pounds of fat will burn fewer calories per day than a woman who weighs 130 pounds but carries only 13 pounds of fat. Note that both women would have the same BMI and the MSJ equation would suggest the same calorie needs for both (assuming equal ages and heights).

We begin by letting $F(t)$ and $L(t)$ be pounds of fat and lean body mass, respectively, at time, t. Since we define lean body mass to be anything other than fat, we can write $W(t) = F(t) + L(t)$, or equivalently, $L(t) = W(t) - F(t)$.

Introducing body composition into our model complicates it in four ways:

1. Fat and lean burn different amounts of calories per pound. We need to know both "burn rates."

2. A normal scale measures only weight, not body fat. We need a way to estimate an individual's body fat.

3. When body weight is gained or lost, some of the weight change is fat, and some of it is lean. We need to adjust the model so that whenever there is a change in weight, some of it comes from a change in fat mass and some from a change in lean.

4. At the beginning of this chapter we assumed that the **energy density** of body weight was 3500 calories per pound. This is a reasonable average value; however, fat and lean actually have different energy densities, and in fact it takes more calories to add/lose a pound of fat than it does a pound of lean. We need to know how many calories are in a pound of fat and how many are in a pound of lean.

We build our new model step-by-step by resolving each of these four issues in turn.

8.4.1   Calorie Burn Rates for Lean and Fat

We have seen ways of estimating REE, the number of calories a body burns at rest, based on considerations like height and weight. Now we take a different approach by estimating REE based on how many calories are burned by a pound of lean body mass and how many are burned by a pound of fat.

This question has been studied by many scientists, and we use the work of Nelson here. In his study Nelson determined that fat mass and lean mass were the most important determinants of REE, and he developed an estimate for REE based solely on the amounts of fat and lean an individual carries (Nelson, Weinsier, Long, & Schutz, 1992). His equation is given by

$$\text{REE} = 1.832 \cdot F + 11.708 \cdot L,$$

where F and L are in pounds and REE is in calories per day We can interpret Nelson's equation to mean that on average a pound of fat burns 1.832 calories per day, while a pound of lean burns 11.708 calories per day. In the next example we show how to use Nelson's formula, and we note how much of a difference body composition can make in REE.

Example 8.10: Kyle and Eric both weigh 156 pounds, but Kyle has 15% body fat and Eric has 30% body fat. Estimate the REE for both Kyle and Eric.

We need to know how many pounds of lean and fat each of the men have. Kyle is 15% fat, so he has $156 \cdot 0.15 = 23.4$ pounds of body fat and $156 - 23.4 = 132.6$ pounds of lean. Similar calculations show Eric has 46.8 pounds of fat and 109.2 pounds of lean. To estimate their REEs, we plug their values for lean and fat into Nelson's equation. For Kyle we get $REE = 1.832 \cdot 23.4 + 11.708 \cdot 132.6 = 1595.3$ calories per day. For Eric we get $REE = 1.832 \cdot 46.8 + 11.708 \cdot 109.2 = 1364.3$ calories per day. By virtue of the fact that he has more lean body mass and less fat, Kyle requires 230 calories more than Eric at rest. $\square$

Nelson's equation for estimating REE is a great start for our new model as it is a very simple equation to use. But it brings us to difficulty number two: how do we know an individual's body fat percentage?

8.4.2 Determining Body Fat Percentage

There are a variety of methods that can be used to estimate body fat percentage, some of which can be done at home. Most methods that can be done easily at home will suffer from some inaccuracy, but as long as the same method is used consistently, they can all be useful as a way of tracking *changes* in body fat %, and that is in many ways more important than a perfectly accurate number.

By far the easiest way to measure body fat is with a **body fat scale**. These scales use **bioelectrical impedance** to measure body composition; some not only report body fat %, but they also measure an individual's body water, muscle, and even bone mass. The scale sends a small electrical current through the body; then based on how much resistance the current encounters, the composition of the body can be calculated. When using a body fat scale, it is important to always weigh in at the same time of day, preferably not directly after exercise or a big meal, in order to get results that are as accurate as possible. Body fat scales are relatively inexpensive and can be purchased for approximately $40–$50.

For those without a body fat scale, a very inexpensive and low-tech do-it-yourself method is available. All that is required is a measuring tape. The idea is to use the measuring tape to take a variety of body measurements and then use a formula based on those dimensions to estimate body fat. Different formulas require that different measurements be taken.

One such method is used by the military and another by author and researcher Covert Bailey. Neither method is 100% accurate, but for a large percentage of people, they will be fairly close. It is probably best to use both methods and think of the results as a range where you can be pretty sure your true percentage lies.

Equations developed by US Navy researchers for assessing body fat in service men and women have been found to be reasonably accurate overall and the best among many existing equations for estimating body fat in young women (Friedl et al., 2001). The Navy's equations are slightly different for men and for women. Men need to measure their height (H), waist at the navel (Wa), and neck (N). Women also need height and neck circumference, but they measure the waist *at its narrowest point*, and they need an additional measurement: the hips at their fullest point (Hp). All measurements should be taken in inches. The Navy's equations are given by

$$(\text{Men})\, \text{body fat}\% = 86.01 \cdot \log(\text{Wa}-N) - 70.041 \cdot \log(H) + 36.76$$

$$(\text{Women})\, \text{body fat}\% = 163.205 \cdot \log(\text{Wa}+\text{Hp}-N) - 97.684 \cdot \log(H) - 78.387.$$

Once measurements have been taken, the individual simply plugs them into the correct formula based on sex. Note that for this formula it will be necessary to use Excel or a calculator capable of computing logarithms. The "log" function is the base-10 logarithm.

As an alternative to using the Navy equations, we can use those developed by Covert Bailey, health researcher and author of the *Fit or Fat* series of books. Bailey developed four different formulas corresponding to different sexes and ages, and his formulas all use a different set of measurements than the Navy's (Bailey C., 2000). Women take measurements of their hips (Hp), thigh (T), and calf (C) at their fullest points and their wrist (Wr) just above the wrist bone. All measurements should be taken in inches. Once all measurements are taken, women select the formula that corresponds to their age category and plug in their measurements. Bailey's equations for women are given by

$$(\text{Women 30 or younger})\, \text{body fat}\% = \text{Hp} + 0.8 \cdot T - 2 \cdot C - \text{Wr}$$

$$(\text{Women over 30})\, \text{body fat}\% = \text{Hp} + T - 2 \cdot C - \text{Wr}.$$

Men take measurements of their hips (Hp) at their fullest point, waist at the navel (Wa), forearm (Fo), and wrist (Wr) just above the bone. Bailey's equations for men are given by

$$(\text{Men 30 or younger})\, \text{body fat}\% = \text{Wa} + 0.5 \cdot \text{Hp} - 3 \cdot \text{Fo} - \text{Wr}$$

$$(\text{Men over 30})\, \text{body fat}\% = \text{Wa} + 0.5 \cdot \text{Hp} - 2.7 \cdot \text{Fo} - \text{Wr}.$$

In the next example we compare the two methods for an over-30 male.

Example 8.11: The author is a 44-year-old male who is 6′ tall and weighs 180 pounds. His measurements are 36″ at the waist, 15.75″ at the neck, 41.25″ at the hips, 12″ at the forearm, and 7″ at the wrist. Estimate his body fat percentage using the Navy's equations and Bailey's equations.

For the Navy's equations we calculate

$$\text{(Men) body fat \%} = 86.01 \cdot \log(\text{Wa} - N) - 70.041 \cdot \log(H) + 36.76$$
$$= 86.01 \cdot \log(36 - 15.75) - 70.041 \cdot \log(72) + 36.76$$
$$= 86.01 \cdot 1.306 - 70.041 \cdot 1.857 + 36.76$$
$$= 19\%.$$

For Bailey's equations we calculate

$$\text{(Men over 30) body fat \%} = \text{Wa} + 0.5 \cdot \text{Hp} - 2.7 \cdot \text{Fo} - \text{Wr}$$
$$= 36 + 0.5 \cdot 41.25 - 2.7 \cdot 12 - 7$$
$$= 17.2\%.$$

The two methods are fairly close in their estimates. $\square$

Equations like these that use body measurements are pretty accurate for a large segment of the population; however, their real benefit is in tracking progress. Whether or not the body fat percentage for an individual is 100% accurate generally matters less than the ability to measure improvement, which these equations do very well.

If it is important to have a truly accurate measurement of body fat %, then underwater weighing is often done. In this method an individual is weighed while fully submerged in a tank of water. Based on the different buoyancies of fat, bone, muscle, etc., an accurate body fat percentage can be calculated based on the weight. Yet another common method is to use skinfold measurements using calipers; this method can yield accurate results, but the person using the calipers must have proper training to avoid measurement errors.

As a final set of body fat equations, we use those developed by Jackson (Jackson et al., 2002) and used by Hall in his development of a model for body weight (Hall et al., 2011). The strength of Jackson's equations is that they require the least amount of information to calculate an estimate for body fat percentage. In fact Jackson's equations require only age (A), weight (W), and height (H), all of which are likely to be known to the user, whereas a measurement like wrist circumference will take effort to find. The function $\ln(x)$ is the natural logarithm and is one of the built-in functions included with Excel. Jackson's equations are given by

$$\text{(Men) body fat \%} = 0.14 \cdot A + 37.31 \cdot \ln\left(703.07\frac{W}{H^2}\right) - 103.94$$
$$\text{(Women) body fat \%} = 0.14 \cdot A + 39.96 \cdot \ln\left(703.07\frac{W}{H^2}\right) - 102.01.$$

Recognizing the expression that appears inside the natural logarithm function allows us to write Jackson's equations as

$$\text{(Men) body fat \%} = 0.14 \cdot A + 37.31 \cdot \ln(\text{BMI}) - 103.94$$
$$\text{(Women) body fat \%} = 0.14 \cdot A + 39.96 \cdot \ln(\text{BMI}) - 102.01.$$

In the next example we show how to use Jackson's equations to estimate body fat.

Example 8.12: Estimate the body fat percentage for the author using Jackson's equations.

The author's BMI is 24.4. Thus according to Jackson's equation, we have

$$(\text{Men}) \text{ body fat} \% = 0.14 \cdot A + 37.31 \cdot \ln(\text{BMI}) - 103.94$$
$$= 0.14 \cdot 44 + 37.31 \cdot \ln(24.4) - 103.94.$$

This gives us a body fat estimate of 21.4% for the author. So far we have three different estimates for the author's body fat percentage: 17.2%, 19%, and 21.4%. As a fourth point of comparison, the author's body fat scale reports a value of 14.4%. This is a fairly wide spread of results and serves to emphasize the point that body fat estimates are not necessarily useful for producing really accurate results, but they can be very useful in tracking progress. ☐

The Jackson equations give us a starting point for body fat %, but note that we can make our model more personally applicable and more accurate if we measure our own body fat percentage using a body fat scale or measuring tape and use that figure in place of the value given by Jackson.

Like BMI and WHR, measurements of body fat percentage are not particularly meaningful without a sense of what levels are considered healthy. The American Council on Exercise guidelines for body fat percentage levels in women and men are presented in Table 8.4.

TABLE 8.4   American Council on Exercise Body Fat Percentage Guidelines

	Women (%)	Men (%)
Essential fat	10–13	2–5
Athletes	14–20	6–13
Fitness	21–24	14–17
Acceptable	25–31	18–24
Obesity	32 or over	25 or over

Source: American Council on Exercise (2014). Reproduced with permission from the American Council on Exercise.

The table reinforces the fact that individuals can carry a wide range of body fat and still be fit and healthy. It also shows that less is not necessarily better when it comes to body fat: it can be dangerous to try to bring body fat levels down below the essential level our bodies need to function properly. According to Table 8.4, the author falls somewhere between fitness and acceptable in terms of body fat.

8.4.3   How Weight Changes Affect Body Composition

Our models of body weight up to now have treated all weight equally as though "a pound is a pound." The reality, though, is more complicated. When someone gains or loses weight, some of that weight will be fat, but some of it will be lean. Even more interesting in recent work by Forbes and Hall is that the proportions of fat and lean gained or lost depend on how much body fat the individual has. In general, the more body fat one has, the higher proportion of any weight change will be fat. If an obese, sedentary person loses a pound of body weight, almost all of that weight will be fat; however, if an Olympic athlete loses a pound of body weight, much of that pound will actually be lean (Forbes, 2000; Hall K. D., 2007).

For our modeling purposes, the big question is how to incorporate the effect of body composition on the composition of weight changes. The key in modeling this body composition effect is to examine how the body "partitions" any calorie deficit or surplus. In other words, when we take in fewer calories than we burn, how does the body compensate for the energy deficit? Broadly speaking, the body compensates by burning some stored fat and burning some lean mass. How much of each is used depends on the amount of fat the individual carries. We call this dependence the **energy partition function**, and different researchers have proposed different formulas for it. The form of the energy partition function we present here was suggested by Forbes in 2000 (Forbes, 2000), though the version we use differs slightly from Forbes' original work. For any caloric excess or deficit, the proportion applied to changes in fat and lean are determined by the energy partition function given by

$$\text{Proportion devoted to fat} = \frac{F}{F + \rho}$$

$$\text{Proportion devoted to lean} = \frac{\rho}{F + \rho}.$$

As before, F represents the amount of body fat in pounds, and we have a new parameter, ρ, which we call the **energy partition parameter**. Note that regardless of the value for ρ, the sum of our two proportions is always one. In his work Forbes found a value of $\rho = 4.41$. We use $\rho = 6$ since this value produces better agreement with Hall's model (Hall et al., 2011).

In the next example we show how the energy partition function works.

Example 8.13: Suppose Max and Paul both weigh 200 pounds. Max is 30% body fat and Paul is 10% body fat. If they each experience a calorie deficit of 500 calories, what proportion of that deficit will be made up for by burning fat?

For Max we know he is carrying $200 \times 30\% = 60$ pounds of fat. Thus the proportion of his energy deficit that is made up by burning fat is $\frac{F}{F + \rho} = \frac{60}{60 + 6} = \frac{60}{66} = 0.909$. So Max will burn $500 \times 0.909 = 454.5$ calories worth of stored fat and the remaining 45.5 calories from stored lean.

Paul on the other hand only has 20 pounds of body fat. Thus the proportion of his energy deficit that is made up by burning fat is $\frac{F}{F + \rho} = \frac{20}{20 + 6} = \frac{20}{26} = 0.769$. So Paul will burn $500 \times 0.769 = 384.5$ calories worth of stored fat and the remaining 115.5 calories from stored lean. □

The final step before we are in a position to construct a model for changes in body composition is to determine the weight changes in fat and lean based on the calories of each as given by the energy partition function.

8.4.4   Energy Content of Fat and Lean

In the previous section we saw that for a given calorie deficit or surplus, some of that delta is due to a change in body fat, and some is due to a change in lean mass. However, knowing how many calories of each are involved does not quite tell us what we want to know, and that is how much the weight of each changes. In this situation we can no longer use the 3500 calorie per pound rule because now we have to ask, "A pound of what?"

It turns out that fat is much more energy dense than lean. In fact from Hall we know that fat contains about 4279.43 calories per pound, while lean contains only 823.38 calories per pound (Hall K. D., 2008). Thus, if we could choose to lose (or gain) 1 pound of pure fat, it would take a calorie deficit (or surplus) of about 4280 calories rather than the standard 3500 calories. Similarly, if we could choose to lose (or gain) 1 pound of pure lean, it would only require a calorie deficit (or surplus) of 823 calories.

The difference in energy densities and the results of Example 8.13 mean that we have a "good news/bad news" situation for those who are obese and seeking to lose weight. On the one hand, the energy partition function tells us that any weight that an obese person loses will be virtually all fat. On the other hand, because any weight loss will be almost all fat, it takes more of a calorie deficit for an obese person to lose a pound than it does for a lean person. We show this with our next example.

Example 8.14: Consider Max and Paul from Example 8.13. Suppose each man incurs a calorie deficit of 4000 calories over the course of a week. Estimate how much fat, lean, and total body weight each man loses during the week.

We assume for the sake of this example that their energy partition proportions do not change during the week. From Example 8.13 we know that 0.909 is the proportion of Max's deficit that will be made up by burning stored fat. Thus he burns $4000 \times 0.909 = 3636$ calories worth of fat. Because fat contains 4279.43 calories per pound, Max burns $\frac{3636}{4279.43} \approx 0.85$ pounds of fat. Similarly Max burns $\frac{364}{823.38} \approx 0.44$ pounds of lean for a total of $0.85 + 0.44 = 1.29$ pounds of body weight.

For Paul, who is much leaner, the situation is different. We know from Example 8.13 that Paul's energy proportion assigned to body fat is 0.769. Thus he burns $4000 \times 0.769 = 3076$ calories worth of fat. Because fat contains 4279.43 calories per pound, Paul burns $\frac{3076}{4279.43} \approx 0.72$ pounds of fat. Similarly Paul burns $\frac{924}{823.38} \approx 1.12$ pounds of lean for a total of $0.72 + 1.22 = 1.94$ pounds of body weight.

These results show that Max loses more fat than Paul, but Paul loses more weight than Max. Thus we have some evidence for the fact that it is harder for obese people to lose weight, and it is because even though they are losing more fat, it takes a larger calorie deficit for them to lose body weight. □

The "3500 calories per pound" assumption makes no distinction between fat and lean, and given the different energy densities of fat and lean, there must be some percentage of each implied by the assumption. We examine the implied proportions in the next example.

Example 8.15: Find the implied percentage of fat and lean in the 3500-calorie assumption.

Since 3500 calories per pound falls between the values for pure fat (4279 calories per pound) and pure lean (823 calories per pound), there must be some percentage of fat and lean that results in 3500 calories per pound. If we let x represent the proportion of fat in our standard pound of body weight, then we will have

$$4279x + 823(1-x) = 3500.$$

Next we solve for x:

$$4279x + 823 - 823x = 3500$$
$$3456x = 2677$$
$$x = \frac{2677}{3456} \approx 0.77.$$

Thus the implied proportion of fat in the standard 3500 calories per pound is 0.77, which means the proportion of lean is 0.23. □

In the next section we finally see how to incorporate body composition into our body weight model.

8.4.5 Section Exercises

1   Calculate how much body fat and how much lean body mass a person has who weighs 205 pounds and is 29% body fat.

2   Joyce and Chloe both weigh 125 pounds, but Joyce has 17% body fat and Chloe has 27% body fat. Estimate the REE for both Joyce and Chloe.

3   Estimate your own body fat percentage using the relevant equation by the Navy.

4   Estimate your own body fat percentage using the relevant equation by Bailey.

5   Estimate your own body fat percentage using the relevant equation by Jackson.

6 *Extension*: Use the Jackson equation to estimate how much weight you have to lose in order to decrease your body fat by 5% points.

7 Estimate your own REE based on the Nelson formula. How does it compare to the MSJ formula for you?

8 Suppose Max weighs 160 pounds and is 28% body fat. If Max arranges a calorie deficit of 400 calories, what proportion of that deficit will be made up for by burning fat and what proportion by burning lean?

9 Suppose Dianne weighs 155 pounds and is 24% body fat. She incurs a calorie deficit of 3500 calories over the course of a week. Estimate how much fat, lean, and total body weight Dianne should expect to lose during the week.

10 Suppose you incur a calorie deficit of 2500 calories over the course of a week. Estimate how much fat, lean, and total body weight you would expect to lose during the week.

11 *Extension*: Create an Excel spreadsheet for calculating body fat percentage based on the Navy equations. Make judicious use of "IF" statements to make the spreadsheet as user friendly as possible.

12 *Extension*: Create an Excel spreadsheet for calculating body fat percentage based on the Covert Bailey equations. Make judicious use of "IF" statements to make the spreadsheet as user friendly as possible.

8.5 THE BODY COMPOSITION MODEL FOR BODY WEIGHT

We now have enough information to construct our new body composition model for body weight. We develop equations for $F(t)$ and $L(t)$ separately and then use the fact that $W(t) = F(t) + L(t)$ to complete the model. First, we use Nelson's equation for REE to estimate our daily energy expenditure:

$$E(t-1) = \lambda_0 \left(1.832 \cdot F(t-1) + 11.708 \cdot L(t-1) \right).$$

Recall that λ_0 is the activity level that represents the amount of physical activity engaged in during a typical day.

Remembering that I_0 represents the constant daily calorie intake, we see that on a given day we have a calorie deficit (or surplus) of

$$I_0 - E(t-1) = I_0 - \lambda_0 \left(1.832 \cdot F(t-1) + 11.708 \cdot L(t-1) \right).$$

The proportion of this difference devoted to fat loss is determined by the Forbes-type energy partition equation $\dfrac{F(t-1)}{F(t-1)+6}$. Furthermore, we know that for every 4279.43 calories devoted to fat loss, we lose a pound of fat. Thus our equation for the amount of body fat in pounds is

$$F(t) = F(t-1) + [I_0 - \lambda_0(1.832 \cdot F(t-1) + 11.708 \cdot L(t-1))] \frac{F(t-1)}{F(t-1)+6} \cdot \frac{1}{4279.43}.$$

In a similar fashion we develop the equation for lean body mass in pounds:

$$L(t) = L(t-1) + [I_0 - \lambda_0(1.832 \cdot F(t-1) + 11.708 \cdot L(t-1))] \frac{6}{F(t-1)+6} \cdot \frac{1}{823.38}.$$

To implement this model in Excel, we have a fair amount of setup to do first. We want the model to be user friendly and as flexible as possible, so we set aside space at the top of the worksheet for the user to enter their body weight, height, age, and sex. The user also must enter an activity level and the daily calorie intake. We allow the user to input a current body fat percentage if known, but if this is not known we have Excel calculate an estimate for body fat percentage using Jackson's equations from Section 8.4.2. Since Jackson's equations use BMI, we get Excel to calculate BMI in a separate cell. The initial setup with the equation for Jackson's body fat estimate displayed is given in Figure 8.8.

	A	B	C	D	E	F	G	H	I	J
1	Body Composition Model									
2										
3	Sex, S =		F		Initial BMI =		25.1			
4	Current Body Weight =		165	(pounds)	Initial Body Fat % =		=IF(C3="M",0.14*C5+37.31*LN(G3)-103.94,0.14*			
5	Age, A =		25	(years)			C5+39.96*LN(G3)-102.01)/100			
6	Height, H =		68	(inches)						
7	Daily Intake, I_0 =		1900	(calories)						
8	Activity Level, λ_0 =		1.375							

FIGURE 8.8   Excel setup for Jackson's body fat equation.

Next we use the initial body fat percentage and current body weight to calculate the initial values for body fat and lean body mass. This initial setup with formula showing is shown in Figure 8.9. Note that body weight is just the sum of fat and lean.

	A	B	C	D
1	Body Composition Model			
2				
3	Sex, S =		F	
4	Current Body Weight =		165	(pounds)
5	Age, A =		25	(years)
6	Height, H =		68	(inches)
7	Daily Intake, I_0 =		1900	(calories)
8	Activity Level, λ_0 =		1.375	
9				
10	t	F(t)	L(t)	W(t)
11	0	=G4*C4	=C4-B11	=B11+C11

FIGURE 8.9   Excel setup for initial values of fat and lean.

Once initial values are set for body fat and lean, we enter the DDS formulas for each. We also add a column to keep track of the user's body fat %, which is just pounds of body fat divided by total body weight. The Excel spreadsheet with the formula for $F(t)$ showing is given in Figure 8.10.

	A	B	C	D	E	F	G	H
1	Body Composition Model							
2								
3	Sex, S =		F		Initial BMI =		25.1	
4	Current Body Weight =		165	(pounds)	Initial Body Fat % =		30.3%	(Jackson)
5	Age, A =		25	(years)				
6	Height, H =		68	(inches)				
7	Daily Intake, I_0 =		1900	(calories)				
8	Activity Level, λ_0 =		1.375					
9								
10	t	F(t)	L(t)	W(t)	Body Fat %			
11	0	49.9	115.1	165.0	30.3%			
12	1	=B11+($C$7-$C$8*(1.832*B11+11.708*C11))*(B11/(B11+6))*(1/4279.43)						

FIGURE 8.10   Excel setup for body composition model.

We finish this section with a computational example.

Example 8.16: Maria is 20 years old, stands $5'6''$ tall, and weighs 155 pounds. She is moderately active and consumes 1900 calories per day. Project Maria's body weight and body fat percentage 1 month from today.

Since we are not given an initial body fat percentage for Maria, we use Jackson's estimate based on her other parameters. We show the screenshot with all parameters entered in Figure 8.11.

	A	B	C	D	E	F	G	H
1	Body Composition Model							
2								
3	Sex, S =		F		Initial BMI =		25.0	
4	Current Body Weight =		155	(pounds)	Initial Body Fat % =		29.4%	(Jackson)
5	Age, A =		20	(years)				
6	Height, H =		66	(inches)				
7	Daily Intake, I_0 =		1900	(calories)				
8	Activity Level, λ_0 =		1.55					
9								
10	t	F(t)	L(t)	W(t)	Body Fat %			
11	0	45.6	109.4	155.0	29.4%			

FIGURE 8.11   Body composition model setup for Example 8.16.

Next we need only drag the model down to 1 month (day 31 in this case). The results are shown in Figure 8.12 with most rows hidden.

We see that Maria's body weight is projected to be 152.8 pounds and her body fat percentage is projected to be 29.0%. Thus the model predicts that a 1900-calorie diet will lead to Maria losing about one half of a pound per week. □

	A	B	C	D	E	F	G
1	Body Composition Model						
2							
3	Sex, S =		F		Initial BMI =		25.0
4	Current Body Weight =	155		(pounds)	Initial Body Fat % =		29.4%
5	Age, A =	20		(years)			
6	Height, H =	66		(inches)			
7	Daily Intake, I_0 =	1900		(calories)			
8	Activity Level, λ_0 =	1.55					
9							
10	t	$F(t)$	$L(t)$	$W(t)$	Body Fat %		
11	0	45.6	109.4	155.0	29.4%		
12	1	45.6	109.3	154.9	29.4%		
41	30	44.4	108.5	152.9	29.0%		
42	31	44.3	108.5	152.8	29.0%		

FIGURE 8.12 Body composition model projections for Example 8.16.

We often want to know what to expect from our body weight over common time periods such as 1 week, 1 month, etc. We can save ourselves a lot of scrolling if we take the time now to display these values in our spreadsheet where they are easy to see. We need only reference the correct cell for each value. Once that is done, these values automatically update, and we avoid having to scroll down in the spreadsheet to find the ones we want. In Figure 8.13 we show the setup for viewing both body weight and body fat percentage predictions over some common time periods.

Results After:	Weight	Body Fat %
1 week	=D18	=E18
1 month	=D41	=E41
6 months	=D193	=E193
1 year	=D376	=E376
5 years	=D1837	=E1837
10 years	=D3663	=E3663

FIGURE 8.13 Excel results for common time periods of interest.

We observe that the projections for year 5 and year 10 for Maria in Example 8.16 are the same. This indicates that the model predicts her weight and body fat % will stabilize at 132.8 pounds and 25.1% body fat if she continues to be moderately active and consume 1900 calories per day.

As with any mathematical model, we should remain somewhat skeptical about model projections that are for so far into the future. In particular for this model, we have neglected the effects of aging, which tend to decrease our energy expenditure over time. Thus the long-term projections for this model are likely to overestimate an individual's calorie needs and hence underestimate an individual's body weight.

Next we analyze long-term projections for body weight by performing an equilibrium analysis.

8.5.1 Equilibrium Analysis

The long-term behavior of our body composition model is more complicated than many of our previous models. For one thing our variable calorie expenditure model allowed us to find an individual's long-term weight based on the initial parameters without having to use Excel. This was because the model turned out to be an affine model and had a stable equilibrium value W^* that was independent of the initial body weight.

For our current model we are not so lucky: for a given initial body weight, we end up with different long-term weights depending on the initial body composition. First we verify this claim with an Excel example.

Example 8.17: Joe and Carl are moderately active, $5'10''$ tall, 25-year-old males who both weigh 180 pounds. Joe has 20% body fat and Carl has 15% body fat. Use the body composition model to predict their long-term weights under a daily diet of 2800 kcal.

To make sure we get both weights to stabilize, we drag the model equations down to a time of 10 years. Plugging in Joe's parameters gives us a projected long-term weight of 188.7 pounds at 21.6% body fat. His results are shown in Figure 8.14.

	A	B	C	D	E	F	G	H	I
1	Body Composition Model								
2									
3	Sex, $S =$		M		Initial BMI =		25.8		
4	Current Body Weight =		180	(pounds)	Initial Body Fat % =		20.0%		
5	Age, $A =$		25	(years)					
6	Height, $H =$		70	(inches)			Results After:	Weight	Body Fat %
7	Daily Intake, $I_0 =$		2800	(calories)			1 week	180.2	20.0%
8	Activity Level, $\lambda_0 =$		1.55				1 month	180.9	20.2%
9							6 months	184.2	20.8%
10	t	$F(t)$	$L(t)$	$W(t)$	Body Fat %		1 year	186.3	21.2%
11	0	36.0	144.0	180.0	20.0%		5 years	188.7	21.6%
12	1	36.0	144.0	180.0	20.0%		10 years	188.7	21.6%

FIGURE 8.14   Joe's body composition projections for Example 8.17.

For Carl we just change the initial body fat percentage to 15% and note that his projected long-term weight is 175.2 at 14.2% body fat. Carl's results are shown in Figure 8.15.

	A	B	C	D	E	F	G	H	I
1	Body Composition Model								
2									
3	Sex, $S =$		M		Initial BMI =		25.8		
4	Current Body Weight =		180	(pounds)	Initial Body Fat % =		15.0%		
5	Age, $A =$		25	(years)					
6	Height, $H =$		70	(inches)			Results After:	Weight	Body Fat %
7	Daily Intake, $I_0 =$		2800	(calories)			1 week	179.8	15.0%
8	Activity Level, $\lambda_0 =$		1.55				1 month	179.4	14.9%
9							6 months	177.3	14.5%
10	t	$F(t)$	$L(t)$	$W(t)$	Body Fat %		1 year	176.1	14.3%
11	0	27.0	153.0	180.0	15.0%		5 years	175.2	14.2%
12	1	27.0	153.0	180.0	15.0%		10 years	175.2	14.2%

FIGURE 8.15   Carl's body composition projections for Example 8.17.

For the same daily diet, Joe is projected to gain almost 9 pounds, while Carl will actually lose almost 5. The differences in body composition and the concomitant differences in calorie expenditure have led to different long-term weights; this is something unaccounted for by our previous models. $\square$

To help us understand this long-term behavior, we begin our usual process of finding equilibrium points by solving for (F^*, L^*) such that

$$F^* = F^* + [I_0 - \lambda_0(1.832 \cdot F^* + 11.708 \cdot L^*)] \frac{F^*}{F^*+6} \cdot \frac{1}{4279.43}$$

$$L^* = L^* + [I_0 - \lambda_0(1.832 \cdot F^* + 11.708 \cdot L^*)] \frac{6}{F^*+6} \cdot \frac{1}{823.38}.$$

Concentrating on F^* first, we see after a little simplifying that we need to solve the equation

$$0 = [I_0 - \lambda_0(1.832 \cdot F^* + 11.708 \cdot L^*)] \frac{F^*}{F^*+6} \cdot \frac{1}{4279.43}.$$

Since a product can only be zero if at least one of its factors is zero, we have two possibilities: either $\frac{F^*}{F^*+6} = 0$, which implies $F^* = 0$ and is not biologically possible, or $I_0 - \lambda_0(1.832 \cdot F^* + 11.708 \cdot L^*) = 0$.

Examining the equation for L^* leads to the same result: the only biologically possible solution is when we have $I_0 - \lambda_0(1.832 \cdot F^* + 11.708 \cdot L^*) = 0$.

Note that intuitively the equilibrium points mean the same thing as for our previous models: body weight will remain stable whenever calories in (I_0) equals calories out $\lambda_0(1.832 \cdot F^* + 11.708 \cdot L^*)$. The difference is that people who start at the same weight but different body compositions end up at different equilibrium points because they have different calorie expenditures. Unfortunately, unlike the variable calorie expenditure model from Section 8.2, we cannot find the long-term point for a particular choice of parameters without using Excel. In our next example we check our equilibrium equation with Joe and Carl from Example 8.17.

Example 8.18: Show that the long-term values for Joe and Carl satisfy the equilibrium equation $I_0 - \lambda_0(1.832 \cdot F^* + 11.708 \cdot L^*) = 0$.

Recall that Joe and Carl started with identical parameters including initial body weight but that they had different body fat percentages. We need to enter the correct parameters for each man and observe the values for fat and lean once they have stabilized. For Joe our Excel work projects long-term amounts of body fat and lean as $F^* = 40.8$ and $L^* = 147.9$. Plugging these values along with $I_0 = 2800$ and $\lambda_0 = 1.55$ into the equilibrium equation yields

$$I_0 - \lambda_0(1.832 \cdot F^* + 11.708 \cdot L^*) = 2800 - 1.55 \cdot (1.832 \cdot 40.80 + 11.708 \cdot 147.91)$$

$$= 2800 - 1.55 \cdot 1806.476$$

$$= 0.$$

For Carl the initial body fat was 15%. His long-term values for fat and lean are $F^* = 24.84$ and $L^* = 150.40$. Plugging these values along with $I_0 = 2800$ and $\lambda_0 = 1.55$ into the equilibrium equation yields

$$I_0 - \lambda_0(1.832 \cdot F^* + 11.708 \cdot L^*) = 2800 - 1.55 \cdot (1.832 \cdot 24.84 + 11.708 \cdot 150.40)$$
$$= 2800 - 1.55 \cdot 1806.39$$
$$= 0.$$

The results for both men are correct to within rounding error. $\square$

The equation $I_0 - \lambda_0(1.832 \cdot F^* + 11.708 \cdot L^*) = 0$ represents a straight line in the F,L-phase plane (see Section 3.2), which we can see more easily once we rewrite the equation in slope–intercept form:

$$L^* = -\frac{1.832}{11.708}F^* + \frac{I_0}{11.708 \cdot \lambda_0}.$$

Unlike previous models where we could expect one, two, or perhaps a few different equilibrium points, this line represents the presence of infinitely many equilibrium points—one for every point on the line.

It is worthwhile to spend some time thinking about the implications of this line for body weight. Notice that the slope of the line, $m = -\frac{1.832}{11.708}$, is constant regardless of the values for all other parameters. On the other hand, the L-intercept, $b = \frac{I_0}{11.708 \cdot \lambda_0}$, is determined by the daily calorie intake and the activity level, both of which are under the control of the individual. Increasing calorie intake raises the line vertically, which corresponds to higher long-term body weights. Lowering calorie intake results in lower long-term body weights. Similarly, increasing the activity level serves to lower the line vertically and thus lowers long-term weight projections. Decreasing activity level has the opposite effect.

Thus in terms of long-term body weight changes, we can set the location of the line we will eventually reach by adjusting our calorie intake and activity level, but the exact location on the line where we will end up is determined by our initial body composition.

8.5.2   Phase Plane Diagram

Whenever we have an entire curve or line of equilibrium values, we have what is called an **invariant manifold**, and creating a phase plane diagram can help us understand how the presence of an invariant manifold affects the long-term behavior of our model.

In the example that follows, we examine a phase plane diagram for the body composition model in the F,L-plane.

Example 8.19: Suppose Julie is a 23-year-old female who is 5′5″ tall, weighs 130 pounds, and is highly active. Create a phase plane diagram that gives body composition projections for Julie for a variety of body compositions assuming a constant daily calorie intake of $I_0 = 1600$.

We know that $A = 23$, $H = 65$, and $\lambda_0 = 1.725$. We also know Julie's initial weight is $W(0) = 130$ pounds. What we do not know is how much body fat Julie has, and we know that this will influence Julie's long-term weight.

To give a graphical representation of the possibilities for Julie, we create a phase plane diagram in the L,F-plane that includes trajectories for Julie's body composition for a variety of initial body compositions. This will give us a global understanding of what to expect for any woman with the same parameters as Julie.

Remembering that $W(t) = L(t) + F(t)$, we can represent all possible initial body compositions for Julie with the equation $L + F = 130$. This gives us a straight line in the L,F-plane that in slope–intercept form is given by $L = -F + 130$. A graph of this line is shown in Figure 8.16.

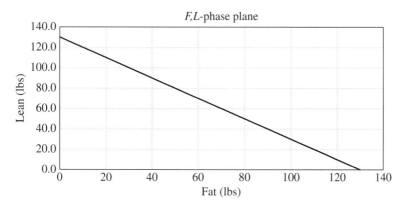

FIGURE 8.16 Possible initial body compositions for a body weight of 130 pounds.

The line divides the plane into two regions. The region above and to the right of the line represents a net weight gain, while the region below and to the left of the line represents a net weight loss.

We know from our equilibrium analysis that Julie's long-term body composition will be somewhere on the equilibrium line in the L,F-plane given by $L = -\dfrac{1.832}{11.708} F + \dfrac{1600}{11.708 \cdot 1.725}$. We plot the equilibrium line along with the initial body composition line in Figure 8.17.

Julie must start somewhere on the initial body composition line and must end up somewhere on the equilibrium line. To get an idea of how the ending point is related to the starting point, we plot trajectories for a variety of initial body compositions and see where they end up. We present such a graph in Figure 8.18.

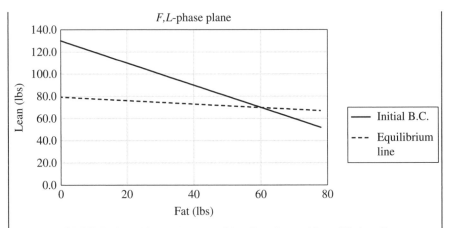

FIGURE 8.17   Initial body composition line along with equilibrium line.

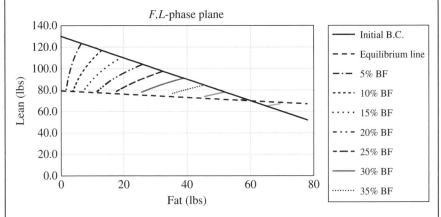

FIGURE 8.18   Phase plane diagram for body composition model.

All trajectories move *toward* the equilibrium line, but the initial body composition determines *where* on that line we end up. Because all trajectories tend toward the line of equilibrium values, we call the line an **attracting invariant manifold** (Chow & Hall, 2008).

By placing our cursor on the Excel graph over the points where the curves meet the equilibrium line, we can see the coordinates for fat and lean. By adding the coordinates together, we can determine the long-term body weight. For example, if Julie starts out with a body fat percentage of 20%, she starts at the point $(F_0, L_0) = (26, 104)$. Her long-term projection is the point on the equilibrium line given by $(F^*, L^*) = (16, 86.6)$. Thus her projected long-term weight is $16 + 86.6 = 102.6$ pounds, and her projected long-term body fat percentage is $\frac{16}{102.6} = 15.6\%$ On the other hand, if Julie's initial body fat percentage is 30%, she will start at the point $(F_0, L_0) = (39, 91)$ and end at the point

$(F^*, L^*) = (32, 84.1)$. Her long-term weight in the second case is projected to be $32 + 84.14 = 112.1$ pounds, while her body fat percentage is projected at $\frac{32}{112.1} = 28.5\%$. This should seem reasonable since we know that lean body mass burns more calories than fat. Thus leaner individuals will have a greater calorie deficit under the same diet than those with more body fat and hence will end up weighing less. We note that the graph shows that if Julie has an initial body fat percentage above about 45%, she will end up gaining weight. □

In the next example we show how we can use the intersection of the initial body composition line and the equilibrium line to our advantage.

Example 8.20: Mark is a sedentary, 5′8″ male who is 40 years old and weighs 195 pounds. Mark knows from careful tracking that his body weight remains constant when he consumes about 2000 calories per day. Determine Mark's body fat percentage.

The idea here is that if Mark's weight is stable, then his current body composition must put him exactly at the intersection of the line $L = -F + 195$ and the line $L = -\frac{1.832}{11.708} F + \frac{2000}{11.708 \cdot 1.2}$. Otherwise, Mark's trajectory would move him off of the initial body composition line and toward the equilibrium line.

Finding this point of intersection is a matter of setting the two equations equal to each other and solving for F. We have

$$-\frac{1.832}{11.708} F + \frac{2000}{11.708 \cdot 1.2} = -F + 195$$

$$0.844F = 195 - 142.353$$

$$F = 62.378.$$

Knowing Mark's amount of initial body fat, we compute his initial body fat percentage to be $\frac{62.378}{195} = 0.320$, or 32%. □

In the next example we consider the implications of the model from an individual perspective.

8.5.3 Individual Weight Trajectories

For an individual with a given set of parameters, our current model projects what body compositions and body weights are possible for that individual in the form of a trajectory in the F,L-plane. Any change in diet or activity level changes where on the trajectory an individual will end up, but the individual will not leave the trajectory. We illustrate this point in the next example.

Example 8.21: Consider Julie from Example 8.19, and assume she currently has an average amount of body fat as determined by Jackson's equation. Graph Julie's individual body composition trajectory and indicate where she will end up for different combinations of diet and activity level.

The graph in Figure 8.19 is Julie's body composition trajectory. It represents all of her possible future body compositions (and hence weights) as projected by the model. Where she ends up on the trajectory depends on her choices regarding diet and activity level.

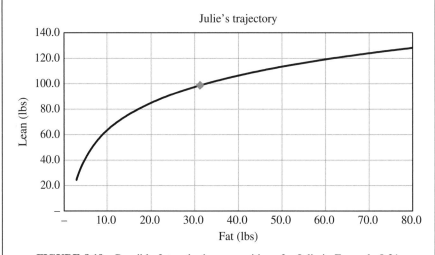

FIGURE 8.19   Possible future body compositions for Julie in Example 8.21.

The diamond in the middle of the graph marks Julie's current body composition. If she consumes more calories than she burns (either by increasing her intake or decreasing her activity level), she will move along the curve up and to the right. This would represent an increase in body fat and weight. If she consumes fewer calories than she burns (either by decreasing her intake or increasing her activity level), she will move along the curve down and to the left. This would represent a decrease in body fat and weight.

In Figure 8.20 we include the equilibrium line corresponding to three different combinations of diet and activity level for Julie. The points where the lines intersect Julie's trajectory are the points where she will end up on her trajectory in each case.

The bottom line represents a daily intake of 1500 calories and a high activity level. The middle line represents a daily intake of 2000 calories and a moderate activity level. The top line represents a daily intake of 2200 calories and a light activity level. □

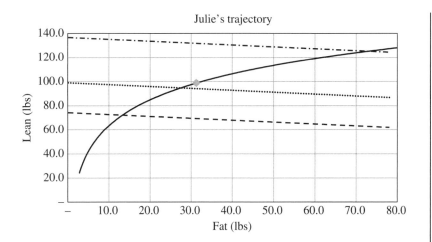

FIGURE 8.20 Long-term body compositions for Julie for three different intake/activity level combinations.

The fact that changes to diet and exercise level do not move us off of our trajectory, just to different points along it, leads to the question, "Is it possible for someone to change their trajectory?"

Unfortunately we cannot tell from the model because the model does not distinguish among different types of activity in the activity level. In particular, the model neglects the effect that a well-designed strength training program can have on body composition. Strength training, especially weight lifting, can build lean muscle mass while limiting fat gain. In this way it has the potential to put us on a new trajectory since for a given weight we will be further up and to the left along the initial body composition line. Then any decrease in daily intake or increase in activity level will lead to lower, leaner body weights in the long-term.

In their 2008 paper "The Dynamics of Human Body Weight Change," Chow and Hall note that all body weight models of the kind we discuss in this chapter fall into one of two types. For a fixed diet and energy expenditure, models either predict an attracting invariant manifold, where there are an infinite number of body compositions possible at equilibrium, or they predict a single stable equilibrium point. Interestingly, it is not yet known which type of model is the correct one for human body weight. The data is insufficient to tell at this time (Chow & Hall, 2008).

8.5.4   Section Exercises

1   Eduardo is 20 years old, stands 5′ 9″ tall, and weighs 165 pounds. He is highly active and consumes 2200 calories per day. Project Eduardo's body weight and body fat percentage 1 week, 1 month, 6 months, 1 year, and 5 years from today.

2   Use the body composition Excel model to show that two individuals of the same weight, height, age, sex, activity level, and daily calorie intake can have very different long-term weight projections.

3   Use a daily calorie intake of your choice and the Excel body composition model to project your own body composition and body weight for 1 week, 1 month, 6 months, 1 year, and 5 years from today.

    a. Run the projections using Jackson's estimate for your initial body fat percentage.

    b. Run the projections using the appropriate Navy equation to estimate your initial body fat percentage.

    c. Run the projections using the appropriate Bailey equation to estimate your initial body fat percentage.

    d. Discuss how much of a difference in long-term weight projection results from using the different body fat estimates.

4   Show that the long-term values for body fat and lean mass for Eduardo in Exercise 1 satisfy the equilibrium equation $I_0 - \lambda_0 \left(1.832 \cdot F^* + 11.708 \cdot L^*\right) = 0$.

5   Show that the long-term values for body fat and lean mass you found for yourself in Exercise 3 satisfy the equilibrium equation $I_0 - \lambda_0 \left(1.832 \cdot F^* + 11.708 \cdot L^*\right) = 0$.

6   Jennifer is a moderately active, $5'4''$ female who is 30 years old and weighs 125 pounds. Jennifer knows from careful tracking that her body weight remains constant when she consumes about 1900 calories per day. Determine Jennifer's body fat percentage.

7   We often use the estimate that 1 pound of body weight contains approximately 3500 calories. Use the Excel body composition model to estimate how many calories would be contained in a 1-pound change of your own body weight.

8   Decide on a personal weight goal. Use the body composition model to determine the required daily calorie intake to reach your goal in:

    a. 1 week

    b. 1 month

    c. 1 year

    d. 5 years

    e. 10 years

9   *Extension*: Set a 2-year weight goal and use the body composition Excel model to find the daily calorie intake necessary for you to achieve your goal. Using this daily calorie intake, answer the following.

    a. What percentage of the long-term weight change takes place in the first month?

    b. What percentage of the long-term weight change takes place in the first 6 months?

    c. What percentage of the long-term weight change takes place in the first year?

10 Repeat Exercises 8 and 9 using a body fat percentage goal rather than a weight goal.

8.6 POINTS-BASED SYSTEMS: THE WEIGHT WATCHERS MODEL

Weight loss is big business in the United States: as of 2010 the weight loss industry in the United States was worth over $60 billion (Sandilands, 2010). One of the largest companies is Weight Watchers International, Inc. The Weight Watchers system has two main components. The first component is a social support structure provided by monthly meetings where subscribers help each other develop healthier eating habits. The second component is a points-based diet tracking system designed to help subscribers lose weight in a healthy way. In this section we analyze the mathematics behind the points-based system, which is now called the PointsPlus® program.

8.6.1 Basic PointsPlus Formula

The PointsPlus program represented an overhaul of the original Weight Watchers Points system. Subscribers are assigned an allotment of **PointsPlus** that they "spend" each day as they consume calories. The number of PointsPlus assigned is based on the subscriber's body weight and weight goal. Some foods cost more in points than others with healthier choices generally costing fewer points to eat than less healthy choices. Fruits and vegetables, for example, cost zero points and so can be consumed freely throughout the day. As subscribers lose weight, their points allotment decreases to reflect an assumed decrease in calorie expenditure.

The PointsPlus system incorporates recent research in nutrition science that indicates some foods are more difficult for the body to process than others. For example, 100 calories of protein is much more difficult for the body to process than 100 calories of fat. Thus a larger percentage of the protein calories will be burned during digestion than will be for fat. This phenomenon is known as the **thermic effect of food** (**TEF**). Roughly speaking, protein is the hardest to digest, followed by carbohydrates and then fat, which is very easy for the body to digest. The theory is that because of this effect, consuming 1000 calories of protein will lead to less weight change than 1000 calories of fat.

In Patent US20100055652, Weight Watchers describes how they assign points to foods based on how much protein, carbohydrate, fat, and fiber they contain (Miller-Kovach et al., 2010). Each type of macronutrient is "discounted" based on how difficult they are for the body to process, and hence the proportion of each type of macronutrient that is actually absorbed by the body is different. The proportions used by Weight Watchers are given in Table 8.5.

Different macronutrients also have different energy densities. One gram of fat, for example, contains 9 calories, while one gram of protein contains only 4. Similarly, one gram of carbohydrate (fiber and nonfiber) contains 4 calories.

TABLE 8.5   Proportions of Macronutrients Assumed to be Absorbed by the Body

Macronutrient	Proportion Affecting Body Weight
Protein	0.80
Carbohydrate (nonfiber)	0.95
Dietary fiber	0.25
Fat	1.0

In order to find the point value for a serving of a particular food, we first find the amount of each of the macronutrients from the nutrition label. Let P be grams of protein, C be grams of carbohydrate, and F be grams of fat per serving. Then the total number of unadjusted calories (TC) in a serving can be found by

$$TC = 4 \cdot P + 4 \cdot C + 9 \cdot F.$$

Dietary fiber is included in total carbohydrates, but the PointsPlus system treats dietary fiber separately. Thus we let DF be grams of dietary fiber and write an equivalent formula for total calories as

$$TC = 4 \cdot P + 4 \cdot (C - DF) + 4 \cdot DF + 9 \cdot F.$$

As we mentioned before not all calories are created equal, and our next step is to calculate total calories adjusted for the TEF using the Weight Watchers parameters in Table 8.2. The equation for total adjusted calories (TAC) in a serving of food is given by

$$TAC = 0.80 \cdot 4 \cdot P + 0.95 \cdot 4 \cdot (C - DF) + 0.25 \cdot 4 \cdot DF + 1.0 \cdot 9 \cdot F$$

Multiplying coefficients together gives us

$$TAC = 3.2 \cdot P + 3.8 \cdot (C - DF) + DF + 9 \cdot F.$$

In our next example we show how to find total calories and total adjusted calories for a serving of peanut butter.

Example 8.22: Find the total calories and total adjusted calories for a serving of peanut butter.

According to the nutrition facts label on a container of Peter Pan creamy peanut butter, one serving of peanut butter is two tablespoons and contains 210 calories. Each serving contains 17 g of fat, 8 g of protein, and 6 g of carbohydrate including 2 g of dietary fiber. First we verify the total calories given on the label:

$$TC = 4 \cdot P + 4 \cdot (C - DF) + 4 \cdot DF + 9 \cdot F$$
$$= 4 \cdot 8 + 4 \cdot (6 - 2) + 4 \cdot 2 + 9 \cdot 17$$
$$= 209.$$

Our calculation gives 209 calories per serving, which agrees with the label within rounding error. Next we calculate the total adjusted calories, that is, the number of calories that will be available to the body after the TEF has been factored in. We get

$$
\begin{aligned}
\text{TAC} &= 32 \cdot P + 3.8 \cdot (C - \text{DF}) + \text{DF} + 9 \cdot F \\
&= 3.2 \cdot 8 + 3.8 \cdot (6 - 2) + 2 + 9 \cdot 17 \\
&= 195.8.
\end{aligned}
$$

Thus the total adjusted calories for one serving of peanut butter is about 196. About 14 calories of the peanut butter will be burned as the body digests it. ☐

Once we have calculated total adjusted calories for a particular food, Weight Watchers assigns a whole number point value by dividing total adjusted calories by a factor that appears to be about 35 and then rounding the result. Thus the Points-Plus value for any food will be approximately equal to

$$
\text{PointsPlus} = \frac{\text{TAC}}{35} = \frac{3.2 \cdot P + 3.8 \cdot (C - \text{DF}) + \text{DF} + 9 \cdot F}{35},
$$

rounded to the nearest whole number. Note that the division by 35 in the formula means that roughly speaking, 1 point is equivalent to about 35 adjusted calories. This is equivalent to 1 point equaling 38.89 calories before adjustment.

Example 8.23: Find the PointsPlus value of a serving of peanut butter.

Since we have already found the total adjusted calories, the PointsPlus value is given by

$$
\text{PointsPlus} = \frac{195.8}{35} = 5.59 \approx 6.
$$

A subscriber who eats one serving of peanut butter would then subtract 6 points from their daily allotment. ☐

E.22   Rounding

With Excel we can create our own PointsPlus calculator. We allow the user to enter the amounts of protein, total carbohydrate, fat, and dietary fiber from any food nutrition label, and then we get Excel to report the equivalent number of PointsPlus for that food. One new command that we will need is Excel's "ROUND" function. The "ROUND" command accepts a number as an input along with the number of decimal places to round to. For example, the command "ROUND(10.53,1)" will return the result "10.5," while the command "ROUND(10.53,0)" will return the result "11." It is this second version we want when computing whole numbers of points. The Excel setup is given in Figure 8.21.

	A	B	C	D	E	F
1	Weight Watchers™ PointsPlus® Values					
2						
3	Protein, P =		8	(grams)		
4	Total Carbohydrate, C =		6	(grams)		
5	Dietary Fiber, DF =		2	(grams)		
6	Fat, F =		17	(grams)		
7						
8	PointsPlus Value =		=ROUND((3.2*C3+3.8*(C4-C5)+C5+9*C6)/35,0)			

FIGURE 8.21   PointsPlus calculator in Excel.

Keeping track of food points is very similar to keeping track of total calories, but it is subtly different. Keeping track of points in this way encourages the user to opt for healthier foods because healthier foods cost less out of their daily target. Fruits and vegetables are obviously privileged in this system because they count as zero points. However, the system also pushes the user to privilege foods high in protein and fiber, which are often healthier than those high in carbohydrates and fat.

Next we see how Weight Watchers determines an individual's daily PointsPlus allotment.

8.6.2   Points Allotments

Under the Weight Watchers system, individuals are assigned a daily PointsPlus allotment based on their calorie needs. Individuals are also given 49 weekly points to use as they see fit throughout the week. If all weekly points are used, this comes to 7 points per day. As discussed in the subsequent section, individuals may also earn extra points by exercising.

Daily PointsPlus allotments are based on an individual's calorie requirements. We have already seen one method for estimating such calorie needs in Section 8.2 where we used the MSJ equation multiplied by the activity level. Weight Watchers uses a similar kind of equation that calculates daily calorie needs (DC) based on age, weight, height, sex, and the assumption of a light activity level as a base. The equations for males and females over 19 years old can be found in the National Academies Press publication *Dietary Reference Intakes: The Essential Guide to Nutrient Requirements* (DRI, 2006) and are as follows:

$$\text{Males}: DC = 864 - 9.72 \cdot A + 7.214 \cdot W + 14.309 \cdot H$$
$$\text{Females}: DC = 387 - 7.31 \cdot A + 5.636 \cdot W + 19.131 \cdot H.$$

As before age is in years, weight is in pounds, and height is in inches.

Next, daily calorie needs are discounted by a factor of 0.90, representing an average value for the TEF for the various macronutrients. The result is adjusted daily calories (ADC) given by the equation

$$ADC = 0.90 \cdot DC.$$

Again, the ADC represents the calories from food not burned during digestion.

Because the goal of most Weight Watchers subscribers is to lose weight, Weight Watchers subtracts the equivalent of about 800 adjusted calories from ADC. This equates to an average weight loss per week of just under 2 pounds.

Using an approximate value of 35 adjusted calories per point, we start with

$$\frac{ADC - 800}{35}$$

points, which we round to the nearest point. Next we subtract 11 from this value, and that will give us our initial total for our daily PointsPlus allotment:

$$\text{Daily PointsPlus} = \frac{ADC - 800}{35} - 11.$$

The subtraction of 11 includes an adjustment of 4 points to compensate for the fact that fruits and vegetables are considered "free" and an adjustment of 7 points to allow the subscriber 49 weekly points to spend at their discretion throughout the week.

There is one final adjustment to make: Weight Watchers places an upper and lower limit on the number of daily points it considers prudent, namely, 71 and 26, respectively. So if our daily points calculation above falls below 26, we would actually use 26, and if the calculation falls above 71, we would use 71. Altogether we have the formula given by

$$\text{Daily PointsPlus} = \min\left\{ \max\left\{ \text{round}\left(\frac{ADC - 800}{35} \right) - 11, 26 \right\}, 71 \right\}.$$

The "max" function accepts two numbers as inputs and returns the larger of the two, the "min" function accepts two numbers as inputs and returns the smaller of the two, and the "round" function rounds to the nearest whole point value.

In the next example we automate the calculation our daily PointsPlus target with Excel.

Example 8.24: Find the daily PointsPlus target for Zack, a 230-pound man who is 5′8″ tall and 24 years old.

Plugging Zack's parameters into the daily calorie needs formula for men gives an estimate of 3263 calories. Then the adjusted daily calorie number will be $ADC = 0.90 \cdot 3263 = 2936.7$. The PointsPlus allotment will therefore be equal to

$$\text{Daily PointsPlus} = \frac{ADC - 800}{35} - 11$$
$$= \frac{2936.7 - 800}{35} - 11$$
$$= 50.04.$$

Since the rounded value, 50, is between 26 and 71, it is Zack's daily PointsPlus allotment.

We check our work by automating the process in Excel. We have done this sort of thing many times before. Here we give the Excel setup with the formula for the PointsPlus allotment displayed in Figure 8.22. ☐

	A	B	C	D	E	F	G
1	Weight Watchers™ PointsPlus® Allotments						
2							
3	Sex, S =		M				
4	Current Body Weight =		230	(pounds)			
5	Age, A =		24	(years)			
6	Height, H =		68	(inches)			
7							
8	Daily Calorie Needs, DC =			3263.0			
9	Adjusted Daily Calorie Needs, ADC =			2936.7			
10	Daily PointsPlus® Allotment =			=MIN(MAX(ROUND((D9-800)/35-11,0),26),71)			

FIGURE 8.22   Calculating daily PointsPlus allotment in Excel.

In addition to the daily allotment for PointsPlus, Weight Watchers allocates 49 weekly points to each subscriber to use when and how they choose throughout the week. These weekly points expire at the end of each week if they are not consumed. Our habit will be to assume that all of these weekly points are consumed so that in effect we can increase the daily points total by 7. Thus in the previous example, if we assume Zack consumes all of his weekly points, then his daily point allotment is in effect 57.

Part of the Weight Watchers system is to encourage physical activity by awarding extra PointsPlus for exercising. We examine how this is done in the next section.

8.6.3   Activity Points

The daily points target assumes a light activity level as a starting point. Subscribers can then earn extra points by exercising. As one exercises, one burns calories, and these calories earn points at a rate of about 1 point for every 77.8 calories burned. Thus in order to calculate activity points, we first need a way of estimating calories burned during exercise.

Weight Watchers distinguishes physical activity by three levels of intensity: low, moderate, and high. As rough guidelines, low intensity exercise is something you could sing while doing, moderate intensity exercise is something you could talk while doing but not sing, and high intensity exercise is something you could only talk halt-ingly while doing if at all. Calorie burn rates are then assigned for each level of inten-sity on a "per pound per minute" basis, reflecting the fact that heavier people generally burn more calories than lighter people for the same activity. Calorie burn rates, based on values provided in DRI (2006), are given by

$$\text{Low intensity} = 0.01804 \text{ calories per pound per min}$$
$$\text{Moderate intensity} = 0.02543 \text{ calories per pound per min}$$
$$\text{High intensity} = 0.06285 \text{ calories per pound per min.}$$

Since we assume every 77.8 calories equals one point, the formulas we will use to calculate activity points earned for each exercise intensity level are as follows:

$$\text{Points low intensity} = 0.000232 \times \text{minutes} \times W$$
$$\text{Points at moderate intensity} = 0.000327 \times \text{minutes} \times W$$
$$\text{Points at high intensity} = 0.000808 \times \text{minutes} \times W.$$

Here "minutes" are the minutes spent exercising, and W is body weight in pounds. We note that in the context of food and daily allotment, 1 point is roughly equivalent to 38.9 calories of food. Since exercise only earns points at half that rate, exercise in the Weight Watchers system should create an additional calorie deficit and should there-fore lead to faster weight loss.

Next we give an example of how to calculate activity points.

Example 8.25: Tonya weighs 150 pounds, and she exercises at moderate inten-sity for one hour. How many activity points does she earn?

Here we just need to plug in Tonya's weight to the formula for moderate inten-sity exercise. We get

$$\text{Points at moderate intensity} = 0.000327 \times \text{minutes} \times W$$
$$= 0.000327 \times 60 \times 150$$
$$= 2,943.$$

After rounding we see that Tonya earns 3 activity points that she can add to her daily PointsPlus allotment. It is a straightforward exercise to include an activity points calculation in our Excel spreadsheet. □

In the next section we are finally ready to set up a dynamic model for tracking weight changes using the Weight Watchers PointsPlus system.

8.6.4 PointsPlus Dynamic Model

The Weight Watchers system is more difficult to implement in Excel than previous models for a couple of reasons. First, the use of points rather than calories means we have to convert back and forth from calories to points. To do so we will use the estimate

$$1\,\text{PointsPlus} \approx 38.9\,\text{calories}.$$

Second, unlike previous models the daily caloric intake I_0 (given in points) is no longer a constant. The way the system works is to lower the PointsPlus allotment and hence the daily calorie intake as weight declines. The idea is to maintain a roughly constant calorie *deficit* rather than a constant calorie *intake*.

We saw in our very first body weight model that any model that involves a constant calorie deficit has no equilibrium and eventually leads to unrealistic body weights. The situation is a little different with the Weight Watchers system for two reasons. One, Weight Watchers sets a minimum number for the daily allotment. Two, once a subscriber reaches their goal weight, they change to a maintenance phase where their daily PointsPlus allotment is reset to a level intended to maintain rather than lose weight.

For now we concentrate on calculating body weight each day. As body weight changes, the daily calories, adjusted daily calories, PointsPlus daily allotment, and activity points all change. Thus they each get their own column. We have already seen how to calculate each of these quantities, so we input the equations into Excel, being careful to reference the body weight column rather than the initial body weight. The formula for *DC* requires an IF statement based on sex, and the formula for activity points requires an IF statement based on exercise intensity.

We assume that all weekly points are consumed, along with the built-in assumption of 4 points for fruits and vegetables. Thus we add 11 points each day to the standard calculation in order to estimate how many points the user will consume without exercise.

To get Excel to calculate body weight, we must make some assumptions. One is the standard estimate that there are 3500 calories in 1 pound. Without considering activity points, our user will consume the equivalent of $38.9 \times (\text{total daily PointsPlus})$ calories per day. Thus the daily calorie deficit will be

$$DC - 38.9 \times (\text{total daily PointsPlus}).$$

When activity points are considered, we must recall that each activity point represents burning 77.8 additional calories. If the activity points are all consumed, that means each activity point results in an additional calorie deficit of 38.9 calories per point. The result is that the daily calorie deficit with activity points included is given by

$$DC - 38.9 \times (\text{total daily PointsPlus}) + 38.9 \times (\text{activity points}).$$

This means that each day our body weight will decrease by

$$\frac{DC - 38.9 \times (\text{total daily PointsPlus}) + 38.9 \times (\text{activity points})}{3500}$$

pounds. The Excel setup with the formula for body weight displayed is given in Figure 8.23.

	A	B	C	D	E	F	G	H
1	Weight Watchers™ PointsPlus® Dynamic Model							
2								
3	Sex, S =		M					
4	Current Body Weight =		230	(pounds)				
5	Age, A =		24	(years)				
6	Height, H =		68	(inches)				
7	Daily Exercise =		30	(minutes)				
8	Exercise Intensity =		2					
9								
10	t		DC	ADC	Daily Pts. Plus	Activity Pts.	Weight	
11	0		3263.0	2936.7	61	2	230	
12	1		3263.0	2936.7	61	2	=F11-(B11-38.9*D11+38.9*E11)/3500	

FIGURE 8.23   Calculating body weight for Weight Watchers PointsPlus system.

Next we give an example of how to use the Weight Watchers PointsPlus Excel model.

Example 8.26: Fred is a new Weight Watchers subscriber. He is $5'8''$ tall, weighs 230 pounds, and is 27 years old. His goal weight is 170. Determine how long it will take him to reach his goal weight if (a) he does not exercise or if (b) he exercises for 30 min each day at moderate intensity.

Once again we assume that Fred uses all of his weekly points along with the 4 points assumed for fruits and vegetables. These assumptions are already built in to the Excel spreadsheet. Next we enter all relevant parameters and drag the model down until we reach Fred's goal weight of 170. The Excel results are presented in Figure 8.24 with most rows hidden. Without exercise it will take Fred about 34 weeks to reach his goal weight, or about 8 months.

	A	B	C	D	E	F
1	Weight Watchers™ PointsPlus® Dynamic Model					
2						
3	Sex, S =		M			
4	Current Body Weight =		230	(pounds)		
5	Age, A =		27	(years)		
6	Height, H =		68	(inches)		
7	Daily Exercise =		0	(minutes)		
8	Exercise Intensity =		2			
9						
10	t		DC	ADC	Daily Pts. Plus	Activity Pts.	Weight
11	0		3233.8	2910.4	60	0	230
12	1		3233.8	2910.4	60	0	229.7
247	236		2803.5	2523.2	49	0	170.1
248	237		2801.7	2521.5	49	0	169.8

FIGURE 8.24   Dynamic PointsPlus model projections for Example 8.26.

If we repeat the calculation but with daily exercise set to 30 min at moderate intensity, we get the results in Figure 8.25. With exercise it will take Fred 31 weeks, or about 7 months, to reach his goal weight. He will also be much fitter. □

	A	B	C	D	E	F
1	Weight Watchers™ PointsPlus® Dynamic Model					
2						
3	Sex, S =		M			
4	Current Body Weight =		230	(pounds)		
5	Age, A =		27	(years)		
6	Height, H =		68	(inches)		
7	Daily Exercise =		30	(minutes)		
8	Exercise Intensity =		2			
9						
10	t	DC	ADC	Daily Pts.Plus	Activity Pts.	Weight
11	0	3233.8	2910.4	60	2	230
12	1	3233.8	2910.4	60	2	229.7
228	217	2803.7	2523.3	49	2	170.1
229	218	2801.6	2521.5	49	2	169.8

FIGURE 8.25 Dynamic PointsPlus model projections for Example 8.26 including daily exercise.

As a final example we examine the necessary PointsPlus allotment for Fred to maintain his goal weight.

Example 8.27: Suppose in the previous example that Fred has reached his goal weight of 170 pounds. Determine how many points Fred should consume per day to maintain his weight. Assume no activity points.

All that is required here is to determine Fred's daily calorie requirement at his goal weight and then divide that requirement by 38.9 to estimate the equivalent number of points. We get

$$DC = 864 - 9.72 \cdot 27 + 7.214 \cdot 170 + 14.309 \cdot 68$$

$$DC = 2801.$$

Thus Fred's maintenance point level would be $\frac{2801}{38.9} = 72$ points per day. This points level does not need to be adjusted for weekly points or for fruits and vegetables. □

The maintenance point level determined by the equations for DC is not the maintenance level recommended by Weight Watchers. The Weight Watchers approach is more of a trial-and-error approach where the subscriber initially ups their daily PointsPlus allotment by 6 points and then checks to see how their weight responds. If they continue to lose weight, then they add another 6 points and check again. Once they are at a point level at which they maintain their weight, they stay at the point level indefinitely. It is not clear how the eventual maintenance levels determined in this way compare to the level we determined in our example. In the exercises the reader is invited to determine how daily exercise impacts the maintenance Points-Plus level.

8.6.5 Section Exercises

1 Use a nutrition facts label of your choice to find the total calories and total adjusted calories for a serving of the food.

2 Use a nutrition facts label of your choice (different from the one in the previous exercise) to find the total calories and total adjusted calories for a serving of the food.

3 Find the PointsPlus value for a serving of the food in Exercise 1.

4 Find the PointsPlus value for a serving of the food in Exercise 2.

5 Find the daily PointsPlus target for Doug, a 200-pound man who is $5'10''$ tall and 22 years old.

6 Find your own daily PointsPlus target.

7 Kara weighs 120 pounds, and she exercises at high intensity for 30 min. How many activity points does she earn?

8 How many activity points would you earn by walking briskly for 1 h?

9 How many minutes of moderate intensity exercise would it take for you to earn 4 activity points?

10 Sophie is a new Weight Watchers subscriber. She is $5'7''$ tall, weighs 190 pounds, and is 23 years old. Her goal weight is 150 pounds. Determine how long it will take her to reach her goal weight if (a) she does not exercise or if (b) she exercises for 30 min each day at high intensity.

11 Set your own weight goal and use the Excel Weight Watchers model to determine how long it will take you to reach it if (a) you do not exercise or if (b) you exercise for a length of time and intensity of your choosing.

12 Suppose in Exercise 10 that Sophie has reached her goal weight of 150 pounds. Determine how many points she should consume per day to maintain her weight. Assume no activity points.

13 Once you meet your goal weight in Exercise 11, determine how many points you should consume per day to maintain your weight. Assume no activity points.

14 Suppose Sophie in Exercise 12 earns 4 activity points each day. What should her new maintenance PointsPlus level be?

15 Suppose you exercise each day for a length of time and intensity of your choosing. What should your new maintenance PointsPlus level be?

16 *Extension*: Using your own parameters, compare the daily calorie needs estimated by the DRI equations in Section 8.6.2 (which assume a light activity level) to

    a. The Mifflin–St. Jeor estimate (with light activity level).

    b. The Harris–Benedict estimate (with light activity level).

    c. The Livingston–Kohlstadt estimate (with light activity level).

    d. The Nelson estimate (with light activity level).

    e. How much difference is there among the methods?

APPENDIX A

THE GEOMETRIC SERIES FORMULA

In this appendix we supply a proof of the geometric series formula from Chapter 1.

Theorem 1.1

Let c be any real number and let x be any real number such that $x \neq 1$. Then for any positive integer n, we have

$$c + cx + cx^2 + cx^3 + \cdots + cx^{n-1} = c\frac{x^n - 1}{x - 1}.$$

Proof: The proof is fairly short and depends on a clever algebraic trick. Consider the sum $c + cx + cx^2 + cx^3 + \cdots + cx^{n-2} + cx^{n-1}$, and call this sum S:

$$S = c + cx + cx^2 + cx^3 + \cdots + cx^{n-2} + cx^{n-1}.$$

Next multiply S by x to get

$$xS = x\left(c + cx + cx^2 + cx^3 + \cdots + cx^{n-2} + cx^{n-1}\right).$$
$$= cx + cx^2 + cx^3 + cx^4 + \cdots + cx^{n-1} + cx^n.$$

Now we subtract S from xS and note that all but two terms cancel

$$xS - S = \left(cx + cx^2 + cx^3 + cx^4 + \cdots + cx^{n-1} + cx^n\right)$$
$$- \left(c + cx + cx^2 + cx^3 + \cdots + cx^{n-2} + cx^{n-1}\right)$$
$$= cx^n - c.$$

Models for Life: An Introduction to Discrete Mathematical Modeling with Microsoft® Office Excel®, First Edition. Jeffrey T. Barton.
© 2016 John Wiley & Sons, Inc. Published 2016 by John Wiley & Sons, Inc.

Factoring out common terms on each side of the equality gives us

$$S(x-1) = c(x^n - 1).$$

We divide through by $x-1$ (which is okay because $x \neq 1$) to get

$$S = c\frac{x^n - 1}{x - 1},$$

and finally we remember what S represents:

$$c + cx + cx^2 + cx^3 + \cdots + cx^{n-2} + cx^{n-1} = c\frac{x^n - 1}{x - 1}. \quad \square$$

APPENDIX B

LANCHESTER'S SQUARE LAW AND THE FRACTIONAL EXCHANGE RATIO

This appendix supplies a proof of Lanchester's square law. The law is based on Lanchester's classic model for combat between two sides, Blue and Red:

$$B(t) = B(t-1) - rR(t-1)$$

$$R(t) = R(t-1) - bB(t-1).$$

The number of Blue and Red units remaining at any time, t, is given by $B(t)$ and $R(t)$ respectively, and the fighting effectiveness is given by b for Blue and r for Red. For our discrete version of Lanchester's model, the square law is given as follows.

Theorem 3.2
Lanchester's Square Law: For the basic Lanchester combat model, the following identity holds for all t:

$$rR(t)^2 - bB(t)^2 = (1-rb)^t \left[rR(0)^2 - bB(0)^2 \right].$$

Proof: We prove the result via induction on the time step t. First we prove the base case for $t = 1$ that $rR(1)^2 - bB(1)^2 = (1-rb)^1 \left[rR(0)^2 - bB(0)^2 \right]$. We begin by using the

Models for Life: An Introduction to Discrete Mathematical Modeling with Microsoft® Office Excel®,
First Edition. Jeffrey T. Barton.
© 2016 John Wiley & Sons, Inc. Published 2016 by John Wiley & Sons, Inc.

DDS to substitute for $R(1)$ and $B(1)$ on the left-hand side of the equation. After expanding and simplifying, we get

$$rR(1)^2 - bB(1)^2 = r[R(0) - bB(0)]^2 - b[B(0) - rR(0)]^2$$

$$= r\left[R(0)^2 - 2bR(0)B(0) + b^2B(0)^2\right]$$

$$- b\left[B(0)^2 - 2rR(0)B(0) + r^2R(0)^2\right]$$

$$= rR(0)^2 - 2rbR(0)B(0) + rb^2B(0)^2$$

$$- bB(0)^2 + 2rbR(0)B(0) - br^2R(0)^2$$

$$= rR(0)^2 + rb^2B(0)^2 - bB(0)^2 - br^2R(0)^2.$$

Next we factor what remains on the right-hand side:

$$rR(0)^2 + rb^2B(0)^2 - bB(0)^2 - br^2R(0)^2$$

$$= rR(0)^2 - bB(0)^2 - rb\left[rR(0)^2 - bB(0)^2\right]$$

$$= (1 - rb)\left[rR(0)^2 - bB(0)^2\right].$$

Thus

$$rR(1)^2 - bB(1)^2 = (1 - rb)\left[rR(0)^2 - bB(0)^2\right],$$

and this completes the proof of the base case.

Next we need to show the induction step that if the result is true for t, then it is also true for $t + 1$. This means we need to show that if

$$rR(t)^2 - bB(t)^2 = (1 - rb)^t\left[rR(0)^2 - bB(0)^2\right],$$

then it is also true that

$$rR(t+1)^2 - bB(t+1)^2 = (1 - rb)^{t+1}\left[rR(0)^2 - bB(0)^2\right].$$

Once again we begin by using the DDS to substitute for $R(t+1)$ and $B(t+1)$ on the left-hand side. Then we simplify. Using identical algebra from the base case gives us

$$rR(t+1)^2 - bB(t+1)^2 = (1-rb)\left[rR(t)^2 - bB(t)^2\right].$$

The last step is to use our induction hypothesis on the right-hand side to yield

$$rR(t+1)^2 - bB(t+1)^2 = (1-rb)\left[rR(t)^2 - bB(t)^2\right]$$

$$= (1-rb)\left[(1-rb)^t\left(rR(0)^2 - bB(0)^2\right)\right]$$

$$= (1-rb)^{t+1}\left[rR(0)^2 - bB(0)^2\right].$$

This completes the proof of the theorem. $\square$

Next we move to the fractional exchange ratio. Recall that for any time, t, the fractional exchange ratio, denoted FER$_t$, is defined to be the ratio of the relative losses for Blue to the relative losses for Red during time step t. We sometimes drop the subscript on the FER and in that case we mean the initial value FER$_1$. For any t we have

$$\text{FER}_t = \frac{rR(t-1)/B(t-1)}{bB(t-1)/R(t-1)} = \frac{rR(t-1)^2}{bB(t-1)^2}.$$

In particular we note $\text{FER} = \frac{rR(0)^2}{bB(0)^2}$. The theorem below shows how important the quantity FER is: it tells us at the outset how the battle will end.

Theorem 3.1
With the FER defined as above, we have the following three cases:

1. If FER < 1, then Blue will win.
2. If FER > 1, then Red will win.
3. If FER $= 1$, then both sides will be put out of action.

Proof: We prove case 1 by assuming FER < 1 and showing that this implies that $\lim_{t\to\infty} R(t) = 0$. We assume throughout that $R(t)$ and $B(t)$ are nonnegative (otherwise the model does not make physical sense).

First we claim that if FER < 1, then FER$_t$ is strictly decreasing in t so that

$$\text{FER}_t < \text{FER}_{t-1} < \cdots < \text{FER} < 1.$$

To see this we first use the Lanchester square law to show that FER$_t < 1$ for all t. If FER < 1, then $rR(0)^2 - bB(0)^2 < 0$, which in turn by the square law implies that

$rR(t-1)^2 - bB(t-1)^2 < 0$. (The term $(1-rb)^{t-1}$ must be positive.) Then we get $rR(t-1)^2 < bB(t-1)^2$ and $\dfrac{rR(t-1)^2}{bB(t-1)^2} = \text{FER}_t < 1$ as required.

Next we show that $\text{FER}_t < 1$ implies that $\text{FER}_{t+1} < \text{FER}_t$. We start with the definition for FER_t and use the Lanchester DDS to make substitutions. We get

$$\text{FER}_{t+1} = \frac{rR(t)^2}{bB(t)^2} = \frac{r[R(t-1) - bB(t-1)]^2}{b[B(t-1) - rR(t-1)]^2}.$$

Next we do some algebra to force FER_t to appear

$$\text{FER}_{t+1} = \frac{r[R(t-1) - bB(t-1)]^2}{b[B(t-1) - rR(t-1)]^2}$$

$$= \frac{rR(t-1)^2 \left[1 - \dfrac{bB(t-1)}{R(t-1)}\right]^2}{bB(t-1)^2 \left[1 - \dfrac{rR(t-1)}{B(t-1)}\right]^2}$$

$$= \text{FER}_t \left[\frac{1 - \dfrac{\sqrt{b}\sqrt{r}\sqrt{b}B(t-1)}{\sqrt{r}R(t-1)}}{1 - \dfrac{\sqrt{r}\sqrt{b}\sqrt{r}R(t-1)}{\sqrt{b}B(t-1)}}\right]^2$$

$$= \text{FER}_t \left[\frac{1 - \dfrac{\sqrt{br}}{\sqrt{\text{FER}_t}}}{1 - \sqrt{br}\sqrt{\text{FER}_t}}\right]^2.$$

The last step is to show that

$$0 < \frac{1 - \dfrac{\sqrt{br}}{\sqrt{\text{FER}_t}}}{1 - \sqrt{br}\sqrt{\text{FER}_t}} < 1.$$

The first part of the inequality follows from our initial assumption that both $R(t)$ and $B(t)$ are nonnegative. To see that the second part of the inequality is satisfied, we note that $\dfrac{1 - \frac{\sqrt{br}}{\sqrt{\text{FER}_t}}}{1 - \sqrt{br}\sqrt{\text{FER}_t}} < 1$ if and only if $1 - \dfrac{\sqrt{br}}{\sqrt{\text{FER}_t}} < 1 - \sqrt{br}\sqrt{\text{FER}_t}$, which in turn is true if and only if $\sqrt{\text{FER}_t} < \dfrac{1}{\sqrt{\text{FER}_t}}$. This last inequality holds by our assumption that

$\text{FER}_t < 1$. Thus $\text{FER}_{t+1} < \text{FER}_t$ whenever $\text{FER}_t < 1$, and this completes the proof that FER_t is strictly decreasing.

Finally we are ready to proceed with the main part of our proof that the number of Red forces tends to zero. Some of the algebra is very similar to what we have already done. First we have

$$R(t) = R(t-1) - bB(t-1)$$

$$= R(t-1) - \frac{\sqrt{b}\sqrt{r}R(t-1)\sqrt{b}B(t-1)}{\sqrt{r}R(t-1)}$$

$$= R(t-1) - \frac{\sqrt{br}R(t-1)}{\sqrt{\text{FER}_t}}.$$

By our previous work and our assumption, we know that $\text{FER}_t < \text{FER} < 1$. Hence $\frac{1}{\sqrt{\text{FER}_t}} > \frac{1}{\sqrt{\text{FER}}}$ and we get

$$R(t) = R(t-1) - \frac{\sqrt{br}R(t-1)}{\sqrt{\text{FER}_t}}$$

$$= R(t-1)\left[1 - \frac{\sqrt{br}}{\sqrt{\text{FER}_t}}\right]$$

$$\le R(t-1)\left[1 - \frac{\sqrt{br}}{\sqrt{\text{FER}}}\right].$$

It follows by the same reasoning that $R(t) \le R(0)\left[1 - \frac{\sqrt{br}}{\sqrt{\text{FER}}}\right]^t$. We note that the right-hand side of this last inequality must go to zero as long as $0 < 1 - \frac{\sqrt{br}}{\sqrt{\text{FER}}} < 1$. This inequality holds as long as $\frac{\sqrt{br}}{\sqrt{\text{FER}}} < 1$, or, equivalently, $br < \text{FER} = \frac{rR(0)^2}{bB(0)^2}$. After some algebra we see that the requirement is equivalent to $bB(0) < R(0)$, which must hold else all of Red's forces are eliminated during the first time step. Thus we have shown that either Red is eliminated during the first time step or $R(t)$ must tend to 0 as the battle goes on. This completes the proof of case 1, and case 2 follows by symmetry.

To complete the proof of the theorem, we assume $\text{FER} = 1$. Since $\text{FER} = 1$, we have $rR(0)^2 = bB(0)^2$, and by Lanchester's square law, we know that for all t we have

$$rR(t)^2 - bB(t)^2 = (1-rb)^t\left[rR(0)^2 - bB(0)^2\right] = 0.$$

Thus for all t we have $rR(t)^2 = bB(t)^2$, or $\sqrt{r}R(t) = \sqrt{b}B(t)$. For the Red side write

$$R(t) = R(t-1) - bB(t-1)$$

$$= R(t-1) - \sqrt{b}\sqrt{b}B(t-1)$$

$$= R(t-1) - \sqrt{b}\sqrt{r}R(t-1)$$

$$= R(t-1)\left(1 - \sqrt{br}\right).$$

For the Blue side similar work gives $B(t) = B(t-1)\left(1 - \sqrt{br}\right)$. Thus both the Red and Blue sides in this case decrease exponentially to zero since $0 < \sqrt{br} < 1$. This completes the proof of the theorem. $\square$

APPENDIX C

DERIVATION OF THE FER = 1 LINE FOR THE HUGHES SALVO MODEL

Recall that for the Hughes salvo model, the fractional exchange ratio is given by

$$\text{FER} = \frac{uR(rR - cB)}{dB(bB - sR)}.$$

If FER = 1, then we have

$$uR(rR - cB) = dB(bB - sR).$$

Multiplying out on both sides and collecting terms on the left give us

$$urR^2 + (ds - uc)BR - dbB^2 = 0.$$

Thus we have a quadratic equation in R that we can solve with the quadratic formula $-b \pm \sqrt{b^2 - 4ac}/2a$, where $a = ur$, $b = (ds - uc)B$, and $c = -dbB^2$. Plugging these values into the quadratic formula and simplifying gives

$$R = \left[\frac{(uc - ds) \pm \sqrt{(uc - ds)^2 + 4urdb}}{2ur} \right] B,$$

Models for Life: An Introduction to Discrete Mathematical Modeling with Microsoft® Office Excel®,
First Edition. Jeffrey T. Barton.
© 2016 John Wiley & Sons, Inc. Published 2016 by John Wiley & Sons, Inc.

where only the positive choice is of interest. So we see that the line that represents FER = 1 is a line through the origin in the B,R-plane with slope equal to

$$\frac{(uc-ds) + \sqrt{(uc-ds)^2 + 4urdb}}{2ur}.$$

APPENDIX D

THE WAITING TIME PRINCIPLE

In this appendix we supply a proof of the Waiting Time Principle that uses some material from second semester calculus and ideas from probability.

Theorem 4.4

Consider a compartment in a discrete dynamical system where a fixed proportion of the compartment leaves each day. Then the average amount of time spent in that compartment (in days) is equal to the reciprocal of the proportion who leave each day. In other words,

$$\text{Average time in compartment} = \frac{1}{\text{proportion leaving each day}}.$$

Equivalently, we write

$$\text{Proportion leaving each day} = \frac{1}{\text{average time spent in compartment}}.$$

Proof: Before we begin work on the Waiting Time Principle directly, we state the infinite sum version of the geometric series formula:

Theorem D.1

Let c be any real number and let x be any real number such that $-1 < x < 1$. Then

$$c + cx + cx^2 + \cdots = c\frac{1}{1-x}.$$

Models for Life: An Introduction to Discrete Mathematical Modeling with Microsoft® Office Excel®,
First Edition. Jeffrey T. Barton.
© 2016 John Wiley & Sons, Inc. Published 2016 by John Wiley & Sons, Inc.

To get an idea why this result is true, recall the finite sum version:

$$c + cx + cx^2 + cx^3 + \cdots + cx^{n-2} + cx^{n-1} = c\frac{x^n - 1}{x - 1}.$$

Now consider what would happen if we let the number of terms, n, get larger and larger. On the left-hand side we are just adding more terms, but on the right-hand side the term x^n gets smaller and smaller because $-1 < x < 1$. In the limit as the number of terms goes to infinity, the term x^n goes to 0. Thus we have the required result:

$$c + cx + cx^2 + \cdots = c\frac{-1}{x - 1} = c\frac{1}{1 - x}.$$

Turning our attention back to the Waiting Time Principle, we let x be the proportion leaving the compartment each day and note that $0 \leq x \leq 1$. The case where $x = 0$ is not interesting because this would mean nothing leaves the compartment each day, and in that case the waiting time would be undefined. At the other extreme if $x = 1$, then the compartment is completely emptied in one day, and the waiting time would therefore be 1 day in which case the result holds. Thus we concentrate on the case where $0 < x < 1$.

We can view x as the probability that a member of the compartment will leave on any given day. Thus the probability of a member leaving on the first day will be x. The probability of a member leaving on the second day, though, will be $(1-x)x$. This is because in order to leave on the second day, the member must not have left on the first day. The probability of leaving on the second day is therefore the probability of *not* leaving on the first day, $1 - x$, times the probability of leaving on the second. Similarly, the probability of leaving on the third day will be the probability of not leaving on day 1 times the probability of not leaving on day 2 times the probability of leaving on day 3: $(1-x)^2 x$. Continuing in this way we know the probability of leaving on day 4 is $(1-x)^3 x$ and on day 5 is $(1-x)^4 x$. We can see the pattern now. The probability of a member leaving on day n is the probability of having not left on the first $(n-1)$ days times the probability of leaving on day n, or $(1-x)^{n-1}x$.

The average waiting time (also known as the **expected value** of the waiting time) is the weighted average of all possible waiting times multiplied by their respective probabilities. Thus the average waiting time is given by

$$1 \cdot x + 2 \cdot (1-x)x + 3 \cdot (1-x)^2 x + 4 \cdot (1-x)^3 x + \cdots.$$

Factoring out the x gives

$$x\left(1 + 2 \cdot (1-x) + 3 \cdot (1-x)^2 + 4 \cdot (1-x)^3 + \cdots\right).$$

The expression inside the parentheses we recognize as the derivative of the expression

$$-(1-x) - (1-x)^2 - (1-x)^3 - (1-x)^4 - \cdots,$$

which we rewrite as

$$-(1-x)\left(1+(1-x)+(1-x)^2+(1-x)^3+\cdots\right).$$

Now the expression inside the parentheses is an infinite geometric series whose sum we know from the infinite geometric series formula:

$$1+(1-x)+(1-x)^2+(1-x)^3+\cdots=\frac{1}{1-(1-x)}=\frac{1}{x}.$$

Working back through our derivation, we write

$$-(1-x)\left(1+(1-x)+(1-x)^2+(1-x)^3+\cdots\right)=-(1-x)\cdot\frac{1}{x}$$

$$=-\frac{1-x}{x}.$$

Thus we can rewrite the expression $1+2\cdot(1-x)+3\cdot(1-x)^2+4\cdot(1-x)^3+\cdots$, which equals

$$\left(-(1-x)-(1-x)^2-(1-x)^3-(1-x)^4-\cdots\right)'=\left(-\frac{1-x}{x}\right)'$$

$$=\left(-\frac{1}{x}+1\right)'$$

$$=\frac{1}{x^2}.$$

Finally we get that the average waiting time will be

$$x\left(1+2\cdot(1-x)+3\cdot(1-x)^2+4\cdot(1-x)^3+\cdots\right)=x\cdot\frac{1}{x^2}$$

$$=\frac{1}{x}.$$

This completes the proof. $\square$

APPENDIX E

CREATING COBWEB DIAGRAMS IN EXCEL

In order to get Excel to produce a cobweb diagram, first we need the reproduction function for our DDS. Here we let $f(x) = x + 0.8\left(1 - \frac{x}{100}\right)x$. Next we set up our Excel spreadsheet to graph the reproduction function along with the line $y = x$ on the same axes. We will also need two additional columns for the cobweb diagram.

The setup is shown in Figure E.1. We set $x = 60$ as our initial population for the cobweb diagram, so the initial point on the cobweb is $(60, 0)$.

What makes the creation of the formulas tricky is that we have to have Excel create the points in the order they are to appear. From $(60, 0)$ we move up to the graph of $f(x)$. This means the next point on our cobweb will be $(60, f(60))$. We get Excel to compute this new point as in Figure E.2.

For our next point we need to move horizontally to the line $y = x$. This means the x-coordinate of the new point will be the y-coordinate of the current point. Since we need to be on the line $y = x$, the y-coordinate of the new point will be the same as the x-coordinate of the new point. In Figure E.3 we show the Excel formulas that make this happen.

We need one more point before copying the Excel formulas down. The next point results from moving vertically to the graph of the reproduction function. This means the new point will have the same x-coordinate as the current point and the

Models for Life: An Introduction to Discrete Mathematical Modeling with Microsoft® Office Excel®,
First Edition. Jeffrey T. Barton.
© 2016 John Wiley & Sons, Inc. Published 2016 by John Wiley & Sons, Inc.

	A	B	C	D	E
1	Cobweb Diagrams				
2					
3	*x*	*f(x)*	*y = x*	Cobweb x	Cobweb y
4	0	0	0	60	0
5	1	1.792	1		
6	2	3.568	2		
7	3	5.328	3		
8	4	7.072	4		

FIGURE E.1   Excel setup for producing a cobweb diagram.

	D	E
1		
2		
3	*Cobweb x*	*Cobweb y*
4	60	0
5	=D4	=D5+0.8*(1-D5/100)*D5

FIGURE E.2   Calculating the second point on the cobweb diagram.

	D	E
1		
2		
3	*Cobweb x*	*Cobweb y*
4	60	0
5	=D4	=D5+0.8*(1-D5/100)*D5
6	=E5	=D6

FIGURE E.3   Moving horizontally to the next point on the cobweb diagram.

new *y*-coordinate will be the reproduction function evaluated at the *x*-coordinate. The Excel version is shown in Figure E.4. The results are shown in Figure E.5.

Finally we are ready to copy the formulas down, but it works a little differently in this case. Here the formulas are based on two rows rather than just one, and we need to preserve the two-row structure. So we select the 2×2 block of cells made up of columns D and E and rows 6 and 7; then we drag the formulas down with the thin cross pointer. Because we are copying two rows at a time, we need to copy down for an even number of rows. Figure E.6 shows the result with formulas displayed.

To get the cobweb on a graph, we first create the graph for the reproduction function and the line $y = x$ in the usual way. Once we have that graph, we add

	D	E
1		
2		
3	*Cobweb x*	*Cobweb y*
4	60	0
5	=D4	=D5+0.8*(1-D5/100)*D5
6	=E5	=D6
7	=E6	=D7+0.8*(1-D7/100)*D7

FIGURE E.4 Moving vertically to the next point on the cobweb diagram.

	A	B	C	D	E
1	Cobweb Diagrams				
2					
3	*x*	*f(x)*	*y = x*	*Cobweb x*	*Cobweb y*
4	0	0	0	60	0
5	1	1.792	1	60	79.2
6	2	3.568	2	79.2	79.2
7	3	5.328	3	79.2	92.37888
8	4	7.072	4		

FIGURE E.5 Resulting point after moving vertically to the reproduction curve.

	D	E
3	*Cobweb x*	*Cobweb y*
4	60	0
5	=D4	=D5+0.8*(1-D5/100)*D5
6	=E5	=D6
7	=E6	=D7+0.8*(1-D7/100)*D7
8	=E7	=D8
9	=E8	=D9+0.8*(1-D9/100)*D9
10	=E9	=D10
11	=E10	=D11+0.8*(1-D11/100)*D11

FIGURE E.6 Result of copying down formulas for cobweb diagram.

the cobwebbing points by first selecting the points we want to add to our graph. Once we do that, we do a copy and paste onto the graph. Specifically, we copy the cobweb points we want to add, then we click somewhere on the graph of the reproduction function. Now from the Paste drop-down menu, we select Paste

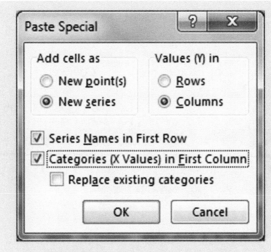

FIGURE E.7   Paste Special dialog box.

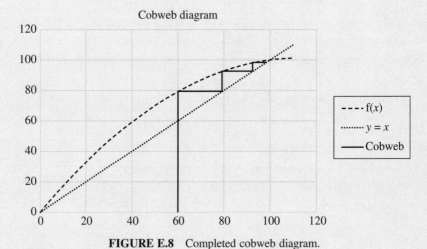

FIGURE E.8   Completed cobweb diagram.

Special. In the dialog box that appears, we check the box for Categories (X Values) in First Column. The completed dialog box is shown in Figure E.7. After clicking "Okay" our graph should be complete. We show the completed graph in Figure E.8.

APPENDIX F

PROPORTION OF TOTAL CREDIT DISTRIBUTED DOES NOT EXCEED 1

We assume there are $N \geq 2$ teams that all play each other once. If teams distribute a just-for-playing credit of $\frac{1}{N^2}$ to teams they defeat and $\frac{1}{L+2}$ to teams that defeat them, then the total proportion of credit distributed by any team is given by

$$\text{Total} = \text{Wins} \times \frac{1}{N^2} + \text{Losses} \times \frac{1}{L+2}.$$

Since wins plus losses must equal the number of games played, $N-1$, we can rewrite the above equality as

$$\text{Total} = \frac{N-1-L}{N^2} + \frac{L}{L+2}.$$

We need to show that

$$\frac{N-1-L}{N^2} + \frac{L}{L+2} \leq 1,$$

or, equivalently, that

$$(L+2)(N-1-L) + LN^2 \leq (L+2)N^2.$$

Models for Life: An Introduction to Discrete Mathematical Modeling with Microsoft® Office Excel®,
First Edition. Jeffrey T. Barton.
© 2016 John Wiley & Sons, Inc. Published 2016 by John Wiley & Sons, Inc.

After multiplying out and collecting terms on the right, we must show that

$$0 \le 2N^2 - LN - 2N + L^2 + 3L + 2.$$

After some algebra we rewrite the expression on the right as

$$N((N-L)+(N-2))+(L+1)(L+2).$$

Since N is at least two and is always greater than the number of possible losses for a team, we see by inspection that the last expression must be positive. This completes the proof. $\square$

BIBLIOGRAPHY

Abernethy, D. R. (1985). Impairment of caffeine clearance by chronic use of low-dose oestrogen-containing oral contraceptives. *European Journal of Clinical Pharmacology, 28*(4), 425–428.

Anderson, R. M. & May, R. M. (1991). *Infectious Diseases of Humans: Dynamics and Control.* Oxford: Oxford Science Publications.

Arciero, P., Goran, M., & Poehlman, E. (1993). Resting metabolic rate is lower in women than in men. *Journal of Applied Physiology, 75*(6), 2514–2520.

Armstrong, M. J. (2013). The salvo combat model with area fire. *Naval Research Logistics, 60*(8), 652–660.

ARUP. (2015). Laboratory Test Directory: Ibuprofen. ARUP Laboratories. Retrieved from: http://ltd.aruplab.com/Tests/Pub/0090176 (accessed on August 19, 2015).

Ashwell, M. & Hsieh, S. D. (2005). Six reasons why the waist-to-height ratio is a rapid and effective global indicator for health risks of obesity and how its use could simplify the international public health message on obesity. *International Journal of Food Sciences and Nutrition, 56*(5), 303–307.

Ashwell, M., Gunn, P., & Gibson, S. (2012). Waist-to-height ratio is a better screening tool than waist circumference and BMI for adult cardiometabolic risk factors: Systematic review and meta-analysis. *Obesity Reviews, 13*(3), 275–286.

Bailey, C. (2000). *The Ultimate Fit or Fat.* New York: Rux Martin/Houghton Mifflin Harcourt.

Bailey, N. T. (1975). *The Mathematical Theory of Infectious Diseases and its Applications, 2nd edition.* New York: Hafner Press.

Bankrate. (2015). Retrieved from www.bankrate.com (accessed on August 19, 2015).

Battle of Trafalgar. (2015). The National Archives. Retrieved from: http://www.nationalarchives.gov.uk/nelson/gallery7/trafalgar.htm (accessed on August 19, 2015).

Birkett, D. J. & Miners, J. O. (1991). Caffeine renal clearance and urine caffeine concentrations during steady state dosing. Implications for monitoring caffeine intake during sports events. *British Journal of Clinical Pharmacology, 31*(4), 405–408.

Bureau of Labor Statistics (BLS). (2012). *Occupational Outlook Handbook.* Washington, DC: BLS.

BLS. (2015). United States Department of Labor, Bureau of Labor Statistics. Retrieved from: www.bls.gov (accessed on August 19, 2015).

Brauer, F. & Castillo-Chavez, C. (2001). *Mathematical Models in Population Biology and Epidemiology.* New York: Springer.

Brin, S. & Page, L. (1998). The Anatomy of a Large-Scale Hypertextual Web Search Engine. Retrieved from: http://ilpubs.stanford.edu:8090/361/ (accessed on August 19, 2015).

Bryant, C. X. Merrill, S., & Green, D. J. (eds). (2014). *American Council on Exercise Personal Trainer Manual,* 5th edition. San Diego, CA: American Council on Exercise.

Car Max. (2015). Car Max. Retrieved from: http://www.carmax.com (accessed on August 19, 2015).

Cars.com. (2015). Retrieved from www.cars.com (accessed on August 19, 2015).

Carter, J., Ackleh, A. S., Leonard, B. P., & Wang, H. (1999). Giant panda population dynamics and bamboo life history: A structured population approach to examining carrying capacity when the prey are semelparous. *Ecological Modelling, 123,* 207–223.

Centers for Disease Control (CDC). (2011). *Body Mass Index: Considerations for Practitioners.* Atlanta, GA: Department of Health and Human Services, Centers for Disease Control.

CDC. (2012). *MMR Vaccine: What You Need to Know.* Atlanta, GA: Centers for Disease Control.

CDC. (2014). Ebola (Ebola Virus Disease) Signs and Symptoms. Centers for Disease Control and Prevention. Retrieved from: http://www.cdc.gov/vhf/ebola/symptoms/index.html (accessed on August 19, 2015).

CDC. (2015a). Effects of Blood Alcohol Concentration (BAC). Centers for Disease Control and Prevention. Retrieved from: http://www.cdc.gov/Motorvehiclesafety/impaired_driving/bac. html (accessed on August 19, 2015).

CDC. (2015b). Update: Ebola Virus Disease Epidemic—West Africa, January 2015. Centers for Disease Control and Prevention. Retrieved from: http://www.cdc.gov/mmwr/preview/ mmwrhtml/mm6404a8.htm (accessed on August 19, 2015).

Consumer Financial Protection Bureau (CFPB). (2013). CARD Act Report. CFPB. Retrieved from: http://files.consumerfinance.gov/f/201309_cfpb_card-act-report.pdf (accessed on August 19, 2015).

CFPB. (2015). Credit Card Agreement Data Base. CFPB. Retrieved from: http://www. consumerfinance.gov/credit-cards/agreements/search/?q=Chase+Bank+Freedom (accessed on August 19, 2015).

Chow, C. C. & Hall, K. D. (2008). The dynamics of human body weight change. *PLOS Computational Biology, 4*(3), e1000045.

Clark, C. W. (1985). *Bioeconomic Modelling and Fisheries Management.* Hoboken, NJ: John Wiley & Sons, Inc.

Clemson Redfern Health Center (CRHC). (2015). Blood Alcohol Concentration (BAC). CRHC. Retrieved from: http://www.clemson.edu/campus-life/campus-services/redfern/ alcohol/bac.html (accessed on August 19, 2015).

Credit CARD Act. (2009). U.S. Government Publishing Office. Retrieved from: http://www. gpo.gov/fdsys/pkg/BILLS-111hr627enr/pdf/BILLS-111hr627enr.pdf (accessed on August 19, 2015).

Curtis, P. D. & Sullivan, K. L. (2001). *White-tailed Deer, Wildlife Damage Management Fact Sheet Series.* Ithaca, NY: Cornell University Cooperative Extension.

Drake, J. & Kramer, A. (2011). Allee effects. *Nature Education Knowledge, 3*(10):2.

DRI. (2006). *Dietary Reference Intakes.* Washington, DC: National Academies Press.

Edmunds. (2015). Retrieved from: www.edmunds.com (accessed on August 19, 2015).

Engel, J. H. (1954). A verification of Lanchester's law. *Journal of the Operations Research Society of America, 2*(2), 163–171.

Food and Agriculture Organization (FAO). (2001). *Human Energy Requirements.* Rome: FAO.

Food and Drug Administration (FDA). (2014). FDA Consumer Advice on Pure Powdered Caffeine. FDA. Retrieved from: http://www.fda.gov/Food/RecallsOutbreaksEmergencies/SafetyAlertsAdvisories/ucm405787.htm (accessed on August 19, 2015).

Forbes, G. B. (2000). Body fat content influences the body composition response to nutrition and exercise. *Annals of the New York Academy of Sciences, 904*(1), 359–365.

Forbes. (2015). The World's Billionaires. Forbes. Retrieved from: http://www.forbes.com/profile/sergey-brin/ (accessed on August 19, 2015).

Frankenfield, D., Roth-Yousey, L., & Compher, C. (2005). Comparison of predictive equations for resting metabolic rate in healthy nonobese and obese adults: A systematic review. *Journal of the American Dietetic Association, 105*(5), 775–789.

Franklin, M. B. (2006). *How Much Is Enough?* Washington, DC: Kiplinger's Personal Finance.

Fredholm, B. B., Bättig, K., Holmén, J., Nehlig, A., and Zvartau, E. E. (1999). Actions of caffeine in the brain with special reference to factors that contribute to its widespread use. *Pharmacological Reviews, 51*(1), 83–133.

Friedl, C. E., Westphal, K. A., Marchitelli, L. J., Patton, J. F., Chumlea, W. C., & Guo, S. S. (2001). Evaluation of anthropometric equations to assess body-composition changes in young women. *The American Journal of Clinical Nutrition, 73*(2), 268–275.

Frontline: The Age of Aids. (2006). Retrieved from: http://www.pbs.org/wgbh/pages/frontline/aids/virus/virus.html (accessed on August 19, 2015).

Gee, G. F. & Hereford, S. G. (1993). *Mississippi Sandhill Cranes.* Washington, DC: National Biological Service and U.S. Fish and Wildlife Service.

Giardinia, E. G. (2015). Cardiovascular effects of caffeine and caffeinated beverages. UpToDate. Retrieved from: http://www.uptodate.com/contents/cardiovascular-effects-of-caffeine-and-caffeinated-beverages (accessed on August 19, 2015).

Giordano, F. R., Fox, W. P., Horton, S. B., & Weir, M. D. (2008). *A First Course in Mathematical Modeling,* 4th edition. Belmont, CA: Cengage Learning.

Goldstein, E. R., Ziegenfuss, T., Kalman, D., Kreider, R., Campbell, B., Wilborn, C., ... Antonio, J. (2010). International society of sports nutrition position stand: Caffeine and performance. *Journal of the International Society of Sports Nutrition, 7*(1), 1–15.

Gunther, K. A. (2003). *Recovery Parameters for Grizzly Bears in the Yellowstone Ecosystem.* Washington, DC: National Park Service.

Gurney, D. (2004). *Cobweb Diagrams with Excel.* Hammond, LA: Southeastern Louisiana University.

Hall, K. D. (2007). Body fat and fat-free mass inter-relationships: Forbes's theory revisited. *British Journal of Nutrition, 97*(6), 1059–1063.

Hall, K. D. (2008). What is the required energy deficit per unit weight loss? *International Journal of Obesity, 32*(3), 573–576.

Hall, K. D., Sacks, G., Chandramohan, D., Wang, Y. C., Gortmaker, S. L., & Swinburn, B. A. (2011). Quantification of the effect of energy imbalance on bodyweight. *The Lancet, 378*(9793), 826–837.

Haugen, H. A., Chan, L.-N., & Li, F. (2007). Indirect calorimetry: A practical guide for clinicians. *Nutrition in Clinical Practice, 22*(4), 377–388.

Healthline. (2015). The Effects of Caffeine on the Body. Healthline. Retrieved from: http://www.healthline.com/health/caffeine-effects-on-body (accessed on August 19, 2015).

Heck, A. (2007). Modeling intake and clearance of alcohol in humans. *The Electronic Journal of Mathematics and Technology, 1*(3), 232–244.

Hughes, W. J. (1986). *Fleet Tactics: Theory and Practice.* Annapolis, MD: Naval Institute Press.

Hughes, W. J. (1993). *The Military Worth of Staying Power.* Monterey, CA: U.S. Naval Postgraduate School.

Hughes, W. J. (1995). *A Salvo Model of Warships in Missile Combat Used to Evaluate Their Staying Power.* Naval Research Logistics. Retrieved from: http://onlinelibrary.wiley.com/journal/10.1002/%28ISSN%291520-6750 (accessed on September 8, 2015).

Institute for Health Metrics (IHME). (2010). Institute for Health Metrics and Evaluation Dallas County Profile. Health Data. Retrieved from: http://www.healthdata.org/sites/default/files/files/county_profiles/US/County_Report_Dallas_County_Texas.pdf (accessed on August 19, 2015).

International Whaling Commission. (n.d.). Retrieved from: http://iwc.int/iwcmain.htm (accessed on August 19, 2015).

Internet Live Stats. (2015). Internet Live Stats. Retrieved from: http://www.internetlivestats.com/ (accessed on August 19, 2015).

Jackson, A., Stanforth, P., Gagnon, J., Rankinen, T., Leon, A., Rao, D., … Wilmore, J. (2002). The effect of sex, age and race on estimating percentage body fat from body mass index: The Heritage Family Study. *International Journal of Obesity and Related Metabolic Disorders, 26*(6), 789–796.

Jannsen, G. M. & Venema, J. F. (1985). Ibuprofen: Plasma concentrations in man. *Journal of International Medical Research, 13*, 68–73.

Karasu, S. R. (2013). College Weight Gain: Debunking the Myth of the 'Freshman 15'. Retrieved from: http://www.psychologytoday.com/blog/the-gravity-weight/201309/college-weight-gain-debunking-the-myth-the-freshman-15 (accessed on August 19, 2015).

Kelley Blue Book. (2015). Retrieved from: www.kbb.com (accessed on August 19, 2015).

Kent, W. (2012). The Pharmacokinetics of Alcohol in Healthy Adults. WebMed Central PHARMACOLOGY. Retrieved from: http://www.webmedcentral.com/article_view/3291 (accessed on August 19, 2015).

Kermack, W. & McKendrick, A. (1927). A contribution to the mathematical theory of epidemics. *Proceedings of the Royal Society, 115*(772), 700–721.

Kiff, L. F., Mesta, R. I., & Wallace, M. P. (1996). *Recovery Plan for the California Condor.* Washington, DC: U.S. Fish and Wildlife Service.

Lanchester, F. W. (1956). Mathematics in warfare. In J. R. Newman (ed.), *The World of Mathematics,* Vol. 4, pp. 2138–2160. London: George Allen & Unwin Ltd.

Langville, A. N. & Meyer, C. D. (2006). *Google's PageRank and Beyond: The Science of Search Engine Rankings.* Princeton, NJ: Princeton University Press.

Langville, A. N. & Meyer, C. D. (2012). *Who's #1? The Science of Rating and Ranking.* Princeton, NJ: Princeton University Press.

Levitt, M. & Levitt, D. (1998). Use of a two-compartment model to assess the pharmacokinetics of human ethanol metabolism. *Alcoholism: Clinical and Experimental Research, 22*(8), 1680–1688.

Lewis, M. J. (1986). Blood alcohol: The concentration time curve and retrospective estimation of level. *Journal of the Forensic Science Society, 26*(2), 95–113.

Lucas, T. W. & Turkes, T. (2004). Fitting Lanchester equations to the battles of Kursk and Ardennes. *Naval Research Logistics, 51*(1), 95–116.

MacKay, N. (2005). Lanchester Combat Models. Retrieved from: http://arxiv.org/abs/math/0606300v1 (accessed on August 19, 2015).

Mahon, C. M. (2007). A Littoral Combat Model for Land-Sea Missile Engagements. Institutional Archive of the Naval Postgraduate School. Retrieved from: https://calhoun.nps.edu/handle/10945/3274 (accessed on August 19, 2015).

Malthus, T. (1798). *An Essay on the Principle of Population.* Aylesbury: Penguin Books.

Marks, V. & Kelly, J. (1973). Absorption of caffeine from tea, coffee, and coca cola. *The Lancet, 301*(7807), 827.

Mattson, D. J., Wright, R. G., Kendall, K. C., & Martinka, C. J. (1993). *Grizzly Bears.* Washington, DC: National Biological Service.

May, R. M. (1976). Models for single populations. In R. M. May (ed.), *Theoretical Ecology: Principles and Applications,* pp. 4–26. Oxford: Blackwell Scientific Publications.

May, R. M. (1978). Mathematical aspects of the dynamics of animal populations. In S. A. Levin (ed.), *Studies in Mathematical Biology,* pp. 317–366. Washington, DC: Mathematical Association of America.

May, R. M. & Oster, G. F. (1976). Bifurcations and dynamic complexity in simple ecological models. *The American Naturalist, 110,* 573–599.

McCullough, D. (1997). Irruptive behavior in ungulates. In W. J. McShea, H. B. Underwood, & J. H. Rappole (eds), *Science of Overabundance: Deer Ecology and Population Management,* pp. 69–98. Washington, DC: Smithsonian Institution Press.

Mifflin, M. D., St Jeor, S. T., Hill, L. A., Scott, B. J., Daugherty, S. A., & Koh, Y. O. (1990). A new predictive equation for resting energy expenditure in healthy individuals. *American Journal of Clinical Nutrition, 51*(2), 241–247.

Miller-Kovach, K., Gerwig, U., Peetz, J., Jacobsohn, C., Frye, W., Rost, S. L., … Halkuff, D. (2010). Processes and systems based on dietary fiber as energy. US patent no.: US2010,00,55,652 A1.

Miyatake, N., Matsumoto, S., Miyachi, M., Fujii, M., & Numata, T. (2007). Relationship between changes in body weight and waist circumference in Japanese. *Environmental Health and Preventive Medicine, 12*(5), 220–223.

Moller, A. P. & Legendre, S. (2001). Allee effect, sexual selection and demographic stochasticity. *Oikos, 92,* 24–34.

Morehouse, C. P. (1946). *The Iwo Jima Operation.* Arlington, VA: USMCR, Historical Division, Headquarters U.S. Marine Corps.

Murray, J. (1993). *Mathematical Biology,* 2nd edition. Berlin: Springer-Verlag.

Nelson, B. (2005). *Infectious Disease Epidemiology: Theory and Practice.* Sudbury, MA: Jones and Bartlett Learning.

Nelson, K. M., Weinsier, R. L., Long, C. L., & Schutz, Y. (1992). Prediction of resting energy expenditure from fat-free mass and fat mass. *The American Journal of Clinical Nutrition, 56*(5), 848–856.

Nelson, H. (1805). Nelson's Trafalgar Memorandum. British Library. Retrieved from: http://www.bl.uk/learning/timeline/item106127.html (accessed on August 19, 2015).

National Highway Traffic Safety Administration (NHTSA). (1994). *Computing a BAC Estimate.* Washington, DC: U.S. Department of Transportation, NHTSA.

Obesity Education Initiative. (1998). *Clinical Guidelines on the Identification, Evaluation, and Treatment of Overweight and Obesity in Adults.* New York: National Institutes of Health.

Ortega, F. B., Lee, D.-C., Katzmarzyk, P. T., Ruiz, J. R., Sui, X., Church, T. S., & Blair, S. N. (2013). The intriguing metabolically healthy but obese phenotype: Cardiovascular prognosis and role of fitness. *European Heart Journal, 34*(5), 389–397.

Palmer, W. L. & Storm, G. L. (1995). White-tailed deer in the northeast. In *Our Living Resources,* pp. 112–115. Washington, DC: National Biological Service.

Phaser. (2015). Phaser Module: Ricker Salmon Model. Phaser. Retrieved from: http://www. phaser.com/modules/students/salmon/salmon.pdf (accessed on August 19, 2015).

Pieters, J., Wedel, M., & Schaafsma, G. (1990). Parameter estimation in a three-compartment model for blood alcohol curves. *Alcohol and Alcoholism, 25*(1), 17–24.

Posterli, B. (n.d). Fast Facts on the Freshman 15. Retrieved from: http://journalism.nyu.edu/ publishing/archives/livewire/health_science/freshman15/ (accessed on August 19, 2015).

Prater, C. (2012). What the New Credit Card Law Means for You. Credit Cards.com. Retrieved from: http://www.creditcards.com/credit-card-news/new-credit-card-law-1282.php (accessed on August 19, 2015).

Population Reference Bureau (PRB). (2014). World Population Data Sheet 2014. PRB. Retrieved from: http://www.prb.org/Publications/Datasheets/2014/2014-world-population-data-sheet/data-sheet.aspx (accessed on August 19, 2015).

Raggett, G. (1982). Modelling the Eyam plague. *Bulletin of the Institute of Mathematics and its Applications, 18,* 221–226.

Ramsay, M. A. (2006). John Snow, MD: Anaesthetist to the Queen of England and pioneer epidemiologist. *Baylor University Medical Center Proceedings, 19*(1), 24–28.

Ricker, W. E. (1954). Stock and recruitment. *Journal of the Fisheries Research Board of Canada, 11*(5), 559–623.

Rutberg, A. T. (1997). The science of deer management: An animal welfare perspective. In W. J. McShea, H. B. Underwood, & J. H. Rappole (eds), *Science of Overabundance: Deer Ecology and Population Management,* pp. 37–54. Washington, DC: Smithsonian Institution Press.

RxList. (2015). Motrin. RxList: The Internet Drug Index. Retrieved from: http://www.rxlist. com/ibuprofen-drug/clinical-pharmacology.htm (accessed on August 19, 2015).

Sandilands, T. (2010). The Financial Side of the Weight Loss Industry. The Houston Chronicle. Retrieved from: http://smallbusiness.chron.com/financial-side-weight-loss-industry-38200. html (accessed on August 19, 2015).

Schwartzlose, R. A., Alheit, J., Bakun, A., Baumgartner, T. R., Cloete, R., Crawford, R. J. M., … Zuzunaga, J. Z. (1999). Worldwide large-scale fluctuations of sardine and anchovy populations. *South African Journal of Marine Science, 21,* 289–347.

Serveheen, C. (1993). *Grizzly Bear Recovery Plan.* Laurel, MD: U.S. Fish and Wildlife Service.

Sibley, F. (1969). *Effects of the Sespe Creek Project on the California condor.* Laurel, MD: U.S. Fish and Wildlife Service.

Sizing Up Your Down Payment. (2005). Bank Rate. Retrieved from: http://www.bankrate.com/ finance/auto/sizing-up-your-down-payment.aspx (accessed on August 19, 2015).

Smith, A. (2002). Effects of caffeine on human behavior. *Food and Chemical Toxicology, 40*(9), 1243–1255.

Spitznagel, E. (1992). Two-Compartment Pharmacokinetic Models, CODEE Newsletter, Fall 1992, Claremont, CA, pp. 2–4.

Spriet, L. L. (1995). Caffeine and performance. *International Journal of Sports Nutrition, 5,* S84–S95.

Spriet, L. L. & Graham, T. E. (2015). Caffeine and Exercise Performance. American College of Sports Medicine. Retrieved from: http://www.acsm.org/docs/current-comments/caffeineandexercise.pdf (accessed on August 19, 2015).

Spruill, W. J., Wade, W. E., DiPiro, J. T., Blouin, R. A., Pruemer, J. M., American Society of Health-System Pharmacists (2014). *Concepts in Clinical Pharmacokinetics,* 6th edition. Bethesda, MD: American Society of Health-System Pharmacists.

Swap-a-Lease. (2015). Retrieved from www.swapalease.com (accessed on August 19, 2015).

Taubenberger, J. & Morens, D. (2006). 1918 influenza: The mother of all pandemics. *Emerging Infectious Diseases, 12*(1), 15–22.

Taylor, J. G. (1983). *Lanchester Models of Warfare.* Arlington, VA: Ketron, Inc.

Thomas, D. M., Ciesla, A., Levine, J. A., Stevens, J. G., & Martin, C. K. (2009). A mathematical model of weight change with adaptation. *Mathematical Biosciences and Engineering, 6*(4), 873–887.

U.S. Army Corps of Engineers (US ACE). (2004). *Adult Fish Counts.* Portland, OR: US ACE.

U.S. Census. (2010). United States Census Bureau. Retrieved from: http://www.census.gov/en.html (accessed on August 19, 2015).

U.S. Fish and Wildlife Service. (1996). *California Condor Recovery Plan,* 3rd revision. Portland, OR: U.S. Fish and Wildlife Service.

U.S. Fish and Wildlife Service. (2011). Mississippi Sandhill Crane National Wildlife Refuge. Retrieved from http://www.fws.gov/mississippisandhillcrane/ (accessed on August 19, 2015).

U.S.A. Today. (2005). U.S.A. Today. Retrieved from: http://usatoday30.usatoday.com/news/washington/2005-11-15-grizzlies_x.htm (accessed on August 19, 2015).

Umulis, D. M., Gurman, N. M., Singh, P., & Fogler, H. S. (2005). A physiologically based model for ethanol and acetaldehyde metabolism in human beings. *Alcohol, 35*(1), 3–12.

Valentine, J. M. & Lohoefener, R. (1991). *Recovery Plan: Mississippi Sandhill Crane,* 3rd revision. Washington, DC: U.S. Fish and Wildlife Service.

Verhulst, P. (1838). Notice sur la loi que la population suit dans son accroissement. *Correspondances mathematiques et physiques, 10*(1938), 113–121.

Wagner, J. & Patel, J. (1972). Variations in absorption and elimination rates of ethyl alcohol in a single subject. *Research Communications in Chemical Pathology and Pharmacology, 4*(1), 61–76.

Wang, H., Li, S., & Pan, W. (2001). Population viability analysis of giant panda (Ailuropoda melanoleuca) in Qinling Mountains. *Beijing da xue xue bao. Zi ran ke xue bao, 38*(6), 756–761.

WebMD. (2015). Brewing Trouble: Coffee-induced Anxiety. WebMD. Retrieved from: http://www.webmd.com/balance/features/brewing-trouble (accessed on August 19, 2015).

World Health Organization (WHO). (2000). *Control of rubella and congenital rubella syndrome (CRS) in developing countries.* Geneva: WHO.

WHO. (2015). Ebola Virus Disease Fact Sheet. WHO. Retrieved from: http://www.who.int/mediacentre/factsheets/fs103/en/ (accessed on August 19, 2015).

Wilbur, S. (1980). Estimating the size and trend of the California condor population 1965–1978. *California Fish and Game, 66,* 40–48.

Wolfram. (2015). Wolfram Alpha. Wolfram Alpha Computational Knowledge Engine. Retrieved from: http://www.wolframalpha.com/ (accessed on August 19, 2015).

Wood, W. (1865). *History of Eyam.* London: Bell and Daldy.

World Wildlife Fund (WWF). (2014). How Many are Left in the Wild? WWF. Retrieved from: http://wwf.panda.org/what_we_do/endangered_species/giant_panda/panda/ how_many_are_left_in_the_wild_population/ (accessed on August 19, 2015).

WWF. (2015). GIANT Panda Social Structure and Breeding. WWF. Retrieved from: http:// wwf.panda.org/what_we_do/endangered_species/giant_panda/panda/ panda_social_structure_and_breeding/ (accessed on August 19, 2015).

Zagorsky, J. L. & Smith, P. K. (2011). The Freshman 15: A critical time for obesity intervention or media myth? *Social Science Quarterly, 92*(5), 1389–1407.

INDEX

Models for Life: An Introduction to Discrete Mathematical Modeling with Microsoft® Office Excel®,
First Edition. Jeffrey T. Barton.
© 2016 John Wiley & Sons, Inc. Published 2016 by John Wiley & Sons, Inc.